# From Hill 92 to Nui Hon Ba

The Journey of Korea Military Academy
Graduates to the Vietnam War

# From Hill 92 to Nui Hon Ba

The Journey of Korea Military Academy Graduates to the Vietnam War

| | |
|---|---|
| **Author** | Chung Youn-son |
| **First Printed** | June 15, 2023 |
| **First Published** | June 23, 2023 |
| **Director** | Koh Yun-sung |
| **Executive Contents Creator** | Shin Sun-ho |
| **Contents Creator** | Jang Hae-jung<br>Roh Jae-eun<br>Byun Da-eun |
| **Designer** | Kim Dae-wook<br>Choi Jae-young |
| **Published by** | HANKUK UNIVERSITY OF FOREIGN STUDIES<br>KNOWLEDGE CONTENTS & PRESS<br>107, Imun-ro, Dongdaemun-gu, Seoul 02450, Korea<br>Tel.) 82-2-2173-2497 Fax.) 82-2-2173-3363<br>E-mail) press@hufs.ac.kr<br>Online) http://press.hufs.ac.kr |
| **Printed by** | K-LAB Tel.) 82-53-583-6885 |
| **ISBN** | 979-11-5901-927-2(03910) |
| **Price** | KRW25,000 $25 |

# FROM HILL 92 TO NUI HON BA

*The Journey of Korea Military Academy Graduates to the Vietnam War*

Chung Youn-son

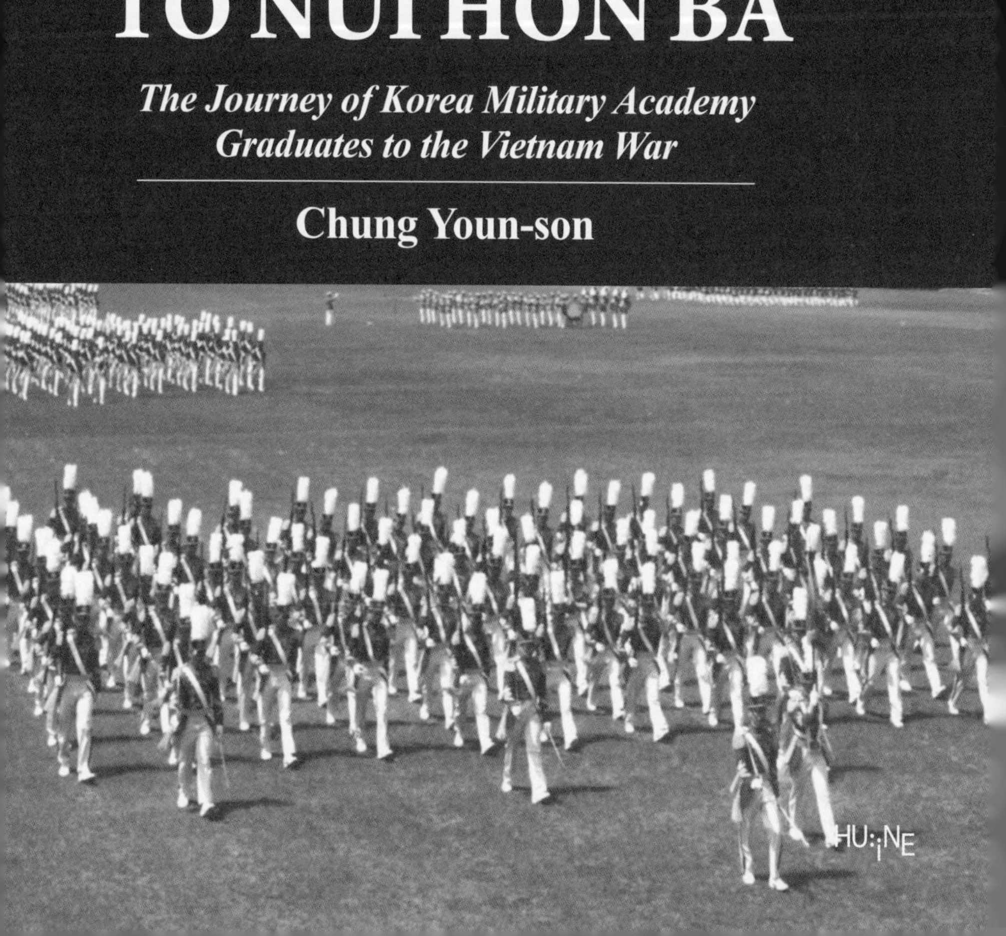

# Acknowledgements

I am grateful to Korea Military Academy Library for allowing me access to its materials and resources concerning my research. The library is noted for having numerous books and articles related to the military and war in Korea, including the Vietnam War. I am also indebted to the Institute for Military History of the Ministry of National Defense and the Office of Records and Information Management of the Republic of Korea Army Headquarters for the necessary data, photos, and documents related to this project. Above all, my thanks go to many KMA graduates who have provided me with their valuable combat experiences and given me constant love and encouragement for this book to be written.

A special thanks to Vietnam veteran Park Jung-ki (KMA Class of 1958), President of the Korea Corporate Members of the Association of the United States Army and the longest-running council member of the International Association of Athletics Federations, for financial support in publishing this book.

# Preface

This book is the war story of the Korea Military Academy (KMA) graduates who fought in the Vietnam War from September 1964 to March 1973, responding to the nation's call to fight. The graduates referred to here are those of the seventeen classes from 1955 (11th Class) to 1971 (27th Class), attending the Republic of Korea's first four-year military institution comparable to the United States Military Academy at West Point. The primary source of this book comes from these graduates' war experiences delivered to me through interviews, memories, testimonies, and many other written documents, including essays, letters, memoirs, and autobiographies. Writing a war story is an act to give order to one's disordered war experience. In this sense, this book aims to summarize the war experiences of all those KMA graduates who fought the war of confusion and thus give meaning to them as a whole.

As a member of the KMA Class of 1970, I interviewed almost all my Vietnam veteran classmates, who generously responded to my request. They provided me with their authentic war stories in the initial stage of conceiving this book. Hundreds of KMA seniors who had been to Vietnam earlier than my classmates told me about their valuable combat experiences in the jungles. Their stories became a part of the book very much. There are also many Vietnam-era KMA veterans whose voices aren't in this book but provided me with tremendous background knowledge about the distinctive aspects of the war.

The title of this book is "From Hill 92 to Nui Hon Ba", which is quite symbolic, suggesting a long journey. Hill 92 is a low hill surrounding the quadrangle of the Corps of Cadets at KMA, like a folding screen to the rear. Hardly a hill to be called a hill, "Hill 92" is a place of memories with countless stories, sweat, and tears for all KMA men. Though the hill's height is 92 meters above sea level, it is the most challenging for KMA graduates and cadets. Even when climbing a mountain outside on a weekend leave, cadets mentally struggle the most at the height of 92 meters. They call it "Devil's height."

On the other hand, "Nui Hon Ba" is a mountain with a height of 1,361 meters. Nui (mountain in Vietnamese) Hon Ba, located in the south-central coastal region of Vietnam, is in the operational area of the Korean White Horse Division. Another mountain in the Tiger Division area, Nui Phu Cat, is far lower than Nui Hon Ba. In this book, I take Nui Hon Ba as a symbol of South Vietnam. The title describes the essence of the book, the perilous journey of KMA graduates, crossing over the ocean to the exotic land of faraway Vietnam.

This book also pays attention to what Korean soldiers and their American counterparts thought of each other in Vietnam while fighting together for eight and a half years. Their close, friendly, but sometimes tensive relationships between the two armies on the battlefield in Vietnam have some implications for their present and future relations in Korea.

I dedicate this book to all the KMA graduates who have been to the war. The regular graduates of KMA are those to be asked for limitless sacrifices for the country. Their participation in the Vietnam War some fifty years ago was not one to seek any personal reward but to come out of the firm conviction that the place where a soldier should be was on the battlefield. Their noble sacrifice significantly contributed to making this country what it is today.

Chung Youn-son
Seoul, Korea
June 2023.

**Acknowledgments** **5**

**Preface** **6**

## TABLE OF CONTENTS

**Introduction** **13**

**PART I**
**INITIATION: INTO THE JUNGLE** **21**

Chapter 1
KMA as the Bastion of National Defense **23**

Chapter 2
The Road to Vietnam **34**

Chapter 3
Why They Went to Vietnam **45**

Chapter 4
Organizing and Training of Combat Troops **55**

Chapter 5
The Voyage: Busan to Quy Nhon **66**

Chapter 6
The Vietnam War: "A Formless War against a Formless Enemy" **75**

**PART II**
**ENGAGEMENT: HOW THEY FOUGHT** **91**

Chapter 1
Class of 1955 to Class of 1957: The Pathfinders **93**

Chapter 2
Class of 1958 and Class of 1959: Into the Fire **116**

Chapter 3
Class of 1960(16th Class): Company Commanders **137**

Chapter 4
Class of 1961 to Class of 1963: In the Midst of Politics **153**

Chapter 5
Class of 1964 and Class of 1965: Platoon Leaders **175**

Chapter 6
Class of 1966 to Class of 1968: In the Footsteps of their Elders **197**

Chapter 7
Class of 1969 to 1971: Toward the End **235**

**PART III**
**CONSIDERATION: ISSUES** **267**

Chapter 1
The Vietnam War as a War of Platoon Leaders **269**

Chapter 2
Combat Motivation: What Makes Them Fight? **278**

Chapter 3
Dehumanization of War **289**

Chapter 4
Appearances and Realities in the Vietnam War **305**

Chapter 5
War and Humanism:
To Humanize an Inhuman War **318**

Chapter 6
The Korean Soldiers Reflected upon the Americans
and the Vietnamese, and Vice Versa **336**

**PART IV**
**RETURN: HOME** 353

Chapter 1
From Fort Hwarang to Dongjak-dong 355

Chapter 2
The War that has Never Ended: The Living and the Dead 370

**Conclusion: Legacies of the Vietnam War** 383

**Notes** 392
**Glossary** 410
**Selected Bibliography** 414
**Index** 418

# Korean Army Locations During the Vietnam War

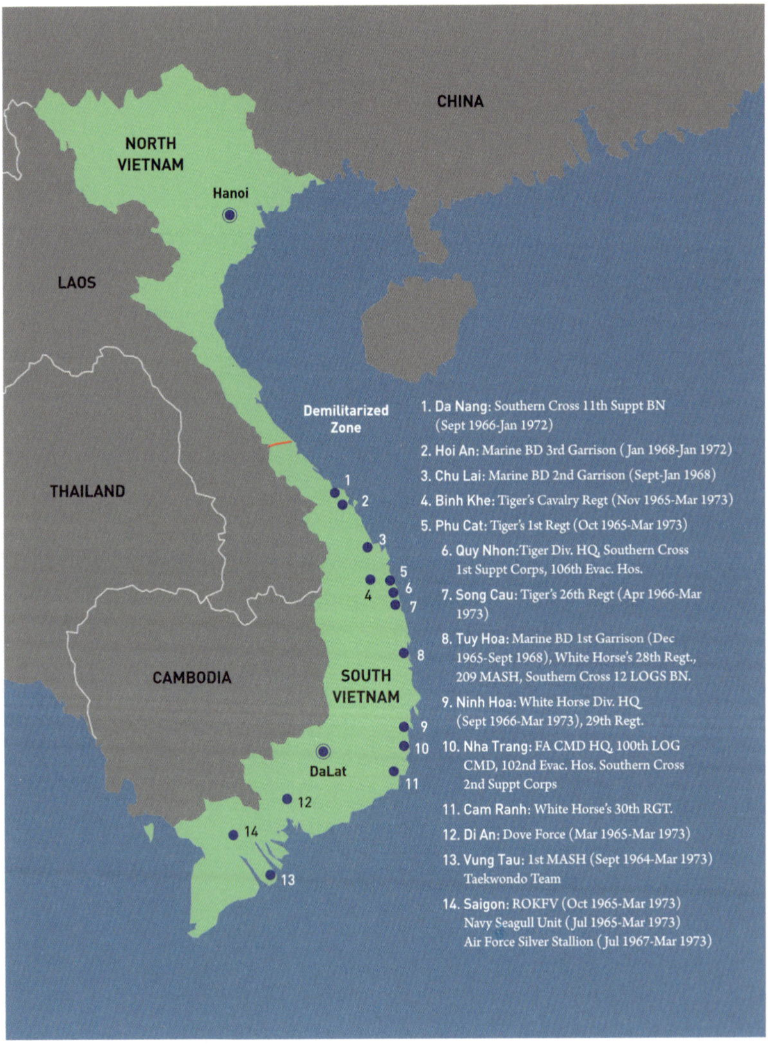


Source: Chae Myung-shin, *The Vietnam War and Me*

# Introduction

The Vietnam War officially ended on April 30, 1975. With the passage of some decades after the war's end, it has almost disappeared from the memory of many Koreans as a forgotten war. But it was not so for the war veterans, especially the Korean Military Academy graduates. From September 1964, when the Korean government began to deploy its troops to Vietnam, to March 1973, when it completely withdrew all of its troops, the Korean army fought alongside the U.S. military and Allied forces for eight and a half years. During and after the war, the numerous combat stories of the Korean military came out in official battle reports, books, novels, research papers, personal memoirs, and many other documents.

However, there is not a single book, in Korean or English, solely devoted to the Korean Military Academy graduates in their entirety. Even though they played crucial roles as chief architects and operators of the Korean army's successful conduct of the war in various positions during its entire period, nobody paid any particular attention to these officers. Thus, I wanted to write about the activities of these KMA graduates apart from the entire Korean Army officers so that international readers, as well as those Allied soldiers, can understand why those KMA men went to Vietnam, how they fought, what they felt, and how they were affected by the war in their long professional career.

This book primarily describes the war experiences of the first seventeen classes from the Class of 1955 to the Class of 1971 of KMA. I call these graduates in this book "the second generation of KMA." The first generation was those commissioned from 1946 (the founding year of KMA) to 1950 (the year KMA was temporarily closed because of the outbreak of the Korean War). This generation of graduates from the first to the tenth class during the five years (1946-1950) was the one whose educational cycle was no more than six months because an unprecedented number of officers were urgently needed to fill the officer corps of the new nation's armed forces, which were being created after the liberation from the Japanese imperial rule in 1945.

As an American Marine veteran of the Vietnam War wrote in his famous memoir *A Rumor of War* (1977), "Every generation is doomed to fight its war,"[1] the first-generation graduates of KMA fought the Korean War (1950-1953), while the second-generation graduates, who had been too young to fight the war, took charge of the Vietnam War. The Class of 1955 (the 11th Class, called such in Korea) was the first four-year class to have graduated with a B.S. degree after the Korean War. The total number of these second-generation graduates from the Class of 1955 to the Class of 1971 was 3,037, of which 2,069 went to Vietnam. Given the total number of 22,339 Army officers committed during the entire war period, the KMA participants were no more than nine percent, yet making a tremendous contribution to the Korean military's successful war effort.[2]

In the sixties and early seventies, the word "Vietnam" was a kind of "albatross" to these second-generation KMA graduates, a metaphor for an onerous burden from which to escape.[3] Indeed, the war was an obsession or "a state of mind"[4] to the graduates. And at the same time, it was also a task for them. Why a burden as well as a task? At that time, the KMA graduates were obligated to participate as regular officers in the overseas war the nation had decided to enter. They had a duty to comply with the nation's call, though their decision still depended upon their own will.

Furthermore, they believed they must experience war as regular Academy graduates pursuing a professional military career. Even so, since the war was a human affair where life and death were at stake, they had an undeniable hesitation as human beings who feared death. This problem worked as a burden on the minds of many graduates. Eventually, about 70% of the total graduates of the seventeen classes voluntarily joined the ranks of Korean troops deployed to Vietnam. They went with the thought that the place where a soldier should be is on the battlefield. While in Vietnam, they never forsook the honor of KMA and the duties that the nation demanded of them.

This book is, in a word, a memoir that sums up the war experiences of all those graduates. With the novelistic tone of a war narrative, this book tries to define their journey to Vietnam as if they were a single protagonist. Although the war stories here are about all the participants of seventeen classes, the actual war experiences taken for each class in this book will only be those of a limited number of graduates. This book is not just a collection of detailed combat descriptions of any individual or class. It is an integrated memoir in which each individual's and class's stories are melted into a whole, thus representing the entire "journey" of the KMA graduates to the Vietnam War.

This book does not discuss complicated international situations, events, the war's chief architects, or the history leading to the Vietnam War. Such events and people--Dien Bien Phu, Robert McNamara, *USS Maddox*, the Tonkin Gulf Incidents, Cold War, Domino Theory, Vietminh, Ho Chi Minh, and other issues of politics--are not this book's concerns. All these matters have been dealt with in many books up to this time. Neither does this book aim to justify Korea's participation in the war or discuss the matter of war's morality. This book tries to describe in an objective way what the KMA graduates saw, did, and felt in Vietnam and what their presence there meant to Korea and its military.

As Tim O'Brien, a renowned American novelist of the Vietnam War, writes, "A true war story is never moral. It does not instruct, nor encourage virtue, nor suggest models of proper human behavior, nor restrain men from doing the things men have always done."[5] I do not want to make this book a moral treatise. Through this book, though, I try to summarize the Vietnam war experiences of the second-generation KMA graduates who responded to the nation's call and speak for them about innocence, patriotism, courage, regret, nostalgia, and the awareness they acquired through the fires of Vietnam in their boisterous youthful years.

The journey structure in this book follows the tripartite vision of war experience in general. Any military expedition, whether in your homeland or abroad, "is governed by a simplified three-part conception of the elements of the process: first, preparation; then execution; and finally critique."[6] In other words, the paradigmatic war memoir comprises three elements: first, the movement to the battlefront; second, the horrific experience of battle; and third, the retirement from the line to a contrasting scene, where there is time and quiet for consideration and reconstruction. This book roughly replicates this process.[7]

This book's three stages are titled in these terms: 1) Initiation, 2) Engagement, and 3) Consideration. But here, one further stage is added with the title of "Return." "Part I Initiation" deals with the KMA as the bastion of national defense, its graduates armed with a sense of mission and honor ready for the war, the processes of leading them to the war, and the aspects of the war. "Part II Engagement" elaborates on how those participating men of each class fought in the jungles of Vietnam. After briefly sketching the history of each class, this part selects less than ten members whose experiences are thought to be different from those of their classmates but representative of each class and describes their individual stories.

And in "Part III Consideration," various controversial issues that arose while conducting the guerrilla war in Vietnam are discussed. Analyzing and evaluating the issues follow methods usually practiced in the criticism of war literature. And finally, "Part IV Return" concerns the homebound journey of the KMA graduates, some dead, but most of them alive. This part focuses on the deceased and bereaved families who miss their loved ones. Though limited to a few graduates, this part describes how the living remember the dead and try to make an effort to sublimate their grief into a meaningful life. This book concludes with a discussion of the legacies of the Vietnam War left to the country and its military, in which the KMA graduates should continue to pursue their careers.

The description of the experiences of the KMA graduates in all these stages is based on their memories. As Professor Viet Thanh Nguyen rightly observes in his outstanding book of memory *Nothing Ever Dies: Vietnam and the Memory of War* (2016), all wars are fought twice, the first time on the battlefield, the second time in memory later years.[8] In other words, the story of a war in which one fought may be a reminder of one's painful experience, occasionally a psychological trauma that he wants to bury forever in the depth of his oblivion. However, as William Faulkner, an American Nobel Prize laureate, writes in his book *Requiem for a Nun* (1951), "The past is never dead. It is not just past."[9] The war memories of the KMA graduates never disappear but reside in their minds. Further, those who experienced combat want their stories to be written so that others may understand the horrible incidents they endured.

Through this book, I want to write down their war stories on their behalf about the Vietnam War, that is, those of the most dramatic overseas expeditionary events in Korean history. I do not want to valorize the experience of the KMA graduates here. Neither do I want to demonize the enemy. As James Webb, an American novelist of the Vietnam War, describes in his widely acclaimed novel *Fields of Fire* (1979), both parties were only "like two scorpions in a jar" to "kill each other."[10] All I desire is that the brutally authentic stories of the KMA graduates and their contribution to the country's war aims shouldn't be forgotten and that due consideration should be given to their dedicated service in Vietnam.

*A Brief Overview of the Vietnam War*

*The Vietnam War or the Second Indochina War (1955-1975)*[11] *dates back to the defeat of the French at the Battle of Dien Bien Phu on May 7, 1954, by the communist-led Viet Minh front, a national independence coalition created by Ho Chi Minh. The negotiations in the Geneva Conference after the year's battle ended the war and recognized Vietnamese independence. The Geneva Accords provisionally divided the country into a northern and a southern zone along the 17th parallel, stipulating general elections scheduled for July 1956 to bring about the unification of Vietnam.*

*The Democratic Republic of Vietnam controlled the northern zone that the Ho Chi Minh regime declared independent in September 1945 and became commonly called North Vietnam, while the southern zone, under the control of the non-communist State of Vietnam, was commonly referred to as South Vietnam. In the South, the anti-communist Ngo Dinh Diem, supported by the U.S., established the Republic of Vietnam in October 1955. Afterward, North Vietnam fiercely criticized the Diem regime for refusing to hold the general election in 1956 aimed at a unified government in Vietnam and forced the South Vietnamese communists to engage in terrorism and guerrilla activities. And North Vietnam forced the National Liberation Front (NLF) to overthrow the Diem regime and wage a revolutionary war to reunify the nation.*

*Meanwhile, when France withdrew its colonial rule from Indochina on April 15, 1956, the United States began actively intervening in Indochina, fearing the domino effect of communist expansion in the region. In particular, the United States was paying more attention to the communization of Vietnam because it had seen China communized and communist attack in Korea a few years before. Then, as the political turmoil in South Vietnam continued, the United States began a military intervention in earnest in the 1960s. The U.S. military intervention, which started with the dispatch of a small military advisory group, established the U.S. Military Assistance Command Vietnam (USMACV) in February 1962 and appointed General Paul D. Harkins as its first commander.*

*Moreover, on August 2 and 4, 1964, the Tonkin Gulf incident occurred when a North Vietnamese torpedo boat attacked the American Navy ship* USS Maddox, *anchored in the Gulf of Tonkin. That led to the United States engaging more directly in the Vietnam War. This incident justified the United States' involvement in Vietnam by gaining Congressional approval and escalated the war in Vietnam.*[12] *Against this move, North Vietnam wanted to collapse the Saigon government before the U.S. ground forces were deployed and sent its regular troops to the south through the Ho Chi Minh Trail, which was opened through Laos and Cambodia, while at the same time expanding support for the guerrillas in the south.*

*Accordingly, the United States bombarded the North and actively deployed ground forces. In April 1964, General William Westmoreland was appointed the commander of the U.S. Forces in Vietnam as the successor to Harkins, and ground forces began to be deployed. In May 1964, President Johnson requested that 25 free countries, including the Republic of Korea, support Vietnam to stop communist aggression in Asia. On March 8, 1965, two battalions of U.S. Marines landed in Da Nang as the first contingent ground troops. After that, the United States drastically increased the number of combat troops. The Korean Army, starting with the non-combat unit in September 1964, deployed two infantry regiments of the Capital Division (Tiger Division) and a Marine Brigade in October 1965. The following year, the Korean government dispatched an additional infantry division (the 9th Division, aka White Horse Division) together with the remaining regiment of the Tiger Division.*

*After years of fighting, the final round of Paris Peace Talks between the United States and North Vietnam started in October 1972, and both countries signed a peace agreement on January 27, 1973. Following the provision that all foreign troops, including the United States and Korea, must withdraw their troops within 60 days of the agreement being signed, the Korean government withdrew its troops entirely by March 23, 1973, thus ending all combat actions in Vietnam. Two years later, on April 30, 1975, South Vietnam's Saigon fell to the People's Army of Vietnam (PAVN) and the Viet Cong. This event marked the end of the Vietnam War, and both Vietnams walked the path of reunifying the country, ultimately becoming the Socialist Republic of Vietnam of today. During the eight years and six months of the war, about 320,000 Korean troops participated.*

# PART I

# INITIATION: INTO THE JUNGLE

Chapter 1

# KMA as the Bastion of National Defense

## A Brief History of KMA

The history of Korea Military Academy goes back to 1946 when it was founded. It was a year after the liberation from Imperial Japan in 1945 and four years before the outbreak of the Korean War in 1950. The Korean War divided the KMA into two different institutions: KMA before the war from 1946 to 1950 and KMA after the war from 1951 to the present. The former KMA was a short-term institution, and the latter was a four-year collegial one. The former's educational cycle ranged from one month to six months for the First to Tenth Classes. The latter started with the Eleventh Class (Class of 1955), who entered the Academy on January 1, 1952, when KMA was normalized as an institution of higher learning, modeled on the U.S. Military Academy at West Point. The five years from August 1945 to June 1950 were a transition period in the formation of the Korean Army and its Military Academy, the most turbulent period ever in the history of 20th-century Korea.

During the colonial period(1910-1945), numerous patriotic young Koreans fled to Manchuria and other parts of China. They fought against Imperial Japan for independence, some as members of the Chinese army and others as members of their military organizations, including the Korean Liberation Army. Several hundred Koreans graduated from the Huangpu Military Academy of China, established by Sun Zhong Shan in 1924, and were commissioned as Chinese army officers and assigned to its units. Many other young Korean men were forcefully mobilized for Imperial Japan's war effort, most as enlisted men and some as officers who graduated from its Army Academy. These Koreans returned home after Japan's surrender and became the human resources for building the national armed forces.

The project of founding the Korean army with all these people from home and abroad was carried out by the U.S. Army Military Government, the then-official ruling body on this land. Its commander was Lieutenant General John R. Hodge, the U.S. Army 24th Corps, who landed at Incheon on September 8, 1945, to disarm the surrendered Japanese forces remaining in the southern half of the Korean Peninsula. General Hodge felt a strong need to create the Korean army in preparation for the official Korean government soon to be established, together with a military institution to produce officers who would become its leaders. And he founded the South Joseon National Defense Constabulary Academy on May 1, 1946, the precursor of the present Korea Military Academy, the bastion of national defense. This date is observed as the founding day of KMA.

## The KMA Graduates: Two Generations

### The graduates before the Korean War: The first generation

The KMA graduates before the Korean War are the first generation. Until the outbreak of the Korean War in 1950, KMA graduated ten classes, starting with the first forty commissioned officers in June 1946.[1] In the tumultuous early years after the liberation, the Academy could not afford to offer a long-term education to these officer candidates because of the pressing need to fill all those army units with young able officers, then in the process of being created on a massive scale. A reason for implementing this policy of short-term education was that most of the candidates were holders of prior military experiences with the Chinese and Japanese armies or the Korean independence army

Forty officer candidates of KMA's First Class and their instructors. They were commissioned on June 15, 1946 after 45 days of training. Until the 9th Class, the education cycle for each was less than six months.(Source: Korea Military Academy)

groups. Thus, the first to the ninth-class candidates were commissioned after education from just 45 days to a maximum of six months. Also, the commissioned ranks varied from second lieutenant to full colonel. Former President Park Chung-hee, a Japanese Imperial Army Academy graduate, is a member of KMA's second class.

However, such a short-term education needed to be lengthened in its duration to foster the elite group of officers for the rapidly growing National Armed Forces. Therefore, in 1949, the Korean Army authorities laid out a new plan to produce regular officers to be educated under the systematic four-year program and also obtained favorable responses from the U.S. Military Advisory Group. Also, they began to discuss producing officers from civilian students rather than military careerists, for these careerists had already been exhausted over time. Thus, from the tenth class in early 1949, KMA was scheduled to start as a four-year institution modeled after the U.S. Military Academy at West Point.

However, the Academy could not realize this plan because of the lack of overall preparation and returned to a short-term one-year program. According to this program, the school accepted 312 students (later called the "Cadet First Class"[2]) on July 1, 1949, followed by the "Cadet Second Class" of 330 on June 2, 1950. Unfortunately, the Korean War broke out on June 25, 1950, just a week before the "Cadet First Class" graduation and three weeks after the "Cadet Second Class" came to the school. Due to the lack of fighting force, the ROK Army War Command decided to put these young cadets into combat and ordered the KMA authorities to form a fighting battalion to deter the invading enemy. So, they organized the 642 cadets into a battalion and committed them to a battle some 20km north of the Academy. Indeed, it was a rare case in world history that officer candidates in training had been put into an actual battle.

The cadet battalion was no match for the well-trained North Korean troops storming down with the overwhelming firepower and tanks ahead. Leaving 151 dead, the cadet troops could only retreat to Daejon in the south, where the Cadet First Class of 223 survivors were commissioned as Tenth Class on July 8, 1950. After the tenth class, KMA could no longer operate as the war intensified, and its closure was inevitable. The Cadet Second Class, who had survived the previous battles, were transferred to Army Integrated School (AIS),[3] established in Busan to rapidly produce junior-level officers badly needed on the war front and were commissioned as second lieutenants on August 15, 1950, after a month's training. The Cadet Second Class, who entered KMA as regular cadets but could not be the men of KMA, remain the most tragic heroes in its history.

Launched as South Joseon National Defense Constabulary Academy four years previously, the Korea Military Academy closed its door on July 8, 1950, with 4,391 graduates from the 1st to the 10th class (May 1946 to July 1950), only two weeks after the war broke out. All first-generation graduates were committed to the war front and dedicated themselves to defending the nation from the attack of the North Korean communists. These officers fought as leaders, commanders, and staff in positions ranging from platoon to division. Among them, 1,220 died in action.[4]

### The graduates after the Korean War: The second generation (Class of 1955 and after)

The Korean War was a watershed event in the history of KMA. Toward the middle of 1951, the fighting was concentrated on the central region around the 38th parallel, and a truce began to be discussed. While the war continued, the Republic of Korea Army authorities considered reopening the closed KMA. Further, there was a consensus that producing regular army officers on a long-term basis should continue despite the immediate need for junior officers for the combat units. Their plan, this time, was to institute the Academy with a four-year course corresponding to that of a civilian college. They submitted the proposal to then-President Rhee Syngman. Aware of the importance of a national academy for fostering elite officers, Rhee immediately approved the plan and sought the assistance of General James A. Van Fleet, then-Commander of the Eighth U.S. Army, who took charge of all combat troops in Korea as a supreme commander of the UN forces. Having been disappointed at the combat performances of Korean troops on the war front, General Van Fleet had also thought of the need to improve the capabilities of Korean officers through education at such a regular military institution, and he immediately accepted the request of Rhee.

Under the direction of General Van Fleet, the Korea Military Assistance Group (KMAG) handled the practical issues regarding the reopening of the Academy and advised its leaders on all matters concerning its operation. This way, Korea Military Academy was reopened on October 30, 1951, at Chinhae near Busan, the southern tip of this peninsula. General Van Fleet was the most significant contributor to establishing the regular collegiate KMA. Brigadier General Ahn Choon-saeng, a relative of the Korean independence patriot Ahn Joong-geun, was appointed its first superintendent.

After some months of preparation under his leadership, KMA selected the first new cadets of 200 in late 1951 and held the historic entrance ceremony for them on January 20, 1952, in the presence of President Rhee Syngman and General Van Fleet. These cadets were members of the Class of 1955 (11th Class), the first class of second-generation graduates. In this book, not all the graduates from 1955 to

The opening parade of the Class of 1955 after their entrance ceremony at Chinhae campus on Jan. 20, 1952 in the presence of President Rhee Syngman, Gen. Matthew Ridgway, Gen. James A. Van Fleet, and many other government dignitaries. (Source: Korea Military Academy)

the present as of 2023 (79th Class) are referred to as second-generation graduates. Only those seventeen classes (1955 to 1971) who graduated under the four-year program and fought the Vietnam War have been dubbed as "the second generation" for convenience to distinguish them from the short-term graduates before the Korean War. No one would have thought then that these graduates of the Class of 1955 would become the pathfinders of all the KMA graduates to be deployed to Vietnam from 1964 to 1973.[5]

Here we need to mention KMA's return to Seoul after the war's end. 1954 was another critical year for the Academy and a step toward the future. Three years after it reopened at Chinhae in 1951, the Academy returned to Taereung[6] in Seoul, its initial site before the war. With the armistice in 1953, KMA returned to Seoul and launched an ambitious project to develop into a better school in terms of system, facilities, curriculum, barracks life, etc. Indeed, most of the practices of today's Academy were founded in that period when it returned in 1954. The historic name *Hwarang-dae* or *Fort Hwarang*,[7] where Hwarangs reside, was designated by President Rhee Syngman at the time. The Eleventh-class cadets who had entered the Academy in 1952 graduated in 1955. The commencement ceremony for the class was held at Hwarang

Parade Ground, constructed in the center of the campus. Since then, the Hwarang Parade Ground has been the spiritual home of all the KMA graduates, where various memorable events are held during their cadet days.

## The Blue White Line

The "Blue White Line," as a continuum, refers to all cadets and graduates who have gone through the Corps of Cadets of Korea Military Academy at Hwarang-dae in Seoul. It is physical and symbolic, similar to the Long Gray Line at West Point. It represents the cadets' formation, whose uniforms are blue and white. Hence the appellation comes from the colors of the cadets' full-dress uniforms. The KMA cadets wore a blue swallow-tailed blouse with a white cross band on their upper body and white trousers and shako for parades and ceremonies. But these days, they wear a blouse with mixed colors of blue, red, and yellow, which the school updated in the late 1970s. Yet, the nom de guerre "Blue White Line" remains a symbol of the KMA and its people. The cadets are proud to belong to the blue white line, a symbol of the elect. It has never failed the people and country in times of war and peace.

As *an ideology* running through all classes of graduates and cadets, the Blue White Line is internalized in their minds and becomes a guidepost in their professional military careers. As an Academy graduate, whoever hears those words is deeply moved by the meaning it conveys and filled with unspeakable pride. All graduates share the same values and ideals that the Blue White Line represents. Even though alumni who get together, elders and juniors, are wide apart in years, they feel a shared identity. They like to talk about the phrase more than the motto—Wisdom, Benevolence, Courage. It comes to their minds as an awakening and shines like a beacon of hope in troubled times. It takes them back on the right track whenever they lose their way and do not know what to do.

The Blue White Line brings all graduates together and allows them to communicate through a mind-to-heart. Those in the line have a

The Blue White Line: Change of command ceremony of the Corps of Cadet's self-governing system in May 1959. (Source: Korea Military Academy)

unique language, behavioral patterns, mindset, and spirit. In a word, these distinctive characteristics can be defined as the Culture of the Blue White Line. This Culture has a typical value that penetrates the mental world of all graduates who have gone through life at KMA. Although they enter the Academy with various motives and backgrounds, they receive the same education and establish the same value system and view of life and country. There is a strict hierarchical order between seniors and juniors. But they share the ideology and values of KMA with a strong sense of solidarity.

Although each class forms a somewhat distinctive culture of its own, it does not stay away from the axis of the blue white line and sustains the shared spirit of KMA. Such a spirit of oneness is blended into the KMA anthem and the Cadet Creed. They all become one when they sing the song and experience the surging emotions inside their hearts. Three lines of the Creed reads: "We give our lives for our country and people; We live in honor and trust, and We choose the difficult path of justice rather than the easy path of injustice." The Cadet Creed has been a guideline for all graduates to pursue their military careers and live by the principle it represents. One can never feel such a sense of sublimity unless he experiences life at KMA.

An example of the oneness of KMA graduates is best shown in its alum association's activities. In 1965, when the Tiger Division, the first dispatching unit to Vietnam, was fighting there, Major Kang Jae-ryun (Class of 1955), the President of the KMA Alumni Association, sent a New Year's message to all the graduates in Vietnam. Emphasizing the significance of the mission all graduates were carrying out, he urged them to do their best for the country so that they might not damage the honor of the Blue White Line specified in the Cadet Creed. The message read as follows:

> The year 1965 is drawing close, and the new year 1966 is about to dawn. At this time of the old year out and the new year in, we wish you all success in war as crusaders of the free world and bulwarks of the ROK Armed Forces fighting in a foreign land far away from home. There is no need to mobilize any adjectives to speak about the significance of your task. We'll not spare any huge but silent applause and support for you all.
>
> We remember the old injunction, 'Our best way is to fight and come back alive courageously, and our worst way is to cowardly fight and come back home dead.' We all had determined this when we sent you off, and you sailed for Vietnam. Let's all do our best with this resolution to achieve our goal. We do wish your sweat and blood will not be shed in vain and yield valuable fruits so that the unification of our country and peace in Vietnam may be accomplished as soon as possible. God bless you all![8]

## The KMA cadet as the Modern Version of Ancient Hwarang

A Hwarang (花郎) is the name of an ancient warrior of the Silla Kingdom, one of the three kingdoms--Silla, Baekje, and Koguryo--having existed in this region of the Korean Peninsula, including Manchuria from circa 57 B.C. to 935 A.D. As an elite group of youths, the Hwarangs of Silla played a critical role in unifying the three kingdoms. A Hwarang, meaning "a handsome young man like a flower," was very strong in *esprit de corps*. A KMA cadet is a modern version of this ancient Hwarang. These Hwarangs had a social, educational, and military function in the politics of the Silla Kingdom. Fort Hwarang is where these young Hwarangs of today are trained and educated. As the bulwarks of the nation's defense, they are the regular officer candidates to man the Korean Army as future leaders and will be the ones to bear the burden of the nation's destiny. Fort Hwarang is called in Korean "Hwarang-dae," a prominent place where Hwarangs reside. The Blue White Line is the proud formation of these modern Hwarangs.

### The Spirit of Hwarang and the "KMA Man"

What is the spirit of Hwarang? Who is the "KMA Man"? What makes him the man he ought to be? First of all, the overarching term referring to all the graduates and cadets of the Academy is "the KMA Man," which is a physical entity and an abstract idea comprising all the values that the Academy represents and cherishes. In a word, the KMA man is the one who embodies the spirit of Hwarang. The spirit of Hwarang is embedded in the "Five Rules to Observe in Life," by which the ancient Hwarangs had to abide. The "Five rules" are:

1. To serve the prince with loyalty.
2. To perform filial piety.
3. To make friends with faith.
4. Not to back down in battle.
5. To be sensible when taking a life.

In the modern sense of the words, they are roughly similar to the core values that the soldiers are required to follow: (1) Devotion to the country and people, (2) Fulfillment of the responsibility to complete the assigned task without giving up, and (3) Mutual respect with trust and confidence. The latent meanings in these lines correspond to those in the Cadet Creed. The KMA Man is educated to be a whole man equipped intellectually, morally, and physically. This idea of being "the KMA Man" is a concept that runs through the minds of all KMA graduates. Sometimes he likes to say that a cadet [KMA Man] is a "Superman."

In his article "The KMA Man and the Cadet Creed," contributed to *The KMA Press* in July 2015, Lieutenant General Park Jong-sun (Class of 1978), then-Superintendent, KMA, emphasized the significance of the Cadet Creed and advised the cadets on what they should do. He said that all the spiritual values of the KMA are summed up in the Cadet Creed. Above all, he noted that about 1,220 graduates sacrificed themselves to defend their country in the Korean War and 36 in the Vietnam War. Referring to the term "the KMA Man," he advised all the graduates to remember and keep the Cadet Creed in mind. According to him, the Creed is a commandment that teaches the KMA Man what to live for and how. He appealed to all cadets' and graduates' hearts and minds for their proper behavior appropriate for the "KMA Man."[9]

The spirit of the KMA Man, an inherited spirit of Hwarang, is the one to be formed through the long period of four years at the Academy. The KMA graduates feel proud that they successfully withstood the harsh life imposed upon them during their Academy days. All the traditions and practices of the Academy have contributed to making the KMA Man a modern Hwarang. The KMA Man likes to say that he is a Hwarang, a member of the *Blue White Line*. Armed with all these ideas, values, and pride, the KMA graduates of the first seventeen classes plunged into the jungles of Vietnam.

# Chapter 2
# The Road to Vietnam

## The Republic of Korea's Entry into the Vietnam War

On May 9, 1964, President Johnson of the United States sent an official letter to 25 allies, including the Republic of Korea, requesting support for South Vietnam. President Park Chung-hee ordered Defense Minister Kim Seong-eun to review the matter. Under the direction of President Park, the Ministry of National Defense organized the "Korean Military Aid Team" and prepared a plan for dispatching a 130-person mobile surgical hospital and a Taekwondo instructor group of 10 officers under the hospital command.[1]

With the approval of some proper government authorities, the First Korean Mobile Army Surgical Hospital (MASH) and Taekwondo instructors boarded the Navy LST on September 11, 1964, departed Busan, and arrived in Saigon on September 22 after ten days of voyage and were stationed at Vung Tau near Saigon. That was the first dispatch of Korean troops to a foreign country in Korea's history. By then, nobody expected that the dispatch would be just the beginning of Korean troops' eight-year-long journey to Vietnam. The Taekwondo instructors were sent to South Vietnam's Army, Navy Academies, and Army Infantry School to teach the martial arts.

However, one thing to note here is that Major General Sohn Hee-sun, JCS, Operations, visited South Vietnam in early January 1965 to discuss the deployment of Korean troops to Vietnam. He met

General William Westmoreland, Commander of USMACV, and South Vietnam's Prime Minister, Nguyen Cao Ky. In the talks with them, he sensed that "there is no will to fight" in the mind of the South Vietnamese military, and "it is the Americans, not the Vietnamese, who are guarding Vietnam as its owners." He recalled later, "He had made up his mind that Korean young men should not make any reckless sacrifices without a just cause in the foreign land."[2]

When the Gulf of Tonkin Incident occurred on August 2, 1964, the Vietnam War unfolded in a new way in which the United States came to the fore. Then, on December 18, 1964, the U.S. government again requested the Korean government to dispatch additional non-combat units. The size of this 2nd dispatch requested by the U.S. was a construction support group of 1,000 troops. Given that they would have to carry out their construction work amid the enemy's constant threat, the Korean government judged that they would need an additional infantry battalion to guard the constructing unit and for self-security. So, it expanded the number of troops even to 2,000. On January 29, 1965, with the consent of the National Assembly, the government established the Korean military aid group in Vietnam and named the group the "Dove Force," a symbol of peace. The aid group was a mixed force composed of an Engineer battalion from the Army and Marine Corps, a security battalion, a transport company, a Navy LST transport squad, and the MASH and Taekwondo instructors previously dispatched. After arriving in Saigon on March 10, 1965, the group moved to the garrison at Di An.[3]

After the Gulf of Tonkin Incident, the United States was mulling over the commitment of ground troops, and its final decision was to send them to Vietnam. Starting with its first ground combat unit landing in Da Nang on March 8, 1965, the U.S. was mired in the longest war in its history. Before and after the commitment of the U.S. ground troops to Vietnam, the deployment of Korean combat units was also a hot issue in Korea. But the Korean government was already making this decision internally, "If the U.S. requests, we will send our combat units." Eventually, the government formalized the dispatch of Korean combat units through the two ROK-US summit meetings in May 1965.[4] This decision was reinforced by South Vietnam's Prime Minister Quat's

Chae Myung-shin (left), CG of the ROKFV, conversing with Gen. William Westmoreland, CG of the USMACV, and Gen. Dwight Beach, CG of the Eighth US Army Korea, in 1966. (Courtesy: Chae Myung-shin and ROK Army Museum)

letter to his Korean counterpart on June 14, 1965, requesting the dispatch of one division-size combat unit. At the official request of the two governments, the Korean Defense Minister announced that Korea would dispatch one infantry combat division (including one Marine Brigade) to Vietnam.

After a serious debate, the National Assembly officially passed the resolution on August 13, 1965, selected the Army's Capital Division (-1) and 2nd Marine Brigade as the first combat units, and appointed Major General Chae Myung-shin (KMA 5th Class) as its commander. Chae had already completed his tour as the division commander at home. After completing a month of training, the Capital Division landed in Vietnam and started taking over the operational area from the U.S. military. The 1st Regiment of the division took over the Area of Responsibility from the South Vietnamese government on November 3. And the Cavalry Regiment from the U.S. 101st Airborne Division and the U.S. 7th Marine Regiment on November 14. From November 15, the Capital Division began to work on crushing the resistance of the Viet Cong and NVA in its TAOR, which spanned 1,200km$^2$ with Quy Nhon at its center.

By the end of 1965, the number of U.S. troops in Vietnam had reached 184,000, but the situation was getting worse. As the casualties of U.S. soldiers increased, the anti-war public opinion within the country spread, and the U.S. requested an increase in the number of Korean combat units as a way of subduing the anti-war opinion in the U.S. while adding to the international character of the Vietnam War. The Command of the Republic of Korea Forces in Vietnam (ROKFV) also needed to bring the Capital Division's 26th Regiment, which had remained in Korea, thus making the Capital Division complete as a standard organization. At the same time, the Korean government already judged that "up to 50,000 strength was not unreasonable" and that there was no difficulty in increasing the number of divisions by one. With its additional deployment of troops to Vietnam, however, the Korean government needed to reap economic benefits and the U.S. firm commitment to Korea's national defense.

In the end, the U.S. promised the Korean government a firm defense commitment and economic aid through the then-U.S. Ambassador to Korea, Winthrop Brown, in the Brown Memorandum. The main contents of Military and Economic Assistance in it were:"support for the modernization of the Korean military," "paying for all equipment and expenses required for dispatching Korean troops to Vietnam," and "the U.S. government's economic support to Korea." On the other hand, members of the U.S. Congress denounced these promises, severely criticizing the U.S. decision as one that brought Korean troops into the American war as mercenaries.[5] This blunt criticism triggered the rise of "the theory of Korean military mercenaries" both in the U.S. and Korea. With the approval of the National Assembly, though, the government selected the Capital Division's remaining 26th Regiment and the 9th Infantry Division for reinforcements. It appointed Major General Lee So-dong (KMA 2nd Class) as the 9th Division Commander. After training for about two months, from June 20, the two units were deployed to Vietnam by October 10, 1966.

## The Korean Combat Troops sent to the Vietnam War

In addition to the articles contained in the Brown Memorandum, the Republic of Korea's decision to join the Vietnam War resulted from various underlying causes: to show the ability of the ROK forces to conduct actual combat to the world and elevate the status of the Republic; to enhance international confidence by helping the friendly country which shared the same ideology; to repay the sacrifices Americans had made during the Korean War; to be a breakthrough for national economic growth and a chance to encourage the sending of the nation's workforce overseas. Public opinion favored the government's decision despite strong opposition to the dispatch of the combat troops. General Chae Myung-shin reflected, "If we hadn't dispatched combat divisions, perhaps the U.S. would have withdrawn its 2nd and 7th Infantry divisions from Korea to deploy them to Vietnam. Then we would have had difficulty defending our country from the aggression of North Korea, whose military power then was far superior to ours."[6]

Thus began the long journey of the Korean military to the Vietnam War for eight and a half years. No one could have imagined the war would be extended for so long as such at that time. On October 20, 1965, the Command of the ROKFV was established in Saigon to command all the Korean units. Total Korean troops deployed to Vietnam throughout the period amounted to 324,864, resulting in 5,099 dead and 10,962 wounded.[7] With the last unit leaving Vietnam on March 23, 1973, the expedition of Korean forces to the war came to an end. During their presence in Vietnam, three combat units (Tiger Division, White Horse Division, and a Marine Brigade) executed as many as 1,175 large unit operations above the battalion level. And they conducted 576,302 small unit operations below the company level, including innumerable "search and destroy" missions and security operations.

While combating Viet Cong and NVA, Korean troops helped build schools, roads, and bridges. Medical units treated tens of thousands of South Vietnamese civilians. During the entire period, four corps-level operations were jointly conducted by Tiger and White Horse divisions. After the Tet Offensive on January 30, 1968, the Allied forces' operations were atrophied due to promoting peace negotiations with the Communist forces. Therefore, the Korean military refrained from large-scale operations and switched to small-unit-oriented operations. During the entire period of its presence, the Korean army took action under the creed called "ROK Army Creed in Vietnam:"[8]

Let's become courageous and fearful Korean soldiers to the enemy.
Let's become polite and kind 'Diahan' to the Vietnamese.
Let's become disciplined and faithful Koreans to the military of our allies.

## The Garrisons and TAORs of Two Combat Divisions

### The Capital Division (Tiger)

The Capital Division's First and Cavalry Regiments set foot in Vietnam on October 22, 1965, as Korea's first expeditionary combat troops. Its remaining 26th Regiment joined them in April next year. The Korean troops were mainly stationed around Route 1, running along the coast from north to south below the 17th parallel. Some prominent regional locations were Quang Tri, Da Nang, and Hoi An, all coastal cities. The USMACV then wanted the Korean forces to be stationed inland up-country. But the Korean military authorities insisted that they place their troops in densely populated areas along Route 1 in the east to unify the military and civil operations. Finally, the USMACV accepted the Korean proposal. All Korean forces were stationed in the regions near the east coast except Dove Force at Di An, 22 km from Saigon to the north.

Quy Nhon, where the Tiger Division was deployed, was a highly concentrated port city with American logistics installations, supply depots, and harbor facilities. Located in the middle part of South Vietnam, Quy Nhon in Binh Dinh Province was a strategic point where Route 1, running north and south along the coast, intersected Route 19, leading to the Cambodian border through An Khe Pass and Pleiku to the west. Before the Tiger Division arrived there, the main forces of VC entrenched themselves on the rocky hills and natural thorny caves in the jungles of Nui Phu Cat. They controlled this entire region, constructing their strategic hamlets in the Go Boi and Phu My plains which were the sources of their food supply. Even Quy Nhon harbor had been in their grip. The Tiger Division built strong tactical bases, set up night ambushes on all passages from Nui Phu Cat to Quy Nhon, destroyed the infiltrating Viet Cong, and protected the vital arteries of

South Vietnam, including Route 1, Route 19, and rice-growing paddies and foothills.

By June 1966, the Tiger Division controlled the entire area north of Quy Nhon to the east of Route 1 and up to the base of Nui Phu Cat. And the Division also extended its control over the area along Route 19, leading to An Khe Pass and bringing peace to this entire region. The Tiger Division mainly carried out all its operations along Route 19 and took charge of the sector, covering 25km from Route 1 to the south of Quy Nhon. The First Regiment was deployed to Phu Cat near the Division headquarters, the Cavalry Regiment to Binh Khe, and the 26th Regiment to Song Cau to the south. The Division took the vast expanse amounting to 1,400 km$^2$ as its TAOR. Throughout its presence in Vietnam, the Tiger Division carried out 517 times above battalion-size operations and as many as over 174,000 small scale ones.[9]

### The 9th Division (White Horse)

Deployed in September 1966, a year later than the Tiger Division, the 9th Division (White Horse) was requested by the USMACV to station in the remote area to the far west bordered by Cambodia. But the ROKFV suggested that the division be located in the coastal area

The arrival of the advance unit of the White Horse Division in Cam Ranh in August 1966 (Source: Institute for Military History, MND)

along Route 1 to the east. The reason was that the distance between the three Korean combat units would be so far apart that their command and control might be difficult and ineffective. After consideration, the U.S. authorities accepted the Korean suggestion, and the White Horse Division was stationed in the coastal area from Tuy Hoa via Nha Trang, leading to Cam Ranh and Phan Rang to the south. Next to the Tiger Division in Quy Nhon to the south, this location was a vast expanse in the region, a total of 270 km in length and 50 km in width. This area was strategically critical because it was highly concentrated with harbors, airfields, and other military installations.

To the west of this region lies Nui Hon Ba (called "Finger Mountain" by Americans), a mountain standing high with a small summit, steep slopes, and local relief of 1,361 meters. The mountain is composed of rocky hills overlooking the entire region and contains numerous natural caves connected like cobwebs to its depth. The enemies sophisticatedly used these natural obstacles and labyrinths to their advantage. In particular, many of the enemy's strong points on the mountain denied the straightforward approach of human beings and any aerial strikes or artillery fire. Those strongholds had to be suppressed one by one only by reconnaissance patrol parties. The 28th Regiment was based in Tuy Hoa, the 29th Regiment in Ninh Hoa with the Division headquarters, and the 30th Regiment in Cam Ranh. At the same time, with the dispatch of the White Horse Division, the ROKFV expanded the Logistics Support Command under the Capital Division into the 100th Logistics Command (aka the Southern Cross Unit) to support combat service for all units of the Korean Army.

In the initial stage, the White Horse Division's primary mission was to clear and eliminate the enemy's strongholds on Nui Hon Ba and the surrounding area west of Route 1. At the same time, it was to protect the Tuy Hoa plains' breadbasket and further enlarge the pacified area for the South Vietnamese government. In a word, the TAOR of the Korean forces in Vietnam extended from Quy Nhon of the Tiger Division to Phan Rang, the southern tip of the White Horse Division, amounting to a total of 400 km of distance. Six months after its arrival, the White Horse Division conducted Operation Ojakgyo Bridge, a corps-level operation, with the Tiger Division for a month from

March 8, 1967. It opened Route 1, the main artery of South Vietnam, from Quy Nhon to Tuy Hoa. The latecomer, White Horse Division, executed a total of 474 large-scale operations and over 211,000 small-scale operations during its stay in Vietnam.[10]

## The KMA Graduates and the Vietnam War

As mentioned above, the second-generation graduates from the Class of 1955 (11th Class)[11] were commissioned after the Korean War. None among this later generation fought the Korean War except for some members of the Class of 1955 and 1956. They were the enlisted men in the field units before being appointed to the newly established KMA in 1951 and 1952. When the Republic of Korea began to deploy its combat troops to the Vietnam War in 1965, the ranks of the Class of 1955 were Majors. Most of them were serving on the level of regimental staff. Of the total 3,037 graduates from 1955 to 1971, 2,069 men, about 70%, joined the ranks of the Korean army's presence in the Vietnam War. (The total number of second-generation entrants from 1955 to 1971 was 3,912, of which 3,037 cadets graduated. The average attrition rate was 22.4%).

The Vietnam War was the war that these second-generation graduates had to fight. These officers gained combat experiences and returned home with many valuable lessons that would serve their military careers and the development of the ROK Army. In the case of graduates of five classes--1962, 1964, 1965, 1966, and 1967--85% of each class's graduates went to Vietnam. Among the Class of 1971, who graduated towards the end of the war, only 22 men had the chance to participate as the last class members of the KMA. Yet the total number of KMA's Vietnam veterans was less than 9% of the 22,339 Army officers who served during the entire war.[12]

For the first three years, from 1965 to 1967, about 500 KMA graduates served in Vietnam. The positions of these men were mainly divisional or regimental staff for the classes of 1955 and 1956, company commanders or battalion staff for those of 1959 and 1960, and company executive officers or platoon leaders below them. The lowest KMA Class fighting in Vietnam in 1967 was the Class of 1965, all second lieutenants, most as platoon leaders. For instance, in the case of the First Regiment of the Tiger Division in 1966, all of its company commanders, staff, and platoon leaders were KMA graduates. This number amounted to about 70% of all of its officers. Also, for the Cavalry Regiment of the Tiger Division, 12 out of 14 company commanders were KMA graduates, and 70% of their platoon leaders were also KMA graduates.

In the early days of Korean deployment to Vietnam, the roles of KMA graduates were crucial in the settlement and operation of the Korean troops.[13] The highest rank of KMA graduates was a colonel

The Number of Vietnam Veterans: Class of 1955 to Class of 1971

| Class | Entrants | Graduates | Viet Vets |
|---|---|---|---|
| 1955 | 200 | 156 | 101 |
| 1956 | 200 | 164 | 94 |
| 1957 | 250 | 175 | 110 |
| 1958 | 243 | 179 | 130 |
| 1959 | 219 | 176 | 126 |
| 1960 | 214 | 178 | 131 |
| 1961 | 203 | 177 | 132 |
| 1962 | 233 | 181 | 153 |
| 1963 | 229 | 205 | 148 |
| 1964 | 218 | 156 | 137 |
| 1965 | 216 | 178 | 145 |
| 1966 | 234 | 181 | 160 |
| 1967 | 235 | 177 | 145 |
| 1968 | 252 | 180 | 115 |
| 1969 | 254 | 203 | 126 |
| 1970 | 256 | 183 | 94 |
| 1971 | 256 | 188 | 22 |
| Total | 3,912 | 3,037 | 2,069 |

Sources: Office of Evaluation, KMA; The Office of ROK Army Records and Information Management.

during the entire war period. Only five officers from the Class of 1955 rose to the rank of full colonel and assumed the positions of regiment commanders, division chiefs of staff, or deputy chief of staff for the command higher than division. They were Chun Doo-hwan, the 29th Regiment commander of the White Horse division, and Park Gap-Yong and Song Sung-han, both Chiefs of Staff of the White Horse division. Park Sung-hwan was the 26th Regiment commander of the Tiger Division. And Lee Seok-myung served as Deputy Chief of Staff, Intelligence of the Field Command of the ROKFV in Nha Trang.

# Chapter 3
# Why They Went to Vietnam

There should be a just cause in war-making. Only a reasonable cause can persuade the people of a nation. Those told to go to war ask, "Why do we have to fight?" "Is the war worthwhile?" Usually, there are two reasons for going to war. One reason is that an individual must comply with the nation's call because their government orders them to fight. When one's nation is under attack from another country, they should be there to defend their own country. Or, when one's nation participates in a foreign war, one can go to war if asked by the government. In this case, the country should properly judge whether its participation benefits the national interest. The war in Vietnam was the latter case. When the Korean government decided to send its troops to Vietnam and designated the Capital Division as the first expeditionary force, most of its organic members of the units were included automatically. However, even in this case, officers were allowed to decide whether to go, depending on their free will.

The KMA graduates knew that the Vietnam War was guerrilla warfare rather than a war in which hundreds of thousands of soldiers had died, such as in World War II or the Korean War. The war didn't appear as dangerous as the past wars they had read about in the books or listened to tales of their elders. As Philip Caputo defined the Vietnam War as "the splendid little war"[1] in his book *A Rumor of War* (1977), the KMA graduates thought it a dangerous but "small war," not unlike how the American soldiers had at first thought about the Korean War.[2] To them, it was such a war as the one they should go through on their path to professional careers. They thought it was essential to gain combat experience as a KMA graduate. They wanted to test the tactics and skills they had learned at the Military Academy in such a small war. Also, they imagined themselves leading their men and charging into the enemy position. There were many ups and downs before making the final decision, but in the end, they joined the KMA ranks going to Vietnam.

## To Have Combat Experience as a Professional soldier

Captain Kim Jin-gyu (Class of 1958, later an Artillery Brigade commander as a one-star general) participated in the war as the 3rd Battery commander of the 61st Artillery Battalion of the Tiger Division. Before making a final decision, he underwent the process of asking some questions to himself about the war's justification. To a Military Academy graduate, real combat experience was essential. The point, however, was a "cause," that is, "Why do I have to go to Vietnam?" and "Is it a just war?" He was caught up with these questions for three whole days.

At last, the desire to experience actual combat won over the doubts about the cause. However, his view of life and death mattered when the desire became a reality. An only son, he had a widowed mother and wife with two children. Indeed, it was regrettable that he had not set up a determined view of life and death as a professional soldier. Any religious belief was not a help to him. He thought about whether to toss a coin for a decision. Finally, he decided to entrust the matter to his destiny and joined the ranks going to Vietnam. For a regular army officer who graduated from the Academy, actual combat experience was more important than any other personal matter. Such a thought prevailed over the conflict in his mind.[3]

Another graduate of the 1958 Class, Major Park Jung-ki (later, President of Korea Electric Power Corporation and Council Member for the International Association of Athletics Federations), joined the ranks of Korean troops' deployment to the Vietnam War with a half-hearted mind. After completing his OAC course at the U.S. Army Air Defense Artillery School in Fort Bliss, Texas (now at Fort Sill, Oklahoma), he served as the Air Defense Artillery

Major Park Jung-ki (center, Class of 1958) after a combat operation in the jungle near Tiger Div. HQ at Quy Nhon in March 1969. (Courtesy: Park Jung-ki)

battery commander in Daegu, Korea. At that time, the Vietnam War was in full swing, and many of his classmates had already been to Vietnam, so he had a sense of duty that he should also go someday to have combat experience as a professional soldier. But, one day, he received a call from Lieutenant Colonel Sohn Young-gil, a 1955 graduate, a three-year senior to him, who was to go to Vietnam as Tiger Division's G-3 Operations, suggesting that he go as his assistant.

Major Park did not think he was the right person for the infantry-oriented division's Operations Office because his branch of service was artillery. Moreover, it was such an unexpected proposal that he wasn't prepared to accept it in his mind. But those things flashed through his head like lightning for a moment: the battlefield, danger, death, and family. And at the same time, something like combat experience, medals, promotions, and honors also began to occupy another corner of his head. He was married and had a wife and two children. The problem was to convince his wife, who never thought her husband had any business with Vietnam as an Air Defense Artillery officer. What made her more ill at ease was that there had not been a word of Vietnam before. She objected to his going. He quarreled with her for a few days. Eventually, his wife was exhausted. Knowing that he was so determined, she agreed. It was a resignation rather than an acceptance. She seemed to think it was her destiny as a soldier's wife. He eventually boarded the airplane leaving for Vietnam in September 1968.[4]

No matter how necessary a war experience for a professional soldier may be, they are a person before being a soldier. Every human being, whoever they are, is supposed to be hesitant in the face of death. This also happened to Captain Park Jong-sik (Class of 1959, later, Chief of the Military Instruction Office as a two-star general). During his four years at the Military Academy, he shouted the Cadet Creed at morning and evening roll calls: "We will give our life for the country and people." As an 18-year-old high school graduate, he did this without thinking deeply about what the line meant. He was still determining whether he was qualified to assume the task embedded in the Creed. He was not mature enough to contemplate a soldier's view of life.

However, when ordered to go to war, he had to contemplate his death seriously as a 29-year-old officer. He knew death always follows an infantry captain who must fight at the forefront. Even Confucius said thousands of years ago, "I still don't know what life is. How much more can we know about death?" referring to death as incomprehensible. Then how could a novice like him understand the depth of life and death? But because he was a soldier going to war, he had to be humble in the face of death. He felt he had no choice but to turn to religion to relieve the fear of death. As human life is believed to depend on the mandate of Heaven, he decided to leave his death entirely to God. He chose to end his three days of contemplation about his death and finally go to Vietnam. Once he decided, he was brave enough to argue that he wanted to be an infantry company commander. After some twists and turns, he ended up the 11th Company commander of the 28th Regiment of the White Horse Division.[5]

At a round-table talk held at KMA one day in early 1967, some officers, who had already returned from Vietnam, shared opinions about their war motives: to have combat experience as professional soldiers. When the Tiger Division was designated as the first combat unit to be deployed, most KMA graduates volunteered. Major Ahn Seung-bong (Class of 1955), Assistant to G-1 Personnel, Tiger

Division, said, "The absence of combat experience was one of the weak points for the KMA graduates because there had not been a war up until the one in Vietnam. So, in consultation with Major General Chae Myung-shin, the Division commander, he included as many KMA graduates as possible in the division officer corps. Captain Lee Kwon-bok (Class of 1960), the 5th Company commander of the Cavalry Regiment of the Tiger Division, said he wanted to command his men on the battlefield and show his bravery to the world.

The same was true of Second Lieutenant Shin Jung (Class of 1964, later a division commander as a two-star general), who had returned home after a one-year tour as platoon leader of the Tiger Division. "As a KMA graduate, I have always wanted combat experience," he said. "When I heard that the government would send our troops to the Vietnam War, I thought it would be an excellent chance to prove myself on the battlefield and volunteered right away."[6] Lieutenant Shin Jung, who got married on his graduation day, served as a platoon leader in the DMZ in Korea before being appointed to the Capital Division deployed to Vietnam. He didn't even notify his bride of his decision, fearing that she would be shocked. Shin informed her of his going after everything was settled.

In a letter to his wife in Vietnam, later published in his memoir *The Wall of Jungles* (1967), Lieutenant Shin explains why he came to Vietnam, "My Darling! I am determined to sacrifice my life for my country and my subordinates fighting in the jungle, waiting for the day to return to their homeland, even though I cannot promise to give them 'tomorrow.'"[7] This remark reminds me of the one made by a U.S. Marine soldier at the Battle of Changjin Reservoir in the Korean War in December 1950. Asked by war correspondent David Duncan about "what is your desire right at this moment," the soldier replied, "Give me tomorrow." "In the coldest weather like this, what kind of desire would be helpful when death was coming any time?" The soldier meant that it would be fortunate enough if only he could stay alive until tomorrow.[8]

In the case of Second Lieutenant Kim Hyung-seok (Class of 1965), an only son of his family, his chief concern was persuade his parents about his decision. The process of his persuading them was dramatic. Upon graduation from KMA in 1965, he was assigned to a division in the DMZ as a platoon leader. One day in early 1966, Vietnam suddenly came to his mind. The Tiger division had landed in Vietnam the previous year and was engaged in fighting. Some of his classmates had joined the Tiger Division and fought in the jungles. Already news of the casualties began to be heard from Vietnam. Among the dead were several of his Academy seniors. However, Kim was not perturbed by the news at all. "They sacrificed themselves for the country as a soldier," he thought, "though their families would grieve over their deaths." And he was incredibly proud of them, wondering how they could courageously confront the horrors of war. While volunteering for the service in Vietnam, he comforted himself that he had no business with death and that nothing would happen to him.

Some officers and men made every effort to have their names dropped from the list of troops to be deployed. But as a professional soldier, Kim had already made up his mind. Yet the point was to inform his parents of his decision and obtain their agreement. Before being transferred to the White Horse Division, he got a week's leave and went to his home in a remote rural village. Rather than the pleasure of going home, he was heavily weighed down with the thought of how to persuade them. His mother was surprised to see him suddenly appear yet filled with all smiles and happiness. He entered the living room and bowed deeply to his parents by kneeling on the floor, as Korean people usually do to their elders after a long absence. Though they did not ask him for any reason for his coming, they vaguely guessed it through his sad facial expression. His mother opened her mouth calmly, "How could you come home when you must be busy with your unit?" He couldn't answer immediately and instead went out to see the places he had used to hang around while growing up. Just at that moment, the sun was setting to the west. When he got to the beach that he used to run around in childhood he could overlook the sea dotted with tiny islands lining up like a folding screen. He felt as if the warm and cozy surroundings embraced him. He said to himself, dominated by nature's splendor, "Will I be able to see this magnificent sunset again here?"

He made it clear why he had to go to Vietnam. Besides the government's official war aims, he had other personal reasons. He wanted to go overseas: to see what the country would be like there, to test his manhood on the battlefield, to recognize his potential as a

KMA graduate, and to contribute to improving his household's poor living conditions.[9] At the time, a junior officer could save $120 monthly from his salary. He came to his house with these personal interests in mind and sat with his parents. After listening to his explanation, his father said, "Well, I understand. But what's important is survival. Isn't it? I'm hearing a platoon leader is the most dangerous." Lieutenant Kim responded, "I'll return safely in just a year with this perfect body of face, hands, and feet." Leaving his seat quietly after a while, his father heaved a groan, "We-ell, be careful." His mother opened her mouth with difficulty, "If you're sure to go, you have to go. What else can we do? Do what you think is right." His parents, who belonged to the generation of the Korean War, were well aware of how cruel and tragic the war was. He tightly held his mother's slender hands and said, "Mom, I'll return safely. Don't worry too much. I promise I won't disappoint you. Your son isn't that weak. You've always had confidence in me, haven't you?"[10]

Second Lieutenant Choi Young-ha (Class of 1966, later a one-star general and Korean ambassador to Uzbekistan) believed that it was "obligatory" to go to the battlefield as a professional soldier. This was the factor that led him to the Vietnam War. But he had a single mother, and he was the only son. He had to live with his old mother, who could not stand alone. Who would care for his mother if he went to Vietnam? That was the thing that weighed on his heart the most. In other words, should he fulfill his duty to go to war as a soldier or stay here as her only son? There was no correct answer. After all, these values--country, duty, soldiering--prevailed over the meanings of personal relationships. But his mother told him to go because she knew her son's convictions too well.[11] The same was true of Lieutenant Noh Nam-seop (Class of 1967, later, a division commander as a one-star general). He decided to go to Vietnam to have combat experience. In his autobiography, he says, "Because a professional soldier needs combat experience, I decided to participate in the Vietnam War." Then his mother gave the following message to her fourth son, who was leaving for the battlefield. "My son! I am proud to hear that you will participate in the Vietnam War. Please take care of yourself. I pray that God will protect you until you return home safely."[12]

Second Lieutenant Paik Nam-hwan (Class of 1969, later a division commander as a one-star general) went to the Tiger Division in June

1970. After graduating from KMA in 1969, he served as a platoon leader in the 2nd Infantry Division in Korea. What led him to Vietnam? He thought that even if he became a general someday, he would not be qualified for it without combat experience. Years later, he recalled, "some intellectuals in Korea agreed to the claim that the soldiers who had been to Vietnam were American mercenaries. I was indeed disappointed when I saw the innocence of the Korean soldiers who responded to the nation's call as being disparaged."[13]

The motives of the Class of 1970 graduates were no different from those of their elders. They also had a powerful sense of duty that, as the country was involved in the war, they had to fight it. Moreover, it was essential for a regular officer who would follow the path of a professional soldier to have combat experience. The battlefield presupposes death, but they did not think they would die. Sigmund Freud once commented about combat psychology in war. To the question, why do men in combat charge toward the enemy position facing the fierce fires, defying danger and fear of death? He answered that they think instinctively, "Nothing happens to me."[14] Indeed, the possibility that they might not return to their homeland did not occur to them even once. They did not know what kind of country South Vietnam was, what language the Vietnamese spoke, or what their staple food was. They didn't even know where South Vietnam was in Southeast Asia. They wanted to test their will and leadership to become professional soldiers.

The motive of First Lieutenant Oh Hyun-koo (Class of 1970, later 1st Corps Commander as a three-star general) was more concrete than that of other classmates. What moved him to Vietnam was a book by Lieutenant Colonel Park Kyung-seok (Cadet Second Class) that he had read during his Academy days. Park, a battalion commander of the First Regiment, the Tiger Division, published a Vietnam War book *Route 19* (1966), in the early phase of the Korean deployment. In the book, based on his combat experiences in Vietnam, he pointed out that even a man who had been courageous under normal circumstances didn't know how to act in an urgent situation during a firefight. Thus, Park concluded that the man who did not get shot on the battlefield would not make a true leader. "Park's conclusion jolted me wide awake," Lieutenant Oh recalled later. He wondered what he would be like in such a situation. Such a question led him to Vietnam. He wanted to test himself on the battlefield. Such an experience, Oh learned from the book, would be an essential part of making one a good leader.

This reason also led First Lieutenant Han Chul-yong (Class of 1970, later the 8th Infantry Division Commander as a two-star general) to Vietnam. Han sums up his motives in two ways: "One was to test my destiny, and the other was to obtain combat experience." Han had a firm belief in his view of life and death. Han says, "I am a fatalist. To live and to die depends upon God's will. If I survive in Vietnam, I'll have a chance to succeed in the military; if not, my life will end there. That's all." Han continues, "When a junior leader who did not experience combat makes a higher-ranking officer like regiment or division commander, will he be able to give an order to his men to fight courageously? Commanding generals of note in all ages and countries were those who had experienced and survived numerous battles, large and small, from being junior leaders. A leader should be equipped with a philosophical belief that his men's lives are precious in combat, as his life is. In so doing, the leader can achieve victory without his men being killed." All of the graduates of the Class of 1970, whose motives were the same as Han's, went to Vietnam, imagining themselves as the figures fiercely charging toward the enemy positions at the head of their men.[15]

Strictly speaking, the Vietnam War was neither a total war nor a war in which the Koreans should bleed to death. This notion was also true of West Point's Class of 1949, who had graduated a year before the Korean War broke out in June 1950. It was the time when they were about to finish their OBC. There was a conflict in their minds about whether they should go to war. To the young officers fresh out of West Point, whose future was full of hope, the war in Korea was a nuisance to evade. Besides, it was not the war to which their country was fully committed but a so-called "limited war," so they were not required to be in it. However, 253 new officers out of 574 graduates came, and 32 were killed.[16] More ignorant of Korea than this class was the Class of 1950, commissioned just a week before the outbreak of war. Just as the KMA graduates didn't know much about Vietnam, the West Point graduates had no idea about Korea, from what kind of food Koreans ate to what language they spoke. However, they learned that their country had decided to go to war, and they were regular officers who had graduated from the USMA. This was all that motivated them to cross the Pacific.

As mentioned earlier, American and Korean generations in the late 1960s had grown up watching the John Wayne movie *Sands of Iwo Jima*. When looking at John Wayne and his men coming to the shore of Iwo Jima, they substituted themselves for those U.S. Marines. The soldiers looked invincible, phoenix-like, and never died. All showed unwavering strength and courage. That's what the KMA graduates had dreamt of while sailing for Vietnam.

When the Korean government started sending across the Tiger Division in 1965, the country was in the grip of Vietnam fever. The people were overwhelmed by national pride that they could help a foreign country by sending troops. All the pop songs then in fashion disappeared from the streets except the military song "Tigers Go!" Almost all schools and workplaces were full of the roaring sounds of army songs. Regarding the martial spirit in "Vietnam," American society was not so different from Korea's. The American version of "Tigers Go!" was the song "Green Berets" by Sergeant Barry Sadler of the U.S. Army Special Forces in 1966. It also became the theme song of the motion picture *Green Berets* starring John Wayne. The patriotic song, was listed No. 1 on Billboard Charts in the U.S. for several months.

Thus, Vietnam was the land of horror to those who knew nothing of war. Yet, the land of adventure called out to challenge them simultaneously. To those who had had no chance of going overseas, Vietnam was a romantic land, far off to the south, with the Southern Cross twinkling during starry nights seen through thickly wooded palm trees. Upon completing some weeks' leave after graduation, most graduates of the 1970 class were sent to their new posts, mainly in the DMZ area on the home front. They forgot Vietnam temporarily, for they could not afford to think about it while adapting to their new environment as newly commissioned officers. However, they gradually began to think about Vietnam again six months later. Although there would be fierce battles, there seemed to be some romances on the night streets of the Vietnamese cities. So, they volunteered for service in Vietnam with a firm conviction that soldiers should be on the battlefield on the one hand and with a romantic curiosity about the foreign country on the other.

# Chapter 4
# Organizing and Training of Combat Troops

## The Capital Division

The ROK Army Headquarters designated the First and Cavalry Regiments of the Capital Division as the first units deployed to Vietnam and decided to reshuffle the existing division completely. It was the prevailing opinion at the time that, no matter how traditionally famed it was, the division should be replaced with elite members and made into a combat-ready unit as it was the first one being dispatched to the battlefield overseas. Thus, the officers above the battalion commander were selected from those who had experienced combat in the Korean War. And the rest of the officers were chosen from those who had already completed their corresponding tours in the homeland.

Taking a few examples of the selected officers, the First Regiment Commander, Colonel Kim Jung-woon (KMA 7th Class), was a hero who had fought in the Korean War as a company commander. Colonel Kim was a man of taciturn personality with a sense of quick judgment and was well-versed in military tactics. The Cavalry Regiment Commander, Colonel Shin Hyun-soo (KMA 10th Class), also fought in the Korean War as a company commander and battalion operations officer of the 32nd Regiment and was a highly decorated officer. A man with a clear-sighted mind, Colonel Shin was a well-informed officer and significantly contributed to developing the Korean armed forces. Both officers graduated from the OBC of the U.S. Army Infantry School at Fort Benning, Georgia. Colonel Kim Chan-bok (KMA 6th Class) was appointed Division Artillery Commander. Well-known as a leader with an iron hand in a velvet glove, he graduated from the U.S. Army Field Artillery School at Fort Sill, Oklahoma.

As for six battalion commanders, Lieutenant Colonels Park Kyung-seok, Kim Yong-jin, and Park Han-young were KMA Cadets Second Class who had come to the Academy as cadets of the first four-year system just one month before the outbreak of the Korean War. As the Academy had been closed, however, they had to be commissioned as second lieutenants from the newly established institution, the Army Integrated School. This school was to mainly produce platoon leaders urgently needed in large numbers during the war. Another three battalion commanders, Lieutenant Colonels Bae Jung-do, Lee Pil-jo,

Officers of the 3rd Bn (later Jae-gu Bn) of 1st Rgt., Tiger Div. From right: Capt. Kwon Joon-taik (Class of 1959, S-2), Lt. Colonel Park Kyung-seok (Cadet Second Class, Bn Cmdr.), Captains Bang Seo-nam (Class of 1959, 12th Co. Cmdr.), Kang Jae-gu (Class of 1960, 10th Co. Cmdr.), Yong Young-il (9th Co. Cmdr.) This picture was taken just two weeks before Capt. Kang died). (Courtesy: Chae Myung-shin)

and Choi Byung-soo were heroes in the Korean War as young student-soldiers at first and later as platoon leaders after being commissioned from the Army Integrated School. They completed the OBC course at the U.S. Army Infantry School and ROK Army War College. These officers had finished their tours as battalion commanders and division staff at home.

Deputy Division Commander Brigadier General Lee Nam-joo (KMA 6th Class) was a war hero in the Korean War who fought as the first battalion commander with the 27th Regiment. As an awardee with multiple decorations, including the U.S. Silver Star, he was the "Soldier among Soldiers." He was also a graduate of OAC of the U.S. Infantry School at Fort Benning. The Division Chief of Staff, Colonel Choi Young-gu (KMA 7th Class), was a company commander with the 21st Regiment in the Korean War. Colonel Choi was a man of reticence and gentleness, yet with a sense of determined responsibility, and he also graduated from the OAC of the U.S. Infantry School. All high-ranking officers above battalion commanders were selected from those who received the best credit in their respective units. These officers, who had been trained at the U.S. military schools, had no problems communicating with their American and South Vietnamese counterparts in English.[1]

The Tiger Division's commanding group was almost wholly reshuffled from top to bottom. In other words, the existing division was almost deactivated for a mission overseas, even from the division commander to a soldier in an infantry platoon. All division officers were those who had volunteered and passed the selection test. Although there was no discrimination in the origin of the commission, the KMA graduates filled almost all the leading positions. Notably, the next higher commander was given the discretion to choose his subordinate leaders. That is, the company commander selected his platoon leaders, and the battalion commander did his company commanders. Indeed, the officer corps of the Tiger Division was composed only of the most elite officers in the history of the Korean Armed Forces.

## The Regular KMA graduates

While almost all the senior officers above the lieutenant colonel of the Tiger Division were short-term KMA graduates, the officers below the battalion were mostly regular KMA graduates. The eldest of the second-generation graduates was the Class of 1955, all with the rank of major. These officers were either Regimental staff or assistants to Division staff. While two operations officers of the First and Cavalry Regiments (Majors Lee Byung-gu and Lee Hyo) were the Class of 1955, all the assistants to the Tiger Division's central staff, G-1, G-2, G-3, and G-4 (Ahn Seung-bong, Park Chan-wook, Kim Ki-taik, and Yang Sung-yoon), were also the graduates of 1955. According to Captain Lee Cheol-ho (Class of 1960), the headquarters company commander of the First Regiment, selecting officers was generally based on 70% KMA graduates and 30% others. According to a direction issued from the higher headquarters, almost all the company commanders and platoon leaders should be KMA graduates. Thus, virtually all company commanders of the two regiments were selected from the graduates of 1958, 1959, and 1960.

For example, 17 of 26 combat company commanders were members of the Class of 1960, and four others were those of 1958 and 1959, amounting to about 80% of all company commanders. The company executive officers were graduates of 1961 and 1962, and all Class of 1964 were appointed platoon leaders. The graduates of 1963 were assigned to the Weapons Platoon leaders, the most senior among the platoon leaders in a company. In the case of the 3rd Battalion, the 1st Regiment, all four company commanders were members of KMA. Three of them (Captains Yong Young-il, Kang Jae-gu, Lee Jae-tae) were from the Class of 1960, and one (Captain Bang Seo-nam) was from the Class of 1959. As for Captain Lee Jae-tae's 11th company, three out of four platoon leaders were KMA graduates, First Lieutenant Bae In-ki (Class of 1963), Second Lieutenant Shin Jung (Class of 1964), and Second Lieutenant Kwon Wu-heon (Class of 1964).

The company commanders of the 9th Division that departed for Vietnam a year later in 1966 were mainly the graduates of 1960 or 1961. And most of the platoon leaders were 1965 graduates.

Starting around 1967, a shortage of junior-grade officers began to arise as many officers were already returning home after a one-year tour. At the same time, since the Vietnam War was unlikely to be so dangerous, many junior officers wanted to go to Vietnam. Therefore, there was some competition for servicing in Vietnam. However, the ROK Army Headquarters prioritized KMA graduates. It even posted an announcement in *Asadal*, the Newsletter of the KMA Alumni Association, notifying that a preference would be given to the junior grade officers of KMA graduates (especially infantry and artillery).[2]

## Enlisted Men

In the case of the enlisted men, the applicants were selected based on their physical condition as well as academic and familial background. Given that, only 20-30% of the personnel from the existing Capital Division were eligible. Additional recruitment had to be made even to the level of reorganization. To this end, the ROK Army Headquarters directed all divisions under the 1st Army Command to organize one company each and send it over to the Capital Division. This reshuffling of the division by sending out more than 9,000 of the existing 13,000 strength and newly accepting 10,000 or more was nothing short of the division being re-activated.

Therefore, when the formation was completed, the current division's personnel accounted for only 30% of the total force, and 70% were soldiers from other divisions. Each company was named after the division to which it originally belonged. For example, the company from the 5th Division became the 5th company, the 6th Division the 6th company, and the company from the 27th Division the Heavy Weapons company in the Tiger Division. And since each company was named after each division from which they came, they did their best in fighting for the division's honor of their origin.[3]

The two divisions accepted a significant number of soldiers from other units. Still, in principle, the soldiers were selected based on their free will. According to Captain Yong Young-il (Class of 1960), the 9th Company commander of the First Regiment, 98% of his soldiers were volunteers, and he or the battalion commander persuaded only 2%. Some soldiers applied out of patriotism, but most men volunteered out of longing for a foreign country, economic reasons, and breaking away from a monotonous life.[4]

The same was true of some American soldiers who had served in the Korean War. William Childress, a veteran of the Korean War who later grew up as a noted poet, felt the excitement only a 19-year-old youth could feel when the Korean War broke out. And the war gave a man who had to work all day, feeding his horses from sunrise to evening, a chance to escape. And he said that those war experiences had changed him and that those changes had given him insight into the people and society in which he needed to live his life.[5] Likewise, such a motive moved many Korean soldiers to volunteer for the Vietnam War.

Despite all these seemingly smooth preparation processes, within the Tiger division was a mood of fear. With newly appointed officers arriving in the chain of command, reorganization started. Outwardly, all soldiers seemed excited about going to a foreign country; however, some things crept in, making the commanders and staff nervous and worried. Most of the officers and men selected were proud that they would go to a foreign country as part of the deployed troops for the first time in the history of Korea. But some soldiers, especially the enlisted men, began to shake a little. They feared they might be killed by the Viet Cong hidden in the jungles or die of malaria and tropical diseases caused by poisonous herbs or mosquitoes. These rumors spread widely among the soldiers. As a result, some 10-20 soldiers began to go AWOL every day. Facing this situation, Major General Chae, the division commander, gave an emotional address to his men while they were in combat training one day:

> Man is to be subject to his destiny. The problem is that he does not know what will happen to him. Does anybody, including you and me, want to go to Vietnam? No, nobody does. But we go because we're destined to go. Man is destined to die. If I am killed by the VC's sniper fire or hand grenade, it will be my fate that I cannot avoid. Let's challenge fate as a man. Don't run away. A man fated to die will die wherever he goes, and a man to live will emerge from the tiger's den. The Viet Cong's first target, I'll always be at the head of you all and

do my best as I have done up to this time. I'll be with you wherever you go. I'll conclude my words by quoting the phrase from Matthew 10:28 in the Bible: "Do not be afraid of those who kill the body but cannot kill the soul. Instead, be afraid of the One who can destroy both soul and body in Hell. Are not two sparrows sold for a penny? Yet not one of them will fall to the ground apart from the will of your Father."[6]

The soldiers cheered when he ended his address by quoting the Bible verse. Another loud applause rang out. Yet some volunteered soldiers said they would not go to Vietnam after meeting with their families. So, Captain Bang Seo-nam (Class of 1959), the 12th Company commander of the 1st Regiment, felt the need to educate their families rather than the soldiers. He explained to their families the nature of the Vietnam War: "The Vietnam War is not such a war like the Korean War. It is a local war, a guerrilla war. I love my subordinates as much as you love your boys. If my men die, I will die, and I will not come back."[7] These words moved many a family and their sons did not desert.

## The Preparatory Combat Training

### The Capital Division

Upon completing the unit organization, Captain Lee Jae-tae's (Class of 1960) 11th Company entered the preparatory training. The problem was that there needed coherence between soldiers from the squad to the company, for most soldiers came from different divisions. The same was true of the officers in the company. However, the thought that they would leave for the war front in a month made them feel united as comrades bound by a common destiny. Yet a rivalry arose between the platoons, making them competitive because their former units differed.

The training began in earnest. Yet nobody knew well about Vietnam's reality and the Viet Cong's tactics. One day in mid-September 1965, a unit of the Cavalry Regiment built a mock Vietnamese village and demonstrated how to catch the VC. But the show was a grab-bag that seldom got to specifics. Although the unit conducted its training for a short period, it grew solid and cohesive. In a

President Park Chung-hee reviewing the Tiger Division troops in the send-off ceremony at an airfield in Seoul in October 1965 with Army Chief of Staff Gen. Kim Yong-bae and MG. Chae Myung-shin, CG of the division. (Source: ROK Army Museum)

month, the division would leave for the battlefield. The last stage of the training was RCT, held as FTX for two nights and three days. When all soldiers completed the entire training in late September, they were determined and full of confidence that they could win in combat.

The Tiger Division completed the training of its First and Cavalry Regiments for adapting to the jungles. And then, on October 12, 1965, the two regiments moved to Seoul and participated in a magnificent send-off ceremony held at Yeoido airfield in the presence of then-President Park Chung-hee and hundreds of thousands of families and friends of the dispatching soldiers and the general public. The reason for holding such a grand send-off ceremony for the troops were to show that we could send troops to a foreign country and that we would never forget the countries that gave us help when we were in need. And it was to show that our national power had grown enough to help another beleaguered country. And also, it was to boost the morale of the troops going abroad to the war front. President Park was moved and proud that the ROK forces would go overseas to help another country for the first time in the nation's history. He said in his address, "You leave for the war front with a holy mission to create a new history. On your long journey, I wish you loving officers and men great success and good fortune."[8]

### The 9th Division

The same was true of another combat division to be deployed. In the case of Captain Park Jong-sik's (Class of 1959) 11th Company of the 28th Regiment, almost all of his company consisted of existing soldiers and recruits freshly out of the Army's boot camp at Nonsan City. Among 180 members of the company, 125 soldiers were recruits from the boot camp. As the Tiger Division had done, this 9th Division received many soldiers from other divisions. Some of those who were supposed to go to the 9th Division refused. In some cases, the drawing of lots was introduced. For example, 30 corporals to be discharged in a year were gathered, and a lottery was held to select three people among them, and those chosen were immensely disappointed at the result. They even beat the ground and wept.

This is Second Lieutenant Kim Hyung-seok's (Class of 1965) story of his transfer to the 9th Division. In June 1966, Lieutenant Kim and his platoon left their 11th Division and reported to the newly reorganized 9th Division stationed in Yang-Pyeong County, about 50 km east of Seoul. Kim's platoon became the 1st platoon, the 6th Company, the 2nd Battalion, the 30th Regiment, the White Horse Division. The original members of the 6th Company, except for a few, were all transferred to other divisions. Three other platoons of the 6th Company came from three different units: the 2nd Platoon from the 13th Regiment of the 11th Division, the 3rd Platoon from the 63rd Regiment of the 21st Division, and the Weapons Platoon from the 77th Regiment of the 27th Division. The men of Kim's platoon here were far better than those of his former platoon in the 11th Division in terms of education and soldierly bearing. The platoon men were all volunteers: one college dropout, five high school grads, seven junior-high, and the rest all educated to elementary level. Most of them came from rural areas and were innocent young men. Kim determined he would make the strongest combat troops out of these men.

The White Horse Division conducted combat training for three months. The programs include firing exercises, individual combat, map reading, field fortification, camouflage, first aid method, field sanitation, laying and clearing land mines and booby traps, etc. He tried to improve their physical power through running and Taekwondo practice using extra hours. Keenly aware of the importance of the training, all soldiers earnestly followed the programs. Sooner or later, they would have to confront the enemy face to face in actual combat, which would not be an exercise or a mock battle. Kim especially emphasized firing practice, for marksmanship of "One Shot One Kill" was badly needed to protect comrades in combat. Small-unit tactics training mainly aimed to instruct the soldiers on searching and destroying the VC in the jungle.

The three-month training ended with the integrated exercise of a night infiltration attack toward the top of a hill over one night and two days. Before leaving for the send-off ceremony, his platoon and other units were allowed to meet with their families and spend some time with them. He paid particular attention to married men by enabling them to have more time with their wives. Corporal Soh, a married man in Kim's platoon, met his wife, who had come from afar with a baby on her back. Though the man and his wife were supposed to stay only overnight, they were lucky to be together for three more days, thanks to the careful consideration of his squad members. Another man from Kim's company was blessed with enjoying a night with his sweetheart, and she was made pregnant. But who in the world knew that the day would be the last reunion of the couple? Some months later, in Vietnam, the man was killed in the battle even without seeing his baby not yet born. Although his flesh was dead, his soul would be reincarnated through his baby. War kills people, but it paradoxically creates new life.

Lieutenant Kim's 30th Regiment moved to Seoul after combat training for three months. On the roads where its trucks were passing, many people gathered, and they could barely pass through. The escorting military police jeeps loudly sounded their sirens to leave the roads open. High school girls in uniforms put their addresses and names on white handkerchiefs, placed small stones, and threw them at passing trucks with notes, "Please, give me a letter," and "Come back alive, not dead." These handkerchiefs brought many consolation letters to the Vietnam front and significantly boosted the soldiers' morale. Because of this connection, some soldiers even had the fortune to get married.

It reminds one of the scenes in Stephen Crane's American Civil War novel, *The Red Badge of Courage* (1895), in which the protagonist is moved and exhilarated by the cheers and coffees given by the crowds on the roadsides. He envisions his heroic action in their joys. Crane writes in the book: "As he basked in the smiles of the girls and was patted and complimented by the old men, he had felt growing within him the strength to do mighty deeds of arms."[9] In Seoul, President Park Chung-hee hosted a grand send-off ceremony for the 30th Regiment held in front of the Capitol Building on October 1, 1966.

From October 1965, when the ROK Army combat units began to be deployed in the Vietnam War, the Tiger Division and White Horse Division were trained at their garrisons and sent to Vietnam. After that, however, it was necessary to train replacements to fill the returning soldiers since the tour of duty for officers and men in Vietnam was one year. As in the U.S. military, the change of troops was not unit by unit but an individual shift in principle. So, on March 15, 1966, the ROK Army Headquarters established the 7th Replacement Training Camp to educate replacement troops at Oeum-ri, Gangwon Province. These troops went to Vietnam monthly, so replenishing troops was not an issue. In addition, there was no deficit in the actual situation because each unit always maintained 102% of the authorized strength in case of lost troops.

# Chapter 5
# The Voyage: Busan to Quy Nhon

Busan, the second-largest city in Korea, was the port of initiation into the war in Vietnam. All the troopships started from this port at the southern tip of the Korean Peninsula. Upon completing the send-off ceremony in Seoul, all the troops went to Busan by train. Also, in Busan, the troops received the warmest farewell ceremony hosted by the city's mayor before departing for Vietnam. Such a splendid farewell ceremony used to be held at the beginning of the war when the Tiger and White Horse Divisions left earlier. As time passed, no grand farewell ceremony was held because the replacement troops departed once a month. The ceremony became increasingly routine, a matter of formality to the general public. No such banners hung on the ship: "Win the war and glorify the country!" "My Lover, please be careful and come back alive from Vietnam!" and "My sweetheart, let's get married on my return." The only people present were some mobilized high school girls, an army brass band, and families and friends. All these people wished for the safety of their kin on the long journey to the war front.

At the time, the Korean troops used American transport ships. Since the U.S. began to commit its forces to Vietnam in 1964, U.S. Army MSTS contracted with the U.S. merchant marine. The service vessels transported the Korean military personnel and equipment to the war front. The *USS General Leroy Eltinge* was one of the service vessels that took the Korean troops to Vietnam. During the Korean War, this ship

The 26th Regiment troops of the Tiger Division departing for Vietnam from the Busan harbor on 15 April 1966 amid the greetings of relatives and well-wishers. (Source: ROK Army Museum)

had been used to transport American soldiers from Seattle, and San Francisco to Japan and Korea.[1] Besides this ship, *USS Barret, Victoria, Gaffey, Darby,* and many other vessels were used to deliver the Korean troops to Vietnam. The size of *General Leroy Eltinge* (12,500 tons) and *Victoria* (13,000 tons) was huge, and the Korean soldiers were amazed at her magnificent figure.

A further surprise was that just 28 sailors were operating *General Leroy Eltinge.* During a week-long voyage, the men who wore uniforms were very few, including the captain, navigator, engineer, storekeeper, etc. The rest were employees for various chores, most Filipinos, with others from Egypt and South American countries. When Captain Lee Jae-tae (Class of 1960) took a course on the "History of the Pacific War" in his cadet days, he learned that the aircraft carrier *USS Enterprise* was 32,000 tons, and the Imperial Japanese Navy Ship *Yamamoto*

*Musashi* was 71,000 tons. Looking at the vast 12,500-ton *General Leroy Eltinge,* he recalled he had imagined how large those ships would have been. The ship still had enough space with the first contingent of 5,000 Korean troops loaded. Such a ship's capacity would be sufficient to carry one division at a time to any place in the world.

Onboard the gigantic transport ship, the KMA graduates were soon lost deep in thought. Myriad images of the past went by in their minds: the ideas of country and people in the Cadet Creed they had recited day and night at roll calls, the images of college girls who had been their partners in group meetings in their cadet days, the faces of their senior cadets who had given them hard times and imposed hazing, the joys and sorrows they had gone through in their life at the Academy, and many others. On the other hand, there were also some anxieties and expectations about the unknown world they would soon face: a vague longing for a strange and curious country, enemies lurking in the jungles, and fear that some of them might be killed. Everybody seemed to be struggling with those things in mind. As the adage, "there is no atheist in the foxhole," suggests, many kept a Bible or Rosary in their belongings.

Above all, there were desperate prayers from families who wished for their sons' safe returns. Second Lieutenant Shin Sang-gil's (Class of 1970) mother, who was there to see him off, quietly shed tears far away from the dock. Already some Christian mothers began to pray for their sons at dawn prayers. And others, like the mother of Second Lieutenant Kim Hyung-seok (Class of 1965), were praying before a bowl of freshly drawn water in the backyards of their country homes every night. This prayer was a kind of shamanic ritual practiced in the traditional society of Korea. Some of the officers cried at the thought of their wives and girlfriends. And others missed their parents, regretting that they did not fulfill their filial piety under the pretense of being tied up in military duty far away on the frontline.

Second Lieutenant Paik Nam-hwan (Class of 1969) vividly recalls the scene at the dock of the Busan harbor. Before boarding the ship, his older sister put soil in a plastic bag and handed it over to him, saying its smell would be effective in seasickness. When combat units went as a group in the early days, the command authorities distributed a loaf of soil the size of laundry soap to each soldier. As time passed, though, rumors about it had already spread, and everyone had prepared. It was said to take a week to go to Vietnam, so Lieutenant Paik even put a paste that would relieve the pain in his navel as the people around him said it would prevent motion sickness. He recalls that the soil was more effective than the paste for seasickness. He had kept the soil in a cabinet and often taken it out to smell the earth.[2]

## "Me Too" menu

The distance from Korea to Vietnam was about 2,200 miles. It took six nights and seven days to get to Vietnam on *General Leroy Eltinge*. During that time, all meals served on the ship were Western food. In the 1960s, Korean people had few opportunities to eat Western food. Of course, soldiers and officers were unfamiliar with the meal and even did not know how to eat it. The first few days were fine, but Western food didn't suit the soldiers' mouths as time passed. Everyone missed kimchi and soybean paste. Some soldiers did not even like the smell of Western food; some episodes related to the food on the ship in the earlier time of passage to Vietnam have been passed down to this day.

Captain Lee Jae-tae (Class of 1960) had two tours to Vietnam. The first time, he was the company commander of the Tiger Division for one year from October 1965. In 1971, he was promoted to Major and went again as a regimental staff. On this second tour, he took the *USS Barret* (13,319 tons). All the food served on the ship was Korean, including rice, kimchi, soy sauce soup, etc. Moreover, as the troops were primarily replacements, their number was small, and there was no problem with food.

The following is what Captain Lee recalls had happened in his earlier time: "There was an officer's mess hall for each different unit. There were tables on which clean tablecloths were laid with forks and knives and a printed menu. Eight people sat at one table. The food served was excellent, but most officers wondered what foods on the menu would be like because Western-style food was unfamiliar. The menu read three categories: appetizer, main dish, and dessert, with an explanation for each. For example, for a grape served as dessert, the explanation was a red grape produced in a valley in California. And for coffee, officers were asked whether the coffee would be black, both sugar and cream or just sugar.

Further, when they chose steak, they were again asked, 'How would you like your steak, well-done, medium-well-done, medium, or rare'? Quite embarrassed, the officers had no idea what to answer. Thus, the safest way was to repeat what the previous officer had ordered (though he wondered whether the officer's choice was right). A crew member received the order from each one. Usually, the officers pointed to food on the menu with their fingers as they did not know its name. When the next officer was asked about his menu, he mumbled hesitantly, 'Me Too.' This same 'Me Too' was repeated throughout the tables. As for the cook in the kitchen, he was comfortable because he had only to make one style of food for everyone. Regardless of whether he liked it, the officer had to eat the food served."[3]

Second Lieutenant Jee Man-won (Class of 1966) had the same experience as Captain Lee. He recalls the scene when he was onboard *USS Victoria* in July 1967. "Looking at the distant sea from the deck, I felt the pleasure of were going through the open ocean. The night sky was gorgeous. The stars descended so closely that you could likely catch them if you stretched out your arm. Countless shooting stars fell on the distant sea, drawing a long curve like a kite's string and moving away without a break. A Filipino employee in a white robe walked around, waving a rattle bell and announcing mealtime. The officer's dining room on the ship was spacious and clean, and there was also a piano. Each table had a menu written in English. It was one he had seen for the first time in his life. The people at the table asked him to explain the menu, but he said he was not familiar with the English menu, either. Yet they trusted him because they saw him converse with the Filipino employees in English. When he decided on a menu and informed the waiter, everyone else started saying 'Me Too' and 'Me Too.' Because of ordering the 'Me Too' menu, the kitchen ran out of chicken and beef in days. On the other hand, some strange foods to Koreans like lamb and turkey were left intact."[4]

This episode originated from the Korean War. During the war, the U.S. military needed to educate Korean officers who showed poor combat performances. So, it had taken hundreds of them to the U.S. Infantry School at Fort Benning, Georgia. An interesting thing happened in the cafeteria. At mealtime, the Korean officers who had been very poor in English lined up in front of the food stand and said, "Me Too," following the IO (Interpretation Officer) at the head who accompanied them." Once the IO had pointed "beefsteak," everyone had followed suit. Confused by the unexpected situation, the kitchen police hurriedly had to prepare the extra food. A problem arose at the next mealtime. The cook, who had experienced from earlier mealtime that the Koreans loved beefsteak, prepared an enormous amount at this time. However, the IO pointed to another food, the "chicken," this time. And then, all the officers who followed did the same. The kitchen cooks were too embarrassed by this news, for they were not ready to cook the chicken that much. They had to throw away a considerable amount of beefsteak. Indeed, the American mess personnel was said to have been deeply stressed every mealtime because of the unpredictable behavior of Korean officers. The kitchen cooks were always alert to what the Korean officers would choose this time. This amusing but sad story has been passed down to this day.

## Arrival in Vietnam

Amidst anxiety and curiosity about the country at war, the Korean combat troops arrived in South Vietnam. However, the first thought that the country would be engulfed in a whirlwind of war changed when the peaceful landscape spread before the soldiers. Second Lieutenant Shin Jung (Class of 1964), who landed in Vietnam as a platoon leader of the first combat unit in October 1965, recalls his first impressions of Vietnam as follows:

Five days after leaving Busan Port, *USS General Leroy Eltinge,* carrying us, dropped anchor at Quy Nhon Port around 6 a.m. It was South Vietnam that he had envisioned in his mind for the past few months. He thought that the whole land of Vietnam would be overwhelmed by the war. To those who saw it for the first time, though, the appearance of Vietnam from the coast was entirely peaceful. The port of Quy Nhon, shrouded in the morning mist, was not like the land of a war-torn country. French-style buildings painted in red and yellow colors and unique Vietnamese houses, and the palm tree forest, a symbol of tropical plants, were seen from afar. Wearing Ao dai, a traditional costume with a Non-La hat on their heads, Vietnamese girls carried bananas in the straw baskets on their shoulders. The appearance of fishing boats with balloons and leisurely catching fish was far from that of the war-stricken front.[5]

A year later, in October 1966, the units of the White Horse Division arrived in Vietnam. Captain Park Jong-sik (Class of 1959), the 11th Company commander of the 28th Regiment, recalls the "beautiful scenery" when he landed at the beach of Nha Trang: "Looking at the sea, I saw an endless white sandy beach of clean and thick sand on which bananas and palm trees were scattered along the coast, swaying in the wind, and the green sea water hit the waves and created white foam, making it truly like a picture. Deep inland, a vast plain spread out, and a dark blue jungle stood out. It had a landscape that had never been seen before, with a large dune close to the airfield lying long and various cactuses over our height growing densely on it. I couldn't believe that this was a bloody battlefield." Indeed, Park had thought that there would be a fierce fight in landing, but what had caught his eyes was a big arch written, "Welcome the White Horse Warriors to Vietnam." Military brass bands played marching songs, and Vietnamese Cong Ais in white Ao dai waited underneath the arch. Park saw many domestic and foreign reporters, cameras, and various reporting equipment.[6]

Captain Lee Jae-tae (Class of 1960), the 11th Company commander of the 1st Regiment of the Tiger Division, had a similar landing experience in Vietnam in October 1965. Yet Lee tried to compare South Vietnam and Korea and grasp what the South Vietnamese people would think of another foreign army stepping on their soil:

After we landed in Quy Nhon, we walked to a public playground where the welcoming ceremony was to be held. When we got there, a lot of people gathered to meet us. The ceremony started when South Vietnam's Prime Minister Nguyen Cao Ky was present. There was a welcoming address, a return address, and presentations of flower wreaths by high school girls. After that, the ceremony ended with our parade.

What was the Vietnamese people's attitude toward Korea? To the Vietnamese, Korea was a country in the east liberated from Japanese colonial rule, a country devastated by war just ten years before, a land of people whose national income was no more than 100 dollars, and one of the most underdeveloped countries. Perhaps all these appellations may have told the Vietnamese people everything they needed to know about Korea. What may have been their first impression of the Korean troops? What might they feel toward us if they saw our troops today? Would they feel any sense of comfort and confidence in such a figure of bravery? Or may they have been disappointed at a poorly equipped division of an underdeveloped country? The formation of Korean troops equipped with those worn-out weapons from World War II might have been a must to see.[7]

The Vietnamese, ruled by foreign powers for quite a long time, could not be very thankful for the appearance of Korean troops in their country. It was just another foreign army that had entered their land. The Vietnamese people passing the streets did not notice the Korean soldiers waving their hands. Captain Kim Jin-gyu (Class of 1958), an Artillery Battery Commander, noted the indifference of the Vietnamese when he landed in Quy Nhon and drove out of the city in a vehicle prepared by the U.S. military. "Not only passers-by but also the residents standing on the side of the road seemed insensitive to our procession. The shops beside the road were crowded with people as if they were unrelated to the war."[8] Captain Park Jong-sik also recalls: "One peculiar thing I felt while stepping on South Vietnamese soil was that the people who welcomed us were not South Vietnamese but American soldiers. Even if the Korean army waved their hands as they passed, they almost turned away. However, when the U.S. soldiers saw us, they greeted us, gestured, and whistled."[9]

As a whole, Korean soldiers were very nervous as they entered the war and feared not knowing when they would die. Even in the peaceful scenery of Vietnam, they were highly wary while drawing the dark image of war. Second Lieutenant Kim Hyung-seok (Class of 1965), a platoon leader of the White Horse Division, describes his landing at

Cam Ranh Bay in this way: "Sailing through Taiwan Strait, East, and South China Seas, the troopship arrived at the port of Cam Ranh in the southeastern region of Vietnam almost a week later. Far behind the Cam Ranh Bay soared high in the sky, the Annamese Mountain range like an enormous back-dropped screen, giving Koreans the impression that Vietnam must be a mountainous country. The thick green jungle forest covering the mountain range gave the Koreans an awe and wonder as if it were a black hole to suck in all forms of life on earth. Kim thought he would conduct firefights for a year, roaming those unfamiliar, rugged, and craggy mountain tops and valleys. He never knew what would happen to him, returning alive or buried somewhere there. When he came to this thought, all of a sudden, he felt enervated, and it got him down. However, there was no way but to confront reality. The old injunction "you'll survive if you try to die" came to his mind"[10]

The same was true of the graduates of 1970. After a week's voyage aboard the passenger ship *Barret* through Taiwan Strait and the South China Sea, they also landed at Quy Nhon and Cam Ranh port cities. What first came into sight was the port full of various-sized ships, the roaring sounds of helicopters, a formation of Phantom fighters disappearing from high in the sky, the far-off sounds of booming guns, and other unfamiliar smells of the beachhead and strange noises. After coming ashore under the burning sun, they realized that they were finally in Vietnam. When they disembarked, they were only met by some officers from higher headquarters. But there were neither welcoming remarks by any South Vietnamese government official nor a crowd of schoolgirls who had placed floral wreaths around the soldiers' necks earlier.

Instead, they were told to be wary of the Viet Cong coming out of the wooded jungles off the road at any time. They all felt like jelly because they feared that they were finally in the war zone. The romantic thoughts they had were gone in an instant. Lieutenant Park Man-joo (Class of 1970, later Chief of Signal Corps as a two-star general) initially felt the reality that he was in the war zone. His first action when he reported to his unit was to submit parts of his nails and hair. It was to take his DNA samples in case something happened. This was a usual practice required of all soldiers entering the war fight.

Chapter 6

# The Vietnam War: "A Formless War against a Formless Enemy"

There would be no other war in history that changed the definition of war and provoked so much controversy about the conduct of war as the Vietnam War. The Vietnam War was a war that had to be fought fiercely, although there was not a clear front, a clear enemy, and a sure target to seize. For this reason, the Vietnam War was dubbed with some other names like the "nasty war," "un-battlefield-like war," the "splendid little war," and "a formless war against a formless enemy." All these appellations converge in a single phrase, guerrilla war.[1] Ultimately, Vietnam was not a war fought on the fronts as was the Korean War or others, where battles ended with clear-cut victories or losses. It was a war fought everywhere and nowhere.

## The Land and the Terrain

Vietnam is located on the eastern Indochinese Peninsula between 8 and 24 degrees of the Northern latitudes. It is a land connected through the mountains of the Annamese Cordillera by the Red River Delta in the north, a flat, roughly triangular region, and the Mekong River Delta in the south, the most extended river basin in the country. Though divided by the 17th parallel into North and South Vietnam in the war era, Vietnam is a fertile land stretched lengthwise from the north to the south along a 3,444 km-long coastal line. And Vietnam's land is mostly hilly and densely forested, with level land covering no more than 20%. Mountains account for 40% of the country's land area, and tropical forests cover around 42%. The coastal line has some good harbors.

Cam Ranh in the south is a port city blessed with a natural environment and the first and foremost strategic base of South Vietnam. North of Cam Ranh is Quy Nhon, which has vital strategic importance for ground operations. This port city of Quy Nhon is the starting point of Route 19, running to the west and intersecting with Route 1, running from north to south. The country's northwestern part bordering Laos and Cambodia is a densely forested mountainous region. Through these natural terrains, the North Vietnamese provided troops, weapons, and other war supplies to Viet Cong in the south and conducted guerrilla warfare in their favor. North Vietnam kept its forces supplied using this so-called Ho Chi Minh Trail, which stretched from North Vietnam to the South.

VC's bicycle supply units moving through jungles and valleys (Source: ROK Army Museum)

Before being assigned to the commander of the Korean forces in Vietnam, Major General Chae Myung-shin was serving as Deputy Chief of Staff, Operations at ROK Army Headquarters in Seoul. After thoroughly researching Vietnam's history and military situation, he concluded that any military forces would have a limit to completely subduing the hostile forces in Vietnam. Some later years, when he was appointed the commanding general of the first combat forces to be deployed to Vietnam, General Chae confessed, "Then, I'll be a commander who plunges into the battlefield even though I know I won't be able to achieve the final victory."[2]

Numerous military commentators in the U.S. had also concurred with Chae's evaluation that the war in Vietnam was not against hostile forces but against Vietnam's land itself, so winning the war would not be possible. No matter how fiercely fighter jets and B-52 bombers poured tens of thousands of tons of high explosives and napalm into the jungles, no change was made except a thin pillar of smoke rising from there. The land itself, which had existed since time immemorial, was unscathed. It was the land itself that resisted. Caputo also pointed to the nature of the hostile land in his book, "it took us all morning to cover the three miles between the landing zone and the village. Four hours to walk three miles, and the company had not once run into significant enemy resistance. It was the land that resisted us, the land, the jungle, and the sun."[3]

All of Vietnam's natural conditions, including mountains and swamps covered with lush tropical jungle, the sweltering heat with an average annual temperature of 34 degrees centigrade, and the ever-changing weather conditions, were limiting factors for the Allied forces. However, more important than the limitations of the natural environment was the spirit of resistance of the South Vietnamese people. They naturally developed a strong sense of nationalism and a spirit of resistance to foreign powers as they had experienced the domination of China and France for a long time. In particular, as the coup d'etat and regime change were repeated in South Vietnam, its people gradually lost trust in their government. And their support for Ho Chi Minh, the leader of North Vietnam, was absolute. They regarded the freedom and independence pursued by Ho Chi Minh as higher ideals than the ideology and system.[4] Support for Ho Chi Minh naturally led them to the Viet Cong. And as time passed, more and more residents supported the Viet Cong. With the support of such residents as the background, the Viet Cong waged a guerrilla war against the incomparable military power of the allied forces, especially the U.S. military.

## The Enemy and the Guerrilla War

The Vietnam War was a guerrilla war or insurgency. In a guerrilla war, knowing who is an enemy and who is not is difficult because a guerrilla wears the same clothes as a civilian during engagements. As the guerrillas mainly reside in the civilian population, most combats are likely against civilians. For this reason, there are lots of innocent civilian casualties. The point was that, as mentioned earlier, the Command of the ROKFV at the time made it a primary policy to protect innocent civilians even if they missed one hundred Viet Cong. That made it more challenging for the Korean troops to execute their operations. As Mao Zedong's dictum about insurgency says, "The guerrilla must move amongst the people as a fish swims in the sea," the guerrillas cannot live away from the people like fish cannot from the sea. In other words, the Viet Cong were among the people, and the people were among the Viet Cong. Thus, the counterinsurgency strategy of Korean troops was to drain the sea and separate the Viet Cong from their support base in the population.

Mao Zedong emphasized the superiority of human resources and mental power over the enemy's economic and military supremacy. As an element of victory or defeat in war, he valued human beings more than material things. Mao Zedong advocated the "Sixteen-Letter Tactics," citing Sun Tzu's *Art of War* in his treatise entitled "A single spark of fire can burn the whole wilderness." Mao's adage is composed of sixteen Chinese characters: 敵進我退 (dí jìn wǒ tuì, When the enemy advances, we retreat), 敵駐我擾 (dí zhù wǒ rǎo, If the enemy is garrisoned, we will go and make a commotion), 敵彼我打 (dí pí wǒ dǎ, If the enemy is tired, we will hit), and 敵退我追 (dí tuì wǒ zhuī, When the enemy retreats, we will attack). Vo Nguyen Giap, the commander of PAVN believed in the theory of Mao's "fish and water" relationship and applied these sixteen-letter tactics to the execution of the guerrilla war in Vietnam.

Captain Park Dong-won (Class of 1958), the 5th Company commander of the First Regiment, analyzed the war in Vietnam and the enemy: "There are no fixed fronts and no uniformed enemies in Vietnam's guerrilla warfare. Those who resist with only weapons are enemies, and those who smile and wave their hands are civilians. But nobody knows when or where they may turn into hostile forces. The area where they wave their hands is the residential area of civilians, and where they protest is the front line. Therefore, it is tough to determine when and where they will fire and fight. Because of this, in ordinary cases, the battle starts when the friendly take damage in the first place."[5]

The defining description of First Lieutenant Bae In-ki (Class of 1963), Weapons Platoon leader of the 11th Company with the First Regiment, best summarizes the unique aspects of the Vietnam War concerning the land, enemy, and nature. Here he minces no words that the Vietnam War was a war with jungles, ambushes, long nights, poisonous insects, and VC.

> When the night curtain begins to swallow the ground, we usually leave the base to perform our duties. It takes a few hours to advance a dangerous distance and settle at a certain point, and soon uninvited guests of the night flock start the first fight, and a tense time of darkness flows. The evil and mentally armed Viet Cong should not be ignored. You cannot clearly distinguish Viet Cong from residents in their appearance. Whether you can or cannot relax in Vietnam does not depend upon the absence or presence of the Viet Cong. It is a battlefield without a frontline. Although he is weak, his movements are agile. Their tiny hands are extraordinary.
>
> We give away the C-ration cans we eat all of, and they pick up the cans and make an anti-personal mine to kill us. In addition, they make an ignition battery to destroy antitank mines by collecting the various batteries used and discarded. And also, they have acquainted themselves with the topography and climate since childhood. They have developed a sense of direction and escape from wetlands, rivers, and mountains more than anyone else. Also, keeping in mind the code of conduct for guerrilla warfare, they practice that the general public is water, and Viet Cong is fish, so fish cannot live out of the water.[6]

And the Viet Cong were mainly short and lean. As they had always been in hiding, exposed to hot weather, and ill-nourished, the Viet Cong mostly had a slight build. Yet they were very familiar with the terrain and the climate, so it was not easy to catch those VCs using their effective hit-and-run tactics. A Korean officer once said, "The

VCs are everywhere and nowhere." In other words, "Viet Cong appears everywhere, whereas they always disappear when you seek them out."[7] Among the VC were some children, girls, vendors, and elders. When these people found a weak point while vending, some abruptly became terrorists throwing hand grenades. Some girls, who seduced soldiers, sometimes became the VC. The Viet Cong were mainly in action at night and only chose points of weakness to make surprise attacks. As Philip Caputo rightly points out, "there was no pattern to the patrols and operations. Without a front, flanks, or rear, we fought *a formless war against a formless enemy* who evaporated like the morning jungle mists, only to materialize in some unexpected place."[8]

Then North Vietnam's Defense Minister, Vo Nguyen Giap, emphasized using people's war tactics, saying, "Mount a powerful attack in an advantageous situation, but retreat in an unfavorable situation." Based on this theory, the VC mainly used ambushing tactics, planting mines and obstacles, or intentionally exposing small troops as decoys. Sometimes they faked running away during the engagements and placed a killing zone on the run, and they turned around abruptly and made a surprise counterattack. The VC's defensive tactics were to hide in underground tunnels or crevices of rocks. They would hold the bamboo-flute-like stick on their lips and stay somewhat longer underwater, only with part of the stick out to the air. These primitive tactics of the VC made the friendly forces frustrating and nerve-racking.

To survive the massive American attacking power--mobility of helicopters, close air support, and artillery fire, the Viet Cong could not afford to maintain an organized retreat and thus had to disperse and run away. And then, they reassembled at a chosen place. The night surprise attack was their primary attacking method. To them, a daytime attack was considered almost taboo. As a flea hopped up in darkness, they attacked a weak spot in the middle of the night and dispersed again quickly. These were the rudimentary tactics of Giap. Another form of attack was an ambush. The VC learned that an ambush was an attack, though it was a means of defense in general. They were forced to use the ambush as an attack, for they could not win over the enemy whose weapons and equipment were far superior to theirs.

The Viet Cong lured the opponent out and attacked. They

intentionally exposed a small group of soldiers and caused them to flee when detected. And they established a killing zone at a proper place and made a surprise attack on the opponent when they fell into that trap. A commander chased the escaping enemy out of a desire for honor and to obtain military achievements and was often counterattacked and failed. The defensive tactics of the Viet Cong were only to hide. Hiding in the crevices of rocks or digging caves between rocks were some of their defensive tactics. Even the Viet Cong stayed for some time underwater, only with a hollow straw in his mouth out in the air to breathe. The VC dug underground caves. They dug deep with some turns inside to avert any hand grenade attack.[9]

In an interview with a Japanese monthly magazine in 1995 for the 20th anniversary of the Vietnam War that had ended in 1975, Vo Nguyen Giap confessed how he had had to confront the mighty U.S. forces. When asked by the reporter about his most cherished thoughts on war, he said:

> The U.S. military was a formidable one. But any formidable opponent had a weakness. The American forces as a whole had no fault, though. So, I made one. They had the best weapons in the world, and the soldiers were one of the most elite forces in the world. Unless you made a weakness [in them], you could not win.
>
> In other words, you must fight where you want, when you want, and your way. I developed a kind of guerilla warfare where the American forces could not use their advanced weaponry. In guerilla warfare, you strike, then retreat. You use the jungle to your advantage. And dig underground trenches to appear like a ghost, then disappear again like the wind. In the jungle, tanks and gunships are useless. It's the same as entering a small room with a long spear, a lance of about four meters. In a small space, a dagger or a short knife is preferable.
>
> The American forces may have been exhausted. They could not exert their great military might. This tiredness becomes the weak point. Do only the opponent's forces get tired? No. The American people tire of the war, the politicians get tired of the war, resulting in an unpopular war. In a democracy, the power of the press is mighty. This is another weakness. The tired Americans don't stay silent. Stop the war, say the American media. The American President also caves into these demands.

And Giap added another crucial point to his story:

> But to be sure, we were also very, very exhausted. It wasn't just the Americans who were tired. Had the Americans endured war for another six months, we would have surrendered and waved the white flag of defeat. Our national power and resources were about to be depleted as well.[10]

Once a specific area was assigned to the Korean troops, they moved its villagers to another improvised center. At first, they found that there were not many men among the villagers. And the Koreans tried to make the women aware that their husbands were fighting for the lost cause in vain. "We know your husbands are out there," said the Korean troop leader, "we'll soon search him out and kill him." In this way, the Koreans persuaded the wives. Afterward, the wives left the accommodation center and brought their husbands from the caves in the neighboring mountains back to their centers.

And then, the villagers outside the refugee center became the objects of being killed, and the Korean troops swept them away. They repeated this work about the area to detect the enemy hiding in the caves or other secret places. According to the Tiger Division commander Major General Ryu Byung-hyun, the successor to General Chae Myung-shin, "It is like the work to dig out potatoes. If we collect the things only on the surface of the field when we harvest the potatoes, it is to collect only the stems and their leaves and not to dig up any potatoes. If you pull up a potato's stem, you will easily dig up five or six potatoes every time you do." He explained thus about the way of searching out the VC.

## The Helicopter War

The Vietnam War was the war of helicopters. Without the helicopters, fighting the war in such a jungle terrain would not have been possible. Then-Defense Secretary Robert McNamara first seized the concept of air mobility in mid-1962. After numerous tests of the deployment of the helicopter, some air assault units of the U.S. Army were activated and committed to the operations in Vietnam. The new UH-I Huey helicopter, the big CH-47 Chinook transport helicopter, and other sisters in rotary-wing aviation were deployed to combat in Vietnam.

The helicopter was an essential war machine in Vietnam. In the Vietnam era, the U.S. Army deployed about 12,000 helicopters in the war, which were committed to providing various support over 2,000,000 times for the combat operations of American, Korean, and South Vietnamese forces. Among the different jobs that the helicopters carried out were: transport of personnel and equipment, an airstrike by gunships, evacuation of casualties by dust-offs, air maneuver, and fire support, and many others. The scenes where a helicopter caught the sight of friendly troops being chased by the enemy in the jungle, picked them up, and swiftly got out of it were frequent occurrences in the movies and in reality. Especially the scores of helicopters flying in formation to the combat areas would have been a spectacular scene if it had not been for the war. War is said to be a fantastic art if only a bullet does not come to me.

Once General Chae Myung-shin described the spectacular sight of helicopters committed to operations: "Operation White Horse 9 began on October 11, 1968, to wipe out the enemies in the Dong Bo area. The 29th and 30th Regiments of the White Horse Division conducted this operation. With 460 helicopters mobilized, the Division deployed 24 companies to attack at once from 06:00 hours, and the attack began with two-sided operations of a ground invasion and airborne maneuver. Fighter jets and 460 helicopters were flying black in the sky. I arrived at Hill 520 early in the morning and observed such an *incredible sight*."[11] Lieutenant Colonel Park Kyung-seok, a battalion commander of the Tiger Division, describes the condition of Vietnam that had made the usage of helicopters inevitable in his book *Route 19*. "To reach the objective, you need to bypass the hostile area. Air mobility is the only means of transportation in an emergency operation, supply, or evacuation. Indeed, the helicopter, which does not need a long airstrip, reigns over the war as its king."[12]

Second Lieutenant Kim Hyung-seok (Class of 1965) once admired the spectacular sight when he saw two infantry regiments flying in helicopters in one location for Operation White Horse 1 on January 29,

A Chinook helicopter carrying 105-mm howitzer to a remote jungle area in Operation Haesanjin 6 in Jan 1969. (Source: Institute for Military History, MND)

1967. Dozens of UH-1 helicopters, which could accommodate one squad each, took off simultaneously and moved to the rugged mountains of Nui Hon Ba. The scene of numerous helicopters flying into the hills covering the sky like a massive flock of crows, reminded him of a scene in a science fiction movie he had seen in his early childhood. In addition, seeing a CH-47 helicopter hanging a 105-mm howitzer and flying away to lay it in remote mountainous areas also aroused awe and wonder. He thought this was a form of modern mobile warfare, and he was fascinated by it, so he felt he had suddenly become a spectator. He was overwhelmed by the U.S. military's state-of-the-art equipment and tremendous resources.[13]

Korean troops heavily depended upon the U.S. Army helicopter battalions in small and large operations. Whenever they went on air assault operations, they requested the U.S. Army units that helicopters be dispatched for transport, strike, evacuation and other reasons. Every Korean regiment appointed a first lieutenant-level officer, commonly called a "helicopter officer," officially a Helicopter Liaison Officer or Operations officer who was an assistant to the G-3 of a regiment. He was in charge of all things concerning the operation of the birds with the U.S. Army helicopter battalion, which was attached to each Korean infantry division. He had to be fluent in English to communicate with the American helicopter pilots.

It was almost impossible for the soldiers to come out of the jungle on foot; if they did, they would have to expect the surprise attack of the VC, who lurked in ambush. In A Rumor of War, Philip Caputo's description shows how much the soldiers greatly depended upon the helicopters. Here he depicts the feeling of soldiers when they watched the planes flying off after they dropped them in the jungle: "Watching the helicopters climb until they were just specks in the sky, some of us felt a momentary but deep longing to go with them back to the small comforts and relative safety of what was called, for lack of a better term, the rear. It was a sense of being stranded on a hostile shore from which there was no certainty of return." Elsewhere, he writes, "When the helicopters flew off, a feeling of abandonment came over us."[14] *Apocalypse Now* (1979), directed by Francis Coppola, is one of the best movies ever produced from the Vietnam War. It is full of the roaring sounds of UH-1H choppers from the opening scene. The sounds dominate the whole stage of the movie and become its background music and, ironically enough, excite nostalgia in many Vietnam veterans' minds.

Since the commitment of the helicopter to the war in 1965, the U.S. army lost over 4,100 aircraft, amounting to the price of one billion dollars. Moreover, over 8,300 men were killed on board the helicopters, including 521 pilots and 3,770 crew members. Among the eight general officers killed in the Vietnam War, five deaths were caused by the incidents relating to the helicopter. According to a U.S. Army officer's report, the aircraft in the war contributed far more gains than losses. On many occasions, the helicopters turned the battles facing defeat into those of victory. General William Westmoreland evaluated that, without the helicopters, the total strength of the allied countries should have been 2,500,000, not 1,500,000, which was the strength at the peak of the war.[15]

One of the critical missions the helicopters had conducted was to evacuate the wounded to hospitals. The pilots proudly said that they evacuated those wounded to the hospitals within 30 to 35 minutes of getting injured. There was a time that one pilot had evacuated as many as 242 wounded men for four days. According to surgeons, 81% of the wounded in Vietnam could survive thanks to the helicopters, compared to 74% in the Korean War and 71% in WWII. Further, in Vietnam, 20% of the wounded were in critical condition, an extremely high number compared to the 2% during the Korean War and the 3%

during WWII. Yet the ratio of recovery in Vietnam was very high due to fast evacuation by helicopter and the courage of its pilots.[16] To the soldiers in action on the ground, the roaring sounds of the helicopters became a source of morale, for they could continue fighting as long as they thought they were being supported by the helicopters.

As far as the helicopters are concerned, the Class of 1970 graduates has some exciting stories. Most of the 1970 class members went to Vietnam first as platoon leaders and were transferred to higher units as staff officers after a certain period. At one time, all six helicopter officers of two divisions (Tiger and White Horse) were members of the Class of 1970, all of whom had already finished their tours as platoon leaders. To name some, First Lieutenants Lim Joong-bok, Lee Seung-gwan, Moon Hee-heun, Cha Byung-gyu, Lee Sung-woo, and Oh Hyun-koo served as helicopter officers.[17] All these officers acquired good knowledge about the usefulness of helicopters in the Vietnam War. The aircraft was a unique means of connecting the world with the bush. Once committed into the jungle by a helicopter, the soldiers felt abandoned and had to stay alive until the helicopter came in to take them out. Thus, the soldiers thought the aircraft was a lifeline connecting the primitive jungle and human society.

Those who had never experienced helicopter operations in Korea had an excellent opportunity to learn about the air assault maneuver by helicopter. The critical point was English, a communication tool with the American pilots. Though incomplete, these officers had a good command of English enough to execute the coordinated operations. While watching the American pilots doing their jobs faithfully, the Class of 1970 helicopter officers were deeply moved by their professional competence and will to accomplish the missions. Some helicopter officers happily remember the events caused by the language barrier. First Lieutenant Lee Sung-woo, the 29th Regiment helicopter officer of the White Horse Division, had to stay up all night, preparing for his first combat mission the following day. The problem was English. For effective coordination with the American pilot in the coming operations, Lee translated all he would say into English and learned it by heart.

When he met with the pilot the following day, he found that the

American did not understand his English. Deeply frustrated, Lee broke out in a cold sweat. He wondered what the problem was. Was it because of the poor English education at the Academy, where he had mainly learned the language? Or was it because of the lack of English competence of the American who could not understand his standard pronunciation? From that time on, however, Lee wrote on a sheet of paper in English what the American pilot had to do at the subsequent operations and showed it to him. And it worked, and he could meet with him without breaking out in a cold sweat. This language problem made him think again about how painful it had been for his elders with poor English competence to fight so many battles in the past Korean War, communicating with their American counterparts.

Perhaps the helicopter officer who had stayed in that position quite a long time and experienced systematic management of the helicopters would be Lieutenant Shin In-sun (Class of 1970), an operations officer in the Cavalry Regiment of the Tiger Division. Shin's story as a helicopter officer shows again how vital the helicopter's role had been in Vietnam. As most areas of operations were in densely forested mountainous regions, at the least scores of helicopters had to be mobilized for all battalion- and regiment-sized operations. A chief command helicopter operated in the air to control all the helicopters committed. This command helicopter was responsible for all helicopters transporting and withdrawing troops to and from the operational area. Throughout the operation, Shin was on board this command helicopter with an American Lieutenant Colonel who was a liaison officer to the Korean units and could observe the whole process from the start to the end. The command helicopter was always the primary target of enemy fire. Sharing his fate with the American colonel, Shin learned valuable knowledge of how to operate the helicopters in such a modern war.

Shin elaborates on an operation he carried out with the colonel on board the command helicopter. To commit the troops to the area of operations, the first thing to do was to strafe and neutralize the LZ. To feel the intensity of the enemy resistance, a veteran pilot flew into the LZ as if his bird's belly would skim over the ground. In other words, the pilot dropped down to treetop level to fly nap-of-the-earth (NOE) on

Successful landing of the 28th Rgt troopers on the top of a hill in Operation Dokkabi 21. (Source: Institute for Military History, MND)

the final approach. Just then, Shin had to throw down the red smoke grenades accurately to mark the LZs. This aimed for the Phantom jets and gunships to clear all the landing zones in advance so that the troops could safely land later. Right then, he had to expect the enemy's rain of bullets. Thus, the danger of being hit on the helicopter's belly or rotor blades by its direct fire always existed.

Sometimes the helicopter pilots feared being hit by the enemy's fire, and without landing, they told troops to jump down to the ground some meters high in the air. That happened to Captain Park Jong-sik (Class of 1959), the 11th Company commander of the White Horse's 28th Regiment. Usually, after moving troops by helicopter, when they reached the landing zone, in some cases, they received fire from the enemy on the ground, so the U.S. pilots did not entirely land on the top of the LZ, and the helicopter crew pushed the soldiers out from two meters in the air. So many soldiers hit a rock, sprained their ankles, and scratched their bodies in thorns, bleeding before even fighting. That's why commanders usually got angry at U.S. pilots and often made the gesture of pointing their guns to kill them. A rumor has it that, once, a regimental commander took off his helmet and hit an American pilot on the head with the helmet.[18]

Tiger Division soldiers landing by rope ladders hung down from the helicopter at the Dong Xuan area during Operation Ojakgyo Bridge in April 1967. (Courtesy: Institute for Military History, MND)

Ultimately, the helicopter was the most effective means of transport and assault in Vietnam. The helicopters were mobilized every time for all "search and destroy" and large-scale operations except for some patrol missions in the nearby villages. Many KMA graduates went through such a unique aspect of modern warfare. That helicopter operations experience became a valuable asset for the future of the Korean armed forces.

# PART II

# ENGAGEMENT: HOW THEY FOUGHT

# Chapter 1
# Class of 1955 to Class of 1957: The Pathfinders

## Class of 1955 (11th Class)

Korea Military Academy reopened as a service academy with a four-year collegial system in Chinhae City at the southern tip of the Korean Peninsula on October 30, 1951, when the war continued. And it accepted the first 200 cadets and held a historic entrance ceremony on January 20, 1952. All domestic and foreign newspapers called it the opening of "ROK West Point." Only 156 cadets graduated with a B.A. degree three and a half years later. And its ceremony was held at the Hwarang parade ground in Seoul on October 4, 1955. The Class of 1955, or the 11th Class (following the short-term 10th Class), became the first four-year class to graduate since the war's end in 1953 and the pathfinders for the fledgling nation's unknown but hopeful future. This class would lead in organizing the combat divisions to be deployed to the Vietnam War ten years later.

The commencement ceremony for this class, held on October 4, 1955, was a touching moment for both the graduating class and all the participants. With a crowd of about 3,000 people on hand, then-President Rhee Syngman gave the graduating class a heartfelt address, "Keeping in mind that the future of this new nation lies on your shoulders, you should make all-out efforts to become the guardian spirit of this country." Following this address, the Eighth U.S. Army Commander General James A. Van Fleet, who had made a significant contribution to founding the Academy, celebrated the cadets with the remarks, "Never forget the words--Duty, Honor, Country—and become fine regular officers with pride."[1] This class dedicated their graduation album to General James A. Van Fleet to whom KMA owes an enormous obligation debt.

The 1955 Class members had a sense of pride and mission as the first graduates of the newly established four-year Military Academy. When they were first assigned to the field units as platoon leaders or staff officers, they had nothing to consider except fulfilling their duty and responsibility. When the first combat unit of the Korean army was to be dispatched to Vietnam in 1965, ten years after their graduation, the members of this class, whose rank was Major, almost all volunteered to join it, but only 101 officers were selected. When the Tiger Division was first deployed to Vietnam in October 1965, the graduates of 1955 took almost all positions on the regimental staff and assistants to the division staff.

After one or two years, they were promoted to Lieutenant Colonel and became battalion commanders or division's principal staff like G-3 Operations of the Tiger and White Horse divisions. Lieutenant Colonel Sohn Young-gil served as the Tiger Division's G-3 Operations, while Lieutenant Colonel Lee Seok-myung did the same for ROK Field Command in Nha Trang. Those promoted to Colonel in 1969 served as White Horse Division's Chief of Staff, like Park Gap-yong, and assumed the posts of regimental commander of the White Horse and Tiger Divisions, respectively, like Chun Doo-hwan (later President of the Republic of Korea) and Park Sung-hwan. Except for these two gentlemen, KMA graduates had yet to take the regiment commander position and beyond in Vietnam.

In the early phase of the Korean deployment, two graduates of 1955, Major Lee Byung-gu and Major Lee Hyo, played vital roles as the operations officers of the First and Cavalry Regiments of the Tiger Division. The most senior KMA graduates, these two officers were chief architects of all operational processes that the two regiments conducted from their homeland to Vietnam.

Major Lee Byung-gu's First Regiment landed in Quy Nhon on 21, October 1965, as the first Korean combat unit. It moved to its location, Phu Cat County, under cover of a U.S. Marine battalion and a brigade of the U.S. 101st Airborne Division. The dominant terrain features of the region were Nui Phu Cat, 800 meters high in the north, and the Highlands, 600 meters in the northwest. To the east was the Quy Nhon Peninsula protruding into the South China Sea, and the central region was scattered irregularly with villages covered by dense tropical forests. Outside the villages spread the rice paddy fields and swamps. The roads within the area were Route 1 and Route 19, one running from north to south and the other from Quy Nhon west to Pleiku, the largest city of the Highlands, and into Cambodia.

The enemy in the region was estimated to be one regular VC battalion (E2B) of Binh Dinh Province and the VC of each village, totaling about one regimental size in strength. Except for the officers above the battalion commanders and high-ranking NCOs, nobody among the Korean troops had experienced actual combat, so everybody worried about what the real war would be like. They knew they had been thrown into a war with no front or rear. For almost a month, the regiment was engaged in constructing the bases of all units under its command. Upon completing its deployment and entrenchment, the First Regiment took over the missions in its TAOR from the U.S. military as of November 15, 1965.

Under the Tiger Division's concept of operations guidelines, Major Lee Byung-gu planned to conduct the operations of his regiment in three stages. In the defensive stage, the first thing to do was to complete the positioning of the units, fortify the tactical bases, conduct positive patrols in their surrounding areas, and construct strongholds in their tactical area of responsibility. In this stage, a company was independent in operating its tactical base, far from its battalion. The basic concept of the fortification was that each base, especially the company tactical base, might stand for over 48 hours when it received the enemy's surprise attack. The second stage, the offensive phase, was to search and destroy the enemy through wide and positive patrol in-depth and secure its tactical area of responsibility. This was the period to conduct company, battalion, and regiment-size operations. And the third stage was the stage of committing reserve forces to expand TAOR and to help the South Vietnamese pacification program within it by carrying out joint mop-up operations with the allied forces.[2]

Some operations in the First Regiment designed by Major Lee Byung-gu include "Battle of Luc Le," November 11, 1965 (Captain Lee Hak-gun, Class of 1960); "Battle of Binh Lam," November 13, 1965; "Operation Advance," November 16, 1965 (Captain Chung Man-gil, Class of 1960), and many others. Operation Advance, in particular, was an airmobile operation about the Quy Nhon Peninsula in the east, the first ever conducted in the history of the Korean army. Major Lee learned a valuable lesson about the air maneuver by helicopter, the most critical form of operation in the Vietnam War. The first stage of operations played a bridging role in improving an individual soldier's combat skills and solidifying the training base.

Another operational brain was Major Lee Hyo, Operations officer of the Cavalry Regiment of the Tiger Division. After landing in Vietnam on November 1, 1965, just ten days later than the First Regiment, the Cavalry Regiment stationed the units under its command at An Khe, Pleiku, Duc Co, and some other places along Route 19 and constructed company tactical bases. Early activities mainly focused on reconnaissance patrols nearby, and its first engagement was protecting harvest conducted by the Second Battalion and the Regiment's Reconnaissance Company. The next operation was Operation Lightning Flash 5 to mop up the VC's control over the area north of River Con.

This operation was carried out in the unique style of the Korean army, different from that of the Americans. If the American troops got fired on the way to attack, they would immediately bombard the enemy with massive firepower and move into its position. But in the process, the VC withdrew or hid in their caves after they delivered fire at the Americans. So, the Americans did not obtain the result that they desired. On the contrary, when the Korean troops got fired on by the enemy, they immediately attempted to counterattack and, at the same time, requested artillery fire to block their escape route. And then they enveloped them and waited, setting up ambushes, even for two or three days until they came out of the hideouts.

The soldiers of the Cavalry Regiment of the Tiger Division are boarding the helicopters to move to an operational area for the second phase of Operation Lightening Flash 66-3 on 27 February 1966. (Source: ROK Army Museum)

As Operations officer of the Cavalry Regiment, Major Lee Hyo established the operational concept of the Korean military in Vietnam. He explained it later in an interview:

After landing in Vietnam, we built a tactical base by battalion and focused on grasping the operational situation while exploring the surrounding area for the first two months. Over the next three months, the Cavalry Regiment moved the 6th and 11th companies to the Viet Cong-controlled area north of River Con, establishing their camps based on the concept of a company tactical base. To this move, the enemy did not launch any significant attacks. As Commanding General Chae judged, expanding the company's base, like a web, has effectively isolated the Viet Cong from the residents.

That the early battles for the Korean troops were unmistakably immature was inevitable. The opponents were guerrillas rooted in the residents who did not form a fixed front. Still, we could not succeed because we applied the traditional warfare method based on American military doctrine. However, we learned various ways to subdue the Viet Cong effectively by analyzing the enemy's tactics and characteristics and developing new tactics through trial and error. We reversed the nighttime behavior that the Viet Cong specialized in and carried out night ambushes, surprises, etc. And as we took over the initiative at night, the residents trusted us. Thus, after the adaptation period of five to six months after landing, we could develop a new countermeasure by analyzing the cause and effect of our engagements and disseminating it to all units, which became a success factor.[3]

Being aware that most of the small-scale operations in Vietnam were mainly ambush operations, Major Lee Hyo intensively researched ambush tactics. In particular, he developed the theory of ambush operations in densely populated areas, one of which was the village where the houses were closely spaced. The idea was well applied to the Vietnam War, for the Viet Cong were based in the village for their living and organization. In other words, the village was the source of their acquisition of food and personnel, and labor force. They also used the village to collect information and propagate it. However small it might be, the village was a crucial terrain feature. This idea of Major Lee was well-received by the U.S. and the South Vietnamese armies. Once, he presented a paper about the theory at a seminar hosted by South Vietnamese 2nd Corps Commander Lieutenant General Vin Lok on 21 September 1966 at Dalat.[4]

His ambush theory was further developed by his classmate Lieutenant Colonel Lee Sang-hoon (later ROK Minister of National Defense), the 1st Battalion Commander of the 26th Regiment, the Tiger Division. Lee applied it to his battalion operations. In an essay written for *Asadal*, the Newsletter of the KMA Alumni Association, in September 1968, Lee emphasized the importance of ambush and explained how to carry out its tactics:

About 80% of all operations conducted by the Korean forces in Vietnam were small unit actions, the core of which was ambush operations. Before the arrival of Korean troops, American and South Vietnamese troops suffered greatly from the enemy's ambush tactics as if it were its specialty. However, the night actions of the Korean forces at present were almost always carried out by ambush operation, and the enemy, who had enjoyed the tactics, was somewhat blocked by the Koreans who chose it as their primary tactic. The importance of the ambush operation was in providing security for the friendly tactical base and working as a means to separate guerrillas from the residents. In other words, the ambush protected the villagers and their properties by blocking the enemy from approaching the village. And it was to provide a chance for the large unit to clear the enemy's main efforts by limiting its activities and fixing it to any single area.[5]

The Tiger Division's TAOR amounted to 3,570 km$^2$, far more extensive than a Corps's in Korea. Within this vast region, the Tiger Division secured Routes 1, 19, and 6B and placed about 640,000 residents under its protection. With these achievements of protecting a large number of people, roads, bridges, and military facilities, this region was known as the most settled one in Vietnam due to the successes of the Tiger's innumerable operations, the ambush operation in particular.

In congratulatory remarks on the publication of *The Crusade of Hwarang Warriors* in 1967, a collection of stories written by KMA Vietnam veterans, General Chae Myung-shin praised the activities of the KMA graduates in this way: "Although the KMA graduates, who occupy most of the junior commanders and mid-level staff of the Korean army in Vietnam, are almost inexperienced in combat, they are usually enthusiastic about unit command and training. They have accomplished many achievements in battle through deliberate planning and bold leadership. I appreciate the education and training of my Alma Mater that raised them for their excellent leadership skills and a strong sense of responsibility. Still, I feel a hidden encouragement from my alums again who are always with me."[6] Although his remarks were not solely directed to the Class of 1955, their significant roles in various levels of office confirmed the expectations toward the KMA graduates in the early phase of the Korean deployment to Vietnam.

The 1955 class members served not only in the combat units but also in support units. Major Kim Young-lin, an Engineer officer, was asked to join the first contingent of Korean troops deployed to Vietnam in March 1965. Judging it the best opportunity to obtain combat experience, he accepted the offer and was assigned to S-3, the Engineer

Battalion of the Dove Force. The Dove Force was a unit organized to contribute to the pacification project in Vietnam. On January 8, 1965, he arrived in Vietnam as a member of its advance party. Major Kim received a briefing on the situation in Vietnam and the pacification plan from an officer of the USMACV. He also met with Commanding General William Westmoreland to tell him about the Dove Force's mission and garrison. Kim played a crucial role in settling the Dove Force in its early deployment phase.

Major Yang Sung-yoon was an assistant to G-4 Logistics, the Tiger Division, who was in charge of logistics support for the division. He explains the logistics support system for the entire Korean army in Vietnam: "As for the logistics support system, the 100th Logistics Command (aka Southern Cross Force) was established to support Korean troops in Vietnam. Under this command, the 101st and the 102nd Logistics Support Corps supported the White Horse Division and Tiger Division, respectively. Food and clothing were provided from Korea. Fuel was unlimited in this area. If you ran out of fuel during long-distance driving, you would get refilled if you entered any Korean or U.S. military bases nearby. Ammunition was supplied so that one company could fight independently for 72 hours. For this reason, a U.S. Army Logistics staff once complained that Koreans spent too much ammunition and continually replenished 6-7 times the standard. Mosquito repellent was provided in only one bottle per person a month. The same was applied to an American soldier. Yet the Koreans always asked more. So, the U.S. military used to make fun of the Koreans, asking more, "Do South Korean soldiers bathe with mosquito repellent?"[7]

Lieutenant Colonel Kim Ki-taik was the first Battalion com-mander among his 1955 classmates. At the death of a battalion commander, General Chae Myung-shin appointed Major Kim Ki-taik, who had served as an assistant to G-3 Operations of the Command of the ROKFV, as the new battalion commander as a replacement. Kim was promoted to a temporary lieutenant colonel as the successor to the late battalion commander. Major Kim was the first among his classmates to earn the rank of lieutenant colonel. The 1955 graduates still had to wait a year to become lieutenant colonels. As the 2nd Battalion commander of the 28th Regiment, the White Horse Division, Kim led Operation Dokkabi (Goblin) 2, conducted around Nui Hon Ba from 25 November to 10 December 1966, a mountain composed of limestone with many natural caves from top to bottom.

In this operation, Kim Ki-taik made an outstanding combat achievement. In recognition of this merit, he was awarded the Choogmoo Order of Military Merit, the third-highest medal of distinction in combat. At the usual time, Kim focused on practicing Taekwondo, bayonet drills, and close combat training in preparation for hand-to-hand fighting. In that operation, his battalion searched and cleared the VC in the caves on Nui Hon Ba, 16 km south of Tuy Hoa. His men under his command performed exceptionally. Especially, Captain Min Pyung-shik (Class of 1960), the 7th Company commander, raided the VC headquarters and weapons arsenal and captured 161 rifles and the flag of the best VC company. This was the most weapons ever seized by a company among all other Korean companies in Vietnam.[8]

Lieutenant Colonel Ahn Jae-seok (later ROK Defense Attaché to Washington as a two-star general) went to Vietnam as the 1st Battalion commander, the 28th Regiment, the White Horse Division, in August 1968. An outstanding cadet in English, he was selected as an English instructor candidate after graduation and transferred to Seoul National University (SNU) as a junior student majoring in English. After graduating with a B.A., he taught English at KMA for two years before going to the field unit. After completing various positions as commander and staff, he was promoted to lieutenant colonel and joined the ranks participating in the Vietnam War. As the battalion commander, he was engaged in large and small operations until he finished his tour of duty in September 1969, including the divisional-size Operations White Horse 9 and 10 and the regimental-size Dokkabi 12 and 13. During his tenure as a battalion commander in the White Horse Division, he gained fame as a courageous combat leader who led most battles to victory."[9]

## Class of 1956 (12th Class)

The date "July 7, 1952" was when history began for the Class of 1956, the second regular class of 200 cadets with the four-year program at KMA. The competitive rate for entrance amounted to 20 to 1. This class did not receive such attention that their elders had. These freshmen had to be prepared to endure being transformed from civilian to soldier. With the pride of being regular cadets like those of West Point, they could survive the beast barracks. These two classes especially had firmer solidarity than any other class as they had to set an example for their juniors. Among 200 entrants, only 164 cadets had the honor of graduating on June 12, 1956.

When the Capital Division was designated as the first unit deployed to Vietnam, its organization had few seats for the 1956 Class. After a year passed, though, when those who finished their Vietnam tours began to return home, they filled their positions as replacements. Among 164 graduates, 94 officers went to Vietnam and followed the path of their seniors. In Vietnam, these class members were initially the regimental staff. When they were promoted to lieutenant colonel, they were also assigned to the posts of battalion commanders like their elders.

The main branches of the battlefield are combat arms such as infantry, artillery, and armor. However, it would be a mistake to think that only combat troops fought alone. When all the combat units completed their deployment to Vietnam, the necessity for logistical support increased significantly, and the Korean army created the 100th Logistics Command. Through this Command, the ROK military could achieve the replacement of the M16 rifle for the old M1, the primary rifle of the soldiers in Vietnam, and at last, it could break away from the outdated army at the level of WWII and make a fresh start as a modern army. According to General Chae Myung-shin, "Major Park Sei-jik (Class of 1956, later Commander of Capital Security Command as a three-star general), an operations officer of the 100th Logistics Command, made an enormous contribution to realizing the project as a liaison officer by cooperating with the U.S. military. He was fluent in English, possessed excellent diplomatic skills, and could conduct missions given to him with boldness and aggressiveness."[10]

His command of English was due to his teaching experience at KMA. Two years after graduating from KMA, Park entered SNU as a junior student to become an instructor at KMA. He returned to his alma mater with a B.A. degree in English. After three years of teaching for mandatory duty, he went out to the field to pursue his career in the professional military. Upon completing his service as a company commander and staff officer in the field units in Korea, he was promoted to a major and went to Vietnam in June 1967. He served for the 100th Logistics Command for a year. After thirty-some years of service in the military, Park retired as a three-star general and successfully held the 1988 Seoul Summer Olympics as the President of its Organizing Committee after taking some ministerial jobs in the government of the Sixth Republic of Korea.

In the early phase of the Korean troops in Vietnam, some of the 1956 class graduates served as regimental staff, mainly like S-2 Intelligence, because their one-year seniors took the S-3 operational positions coveted by most field officers. Major Kim Joi-soo (Class of 1956) was S-2 Intelligence of the 1st Regiment, the Tiger Division, in 1965. Major Kim Moo-kyung (Class of 1956) went to Vietnam in 1966 as S-2 Intelligence and later was transferred to S-3 Operations of the 26th Regiment, the Tiger Division. Also, Major Han Il-soo (Class of 1956) was S-2 of the 28th Regiment of the White Horse Division. These officers gathered information about the enemy activities, interrogated the captured enemies, and distributed the produced intelligence data to the subordinate operating units. In Operation Fierce Tiger 6, conducted on October 9, 1966, Captain Yong Young-il (Class of 1960), the 9th Company commander of the 1st Regiment, was able to achieve an excellent combat result on the hilltop of Nui Ba, thanks to Major Kim Joi-soo who had provided a piece of important information on the Viet Cong company secluded there.

Major Kim Moo-kyung contributed an essay titled "Parts of the Vietnam War Seen by an Intelligence Officer of the Tiger Division" to *The Crusade of Hwarang Warriors* in 1967.[11] The following is part of his description of the Viet Cong and Operation Fierce Tiger 6, where he participated as an intelligence officer with the 26th Regiment of the Tiger Division, in April 1966.

First of all, the Vietnam War is unique because of the Viet Cong. The Viet Cong are active among the villagers, and it isn't easy to think of a Viet Cong separate from the residents. The organization of the Viet Cong is closely related to the size of administrative districts: battalion-level units in provinces, company-level units in counties, and platoon-level units in towns. They form a direct relationship with each organizational unit and take control of the residents. Also, they become agents of the NVA regular army, a spy network, and a tool for collecting taxes. They don't have combatant clothes, so it's hard to distinguish them from the villagers because they pretend to be villagers.

The Viet Cong's weakest point is undoubtedly the supply. Because their supply routes are cut off by friendly aircraft and artillery, they cannot perform exposed supply operations under any circumstances. The only way to solve the problem is only through local procurement. So, taxes are collected from residents, agricultural products are confiscated, and labor is mobilized to produce military supplies. The Viet Cong's equipment is mainly armed with China-made weapons. However, more than half of one squad does not have firearms. Once, a gun was captured from Nui Phu Cat during Operation Fierce Tiger 6; it was a wooden gun. It is common sense in the Vietnam War that "you should open any C ration cans and bury their leftovers during the operation or at any time." The reason is that the enemies try to find food and make booby traps using empty cans.

The Viet Cong attaches great importance to propaganda. The reason is that the Viet Cong lives among the residents of a village. That is why the enemy focuses on strengthening propaganda and instilling political ideology. Each household is forced to have a flag that symbolizes the Viet Cong and hang a picture of Ho Chi Minh to honor him as an eternal hero. The Viet Cong are doing their best to incite hostility toward the United States and its allies. Their propaganda is in vain, though. Since the Korean troops arrived here, innumerable propaganda leaflets were discovered within the TAOR of the Tiger Division, and the contents were bizarre. However, our soldiers were never swayed by the Viet Cong propaganda. As the ROK military continued to engage in civil affairs activities in its garrison and the pacified area, we could see that many residents resented their deception and rejoiced in getting out of the VC's control.[12]

## Major Kim Moo-kyung participated in Operation Fierce Tiger 6

as an intelligence officer, executed from September 23 to November 9, 1966. It was a 48-day operation, the longest ever conducted by the Korean troops up to that time. The operation focused on Nui Phu Cat and its surrounding areas, known as the home of the E2B battalion, the main force of the Viet Cong in Binh Dinh Province. According to the information, the Viet Cong were struck hard in Operation Fierce Tiger 5 across the Go Boi and Phu Mi plains but achieved significant rebuilding of its power. Nui Pu Cat was an independent mountain located 35 km north of Quy Nhon, with an overall width of 350 $km^2$. For the past 20 years, this mountain had been the enemy's hideout, training center, and stronghold of various facilities, including a hospital.

Two battalions from the 1st and 26th Regiments participated in this operation as the main efforts. And included were some units of the Cavalry Regiment and the Division Reconnaissance Company. The operation results were equivalent to the records accumulated by the Korean army in Vietnam for a year. This operation yielded 1,161 enemies killed and 518 prisoners and captured 43 crew-served weapons, 454 small arms, and 963 grenades. Four infantry battalions were deployed in the first stage, but two more were sent as reinforcements in the latter stage, totaling six battalions engaged in the operation. Major Kim Moo-kyung made a significant contribution to collecting and analyzing information about the enemy as a regimental intelligence officer.[13]

The artillery firepower in Vietnam was enormously influential, and it is not an exaggeration to say that it dominated the battlefields. All around the company's tactical bases or primary objectives to attack, there were numbered "fire concentration points," that is, the pre-planned targets on which fire was delivered when requested. Thus, if a forward observer asked for fire on any such points, artillery shells came down on them in no time. When a large-scale operation was carried out in the jungles, enormous artillery fire was delivered accurately to the requested targets. Artillery fire support during offensive and defensive operations by infantry was essential and caused the most severe damage to the enemy, either physically or psychologically.

Three 105-mm howitzer battalions and one 155-mm howitzer battalion provided artillery fire support to the Tiger Division in the

An Artillery Battalion of the Tiger Division in action. (Source: Institute for Military History, MND)

early phase of the Korean deployment. The 60th, 61st, and 10th 105-mm Artillery Battalions directly supported the 1st Regiment, the Cavalry Regiment, and the 26th Regiment, respectively, and the 628th 155-mm Artillery Battalion provided general support to the Division. As S-3 Operations Officer of the 628th Artillery Battalion, Major Park Tae-deok (Class of 1956) contributed significantly to establishing the early artillery operation plan. In *The Crusade of Hwarang Warriors* in 1967, Major Park elaborates on how the artillery firepower was operated in the early phase of the Korean troop deployment to Vietnam.

In phase one of the defensive operation, each battalion selected the area to give fire support for each regiment and prepared for self-defense after occupying it. And to facilitate the concentration of firepower in any direction within the area of operation, the artillery positions were circularly arranged to be suitable for firing 6,400 mm.

The second stage of the operation was the sweeping stage, which was to annihilate the enemy through extensive and in-depth reconnaissance and search. During this period, the artillery delivered suppression fire at the helicopter landing zones for preparation fire when an infantry company or battalion attacked. It minimized the damage to the friendly forces by concentrating fire at the already planned point.

In the third stage of operation, which supports the pacification task of the South Vietnamese government by destroying the enemy's main force, the artillery mobilized all the available firepower of the division artillery. In addition, in division-size operations, i.e., Operation Fierce Tiger 5 and Operation Fierce Tiger 6, the Korean troops received fire support from the U.S. Artillery. In this stage, the concept of fire support usually selects enough concentration points on or in the target or area in advance as the maneuver concept changes. In this way, the maneuver force could act under the cooperative support of the close artillery firepower. In particular, this artillery firepower suppressed the enemy's stubborn resistance, blocking the friendly maneuver force's advance. And in exposed fields or open areas, the artillery fired smoke screens up to 100 meters in front of the maneuver unit to obscure enemy observation and visibility.[14]

Lieutenant Colonel Choi Ung was one of the few among his 1956 classmates who took the position of battalion commander in Vietnam. He served as the 3rd Battalion commander of the White Horse Division's 28th Regiment from April 1969 to February 1970. During the period, he led his battalion to Operation White Horse 10 and Operations Dokkabi (Goblin) 13-17, and he recalls Operation Dokkabi 14 as the most memorable. At the beginning of the operation, his battalion started to penetrate on foot from the outpost to the designated PZ. The distance traveled on the road was just 10 km on the map, but it took as many as eight hours to cover. As he did not use helicopters to deceive the enemy, the soldiers were excessively consumed, even before starting the operation, which caused great difficulties in completing the operation.

This operation taught him some valuable lessons: first, when the troops with superior firepower attempted to conduct an operation against the enemy to wage guerrilla tactics, improving each soldier's combat ability and "one-shot one-kill" marksmanship was more urgent than anything else to defeat the enemy effectively; second, both sides were limited in observation and visibility in the jungle operation;

and third, the dense forest reduced the effectiveness of firepower. Overgrown thorn bushes became significant obstacles to employing and maneuvering search formations. Therefore, it was necessary to consider all these problems and study how to overcome them.

Lieutenant Colonel Choi was known as a commander who did considerable research on the operation in the jungle and suggested some theories regarding them. For example, it took a lot of work to maintain the line formation in the dense forest despite the emphasis on its importance. So, conducting a search operation in a column formation in the jungle was more effective. Yet the formation selection depended upon the enemy situation and terrain to achieve the maximum effect of rapid encirclement or firepower. One method of jungle search devised by Choi was to use dogs. Each company base raised a lot of dogs, and if you trained these dogs properly and brought them to the jungle search, you would get help to find the enemy's hiding places and facilities by using the dogs that were sensitive to smells and sounds.

As his artillery classmate Major Park Tae-deok mentioned, Lieutenant Colonel Choi also realized the importance of artillery firepower in the Vietnam War and was keenly aware that close cooperation between infantry and artillery was necessary. When infantry units conducted an attacking operation under the observation of the enemy, they could reduce damage if they got adequate artillery firepower. In particular, during Operation Dokkabi 14, the 12th Company on D+8 made a successful attack under the smokescreen of the artillery. And it could perform an effective movement by obstructing the enemy's visibility without revealing its operational intentions. Each company FO could concentrate firepower at the desired time and place during this operation by carrying out proactive fire adjustments.[15]

## Class of 1957 (13th Class)

Two hundred and fifty men entered KMA as the Class of 1957 on July 1, 1953, at the Chinhae campus. Of these entrants, 175 had a graduation ceremony at Taereung Campus in Seoul on June 12, 1957. The attrition rate was 30%, the highest among all the graduating classes. During the whole period of the Vietnam War, 110 of them went to the war. When the first combat unit was deployed to Vietnam, they mainly participated as a company commander, battalion staff member, and military history officer to record combat activities. Toward the war's end, some were assigned to battalion commanders upon being promoted to lieutenant colonel. There were few berths for the 1957 class members to take in the first deploying units. They had to wait one or two more years to assume the higher positions in the regiment or division. The Class of 1957 was the sandwich, an odd class of people, between being eligible for regimental staff and company-level positions in the early phase of the Korean dispatch.

Perhaps Captain Kim Dae-sik is the one who fought the severest battle among his classmates. As the 1st Company commander of the Cavalry Regiment, the Tiger Division, he landed in Quy Nhon on November 3, 1965, and moved to its garrison in the rain for a whole day. Perhaps, for this reason, he began to develop a high fever, which lasted for three days at 40 degrees, and he was hospitalized in a state of unconsciousness. Even before encountering the VC, he was fallen ill from contracting malaria in Vietnam. He eventually ended up in the hospital for 45 days. On January 1, 1966, he was discharged from the 1st Company commander. But he was immediately reinstated to his former position. The 1st Company commander Captain Park, who had replaced him, was killed by a mine explosion with his Weapons Platoon leader First Lieutenant Kim Hyun-young (Class of 1963).[16] At the request of the soldiers, Captain Kim returned to his initial unit, the 1st Company, whose morale was so low because its company commander and platoon leader were killed in a single attack.

Asked about any unforgettable events during his tour, Captain Kim Dae-sik mentioned two incidents: one was the ambush operation he won by killing ten enemies and capturing some weapons without sacrificing his men. And the other was Operation Lightning Flash 66-3, in which he took ten casualties among his men and had to retreat with those dead bodies in the middle of deep night. Captain Kim contributed the details of the latter operation to *The Crusade of Hwarang*

*Warriors* for a future lesson. The following is the summary of the article he wrote:

> Operation Lightning Flash 66-3 was carried out north of River Con in Bin Khe County for three days from February 25 to 28, 1966. The operation's objective was to search and destroy the enemies within the area. At that time, the 1st Company was assigned to advance to the point set by the Cavalry Regiment and block the enemy's escape route that the regiment had suppressed. My 1st Company killed seven VCs in this operation but suffered some losses, three dead and seven wounded. I could not forgive myself for losing my men as their commander. I felt ashamed as I thought of my poor fighting competence as a leader.
>
> As I watched each platoon returning to the company's assembly area after the battle, I unwittingly shed tears in front of my men. I showed my soldiers what I should not have shown as a commander in war. I couldn't hold back my tears when I looked at the ten casualties a day, including a fallen comrade on a stretcher made out of bamboo, a maimed soldier whose buttocks had been ripped off by a mortar shell, a platoon sergeant whose leg was pierced by a bullet. Indeed, I paid a high price as a commander with no combat experience in this operation. I learned valuable lessons from this harrowing experience in pursuing my future career as a professional soldier.[17]

The 1957 Class had several officers who served as Chiefs of the Department of War History. Their jobs were to directly observe the combat actions of the units and compose the reports to be delivered to the higher headquarters. While working in that capacity for the White Horse Division, Major Kwon Yong-soo and Major Koo Ja-yoon directly participated in many combat operations. Two crucial points for an officer of military history were to check whether a commander had a keen insight into the war's situation and the leadership ability to perform combat successfully. Above all, though, they were to write reports objectively because their reports could be a valuable reference for later uses. Major Kwon's first action report was about Operation White Horse 1, the first division-level operation since it landed in Vietnam. And Major Koo's report was about the regiment-level Operation Dokkabi (Goblin) 6 conducted by the 28th Regiment on

5-7 April 1968. Major Koo had been an instructor of Military History at KMA before coming to Vietnam. The following is part of the combat report composed by Major Kwon, the White Horse Division's military history officer:

Since the takeover of the present TAOR from the U.S. Army on October 22, 1966, the White Horse Division has opened Route 1 and conducted regimental-level operations. Entering 1967, however, the division planned to carry out the mopping-up operations in its TAOR and further expand it. Operation White Horse 1 was the first attempt to accomplish the scheduled task of the year. It was to search and destroy the Viet Cong's Khan Hoa headquarters, its subordinate units, and part of the NVA 18B Regiment, all based on Nui Hon Ba and its surrounding hills west of Route 1 between Ninh Hoa and Nha Trang. This operation would eliminate the enemy threat and protect the residents in the TAOR.

The operation period spanned 36 days, from January 29 to March 5, 1967. The Area of Operation, Nui Hon Ba and its surrounding hills and Cao Valley was thickly wooded jungles and scattered with numerous natural caves between rocks. Excepting some lowlands at the bottom of Cao Valley infantry units' ground maneuvers and helicopters' landing were almost impossible. The enemy situation in the AO was identified as Viet Cong Khan Hoa Province Headquarters and its five subordinate companies and NVA 18B Regiment Headquarters and its subordinate 7th and 8th Battalions, amounting to about 1,700 men.

As the division's 28th Regiment was conducting Operation Horse Head 1 around the hills south of Tuy Hoa, the strength to be committed to this operation was limited. So, for the defense of TAOR, the minimum strength remained at the company tactical bases. A total of four infantry battalions and five artillery batteries were committed to this operation, two battalions each from the 29th and 30th Regiments. And the operating units were supported by the U.S. Air Force's Ark-light (Cross Ocean Bombing by B-52 bombers coming from Guam), one 105-mm howitzer battery, and one combined heavy artillery battery composed of 8-inch howitzers and 175-mm guns from the U.S. 1st Field Army. And the helicopters for maneuvering were provided by some helicopter companies under the U.S. 17th Aviation Corps.

In order to destroy the enemy itself, the regiment set three objectives to defeat: first, the 8th Battalion of 18B Regiment north of the AO; second, the Viet Cong Khan Hoa Province Headquarters in the middle of the AO; and third, the 7th Battalion of 18B Regiment south of the AO. The operation was planned to be executed in three phases so that the four battalions' overwhelming strength and firepower could be concentrated on each objective. The first phase was the air assault maneuver by helicopters to strike a heavy blow on the heart of the enemy by a vertical envelopment and to disrupt the enemy's command and control. The second and third phases were to defeat in detail the enemy forces at staggered times while some of our troops were blocking them from their escape.

The results of the operation were enormous: (1) The division destroyed the Viet Cong's Khan Hoa headquarters, (2) It inflicted about 25% loss on the NVA's 8th Battalion, 18B Regiment, and (3) The division had confidence in the battles of the densely forested jungles by applying the ranger training they had received at home to the actual combat in Vietnam. One regretful thing was that, although we had contact with almost all the enemy units we had expected during the operation, we could not contact the 7th Battalion, 18B Regiment. The significant achievements that the operation had made were 393 enemy KIAs, 8 POWs, one defector, and 305 various weapons, including 81-mm mortars, crew-served weapons, and anti-aircraft weapons. Our casualties were 19 KIAs and 35 WIAs, and the KIA ratio was 1:21.[18]

Another military history officer, Major Koo Ja-yoon, participated in Operation Dokkabi 6 conducted in My Thanh and Phu Loc villages 6 km northwest of Tuy Hoa on 5-7 April 1968. He was Chief of the Military History Department of the 28th Regiment of the White Horse Division. This is the combat report of the operation by Koo:

The enemy left their strongholds on Mt. Ong La in the area 15 km northwest of My Thanh village, arrived at their destination at midnight on April 5, and was entrenched there for defense. Their infiltration was to destroy the 3rd Battalion, the 47th Regiment of the South Vietnamese army, 3 km west of the village and expand its control area. And also, it was to boost the morale of the Viet Cong in the area and solve their food problem by acquiring it from the village. The enemy proved to be the elements of the NVA's 95th Regiment and the village guerrillas totaling up to about 270.

The soldiers of the 28th Rgt charging into the enemy positions in Operation Dokkabi 19 in Oct. 1970. (Source: Institute for Military History, MND)

As this area belonged to the South Vietnamese army, the operation could be conducted only at its request. Some part of the 3rd Battalion of the South Vietnamese army was engaged with the enemy but defeated. So, the South Vietnamese regiment commander requested the 28th Regiment of the White Horse Division to support his operation. Immediately the 28th Regiment accepted the South Vietnamese request and searched the area in question. Major Kim Kye-doo (Class of 1956), S-2 Intelligence, confirmed the presence of the enemy in the area of My Thanh and Phu Loc villages by helicopter, and Major Chung Yong-woo (Class of 1956), S-3 Operations, composed the operational plans based on the enemy situation.

Following the artillery prep fire and aerial bombing, the regiment was to form the enveloping net with the overwhelming troops, block the enemy's likely escaping route, and destroy the enemy within the area.

From 3:00 p.m. on April 5, the troops began air maneuvering by helicopter and created the enveloping net after four hours, followed by night positioning. The regiment carefully enveloped the small area with a diameter of about 2.2 km and a perimeter of 6 km with about 900 men from seven companies from the 1st and 2nd Battalions under Lieutenant Colonel Han Il-soo (Class of 1955). Before forming the enveloping net, the artillery battalion fired 3,000 projectiles, plus aerial bombing and gunship strikes without lifting.

D-day: Completely enveloped by the friendly forces, the enemy judged they could no longer stand and began withdrawing around 7 o'clock after dark. The first encounter with the enemy was made at the front of the 6th and 8th Companies from 7:45 p.m. Around 20 enemy soldiers tried to break through the friendly forces' enveloping net but failed. At night, the enemy tried desperately to find the breakthrough point of the friendly's envelopment net.

So, they divided their units into 30-50 men, attacking with available arms like AK-47 rifles, machine guns, mortars, and B-40 rockets. However, the enemy fell into a trap set up by the regiment, and their attempt failed. That night, the regiment killed 64 enemies and captured about 30 weapons.

D+1: The main attacking operation was conducted in the daytime on D+1 day. From 10 a.m., the regiment changed its course into the offensive, and the search and destroy peaked around noon. The 6th and 8th Companies conducted precision searches for the village houses and killed the enemy hiding in the ceiling. When the daylight attack ended at seven o'clock, the regiment killed 104 enemies and captured 16 men and 50 weapons. However, the regiment also took five dead and 17 injured.

D+2: From 9 a.m., the regiment restarted the search operation for the houses they had missed the other day. When the search ended at 4 p.m., the enemy dead were 51. What was special that day was that the friendly troops captured alive the executive officer of the enemy's 2nd Company, the 4th Battalion. They took him to the village on board the APC and had him announce through microphone inducement of his men to defect. The result was that only two men came over to this side. According to the investigation of a POW, almost no enemy remained because of being killed or captured. It was a rare case in the Korean army in Vietnam that the indoctrinated communist had his men defect to the opposite faction.

Operation Dokkabi 6 was one of the best operations in which the Korean unit achieved the maximum result within the shortest period among the regiment-size operations. Then what were the reasons for the success of the operation? First, based on the exact enemy situation, the regiment rapidly concentrated the troops in the appropriate space and formed the complete envelopment net. Second, it was the concentration of firepower by 105-mm and 155-mm howitzers, air bombing by 28 jet fighters, and four gunships. These fire powers neutralized the enemy, and the infantry launched the attack. Third, it was because of the individual soldier's proficient combat skills. Adept as the individual soldiers have become through the incessant searching operations by company or battalion, they came to have confidence in village operation.[19]

Some of the 1957 graduates took the position of battalion commander in the later years of the Vietnam War. Lieutenant Colonel Paik Myung-yoon was one of them. After completing his tour as the

3rd Battalion commander of the 1st Airborne Brigade in Korea, he went to Vietnam as the 1st Battalion commander of the 29th Regiment, the White Horse Division, in December 1970. His battalion's mission was to guard Route 1 around Vung Ro Bay near Ninh Hoa. During his tenure, his battalion conducted Operations Vat 25, 26, and Eagle 71-1. His immediate superior was Colonel Chun Doo-hwan, whose slogan for the regiment was "to build the best unit in the world."[20]

# Chapter 2
# Class of 1958 and Class of 1959: Into the Fire

## Class of 1958 (14th Class)

The Class of 1958 took the entrance exam at the Chinhae campus, but the entrance ceremony was held at the Taereung campus in Seoul on July 1, 1954. This class makes a date—July 1, 1954—the day of their birth. At first, 243 cadets enrolled, and only 179 graduated on June 16, 1958. Seven years later, in 1965, many graduates of this class joined the Korean army to be deployed to the Vietnam War, and as many as 130 graduates participated in the war. The officers who initially joined the Tiger Division were mostly company commanders and battalion staff and later served as the regimental staff. Those who graduated in 1958 and later were short of rising to the position of battalion commander in Vietnam.

At the Corps of Cadets at KMA, the new entrants are often compared to "lion cubs." In other words, when a lion gives birth to cubs, it takes them to a cliff and drops them below, and it is said that the lion only raises the cubs that survive the height. This story is often compared to the newly admitted students. In actuality, "The Ritual of Lions' Den" began to be held for the recruits from this class. On the first day they enter, the recruits are led into the formation of their seniors, symbolically "lions' den," at the quadrangle of the Cadets' Corps and pass through it. And then, they are officially welcomed by the address of the Captain of the Cadets' Corps (today Regiment Commander Cadet). After this ritual, these recruits enter the first five-week basic military training (aka beast barracks), probably the harshest experience they have ever had. Only those who have completed this training are officially accepted as cadets. The graduates of 1958 likened their early selves to "lion cubs" who survived falling off a cliff. Seven years later, since they had completed their schooling at KMA, they had a chance to see the war for the first time.

Captain Park Dong-won (Class of 1958) joined the Tiger Division when it was designated as the division deployed to Vietnam in 1965. As the 5th Company commander of the 1st Regiment, he fought in Operations of Flying Tiger 1, 2, 3, 5, and 6 and Fierce Tiger 5. He was awarded multiple medals for his meritorious achievements in combat--Choongmoo and Eulji medals, and South Vietnam medals, including Gold Star and 1st degree Distinguished Service Medal. When he graduated from KMA, Park was awarded the "Model Hwarang," one of the two most honorable prizes given to the best graduate who had shown the most outstanding and exemplary leadership with upright character among his classmates during the cadet days. This award is the one most coveted by all cadets. (Another prize is the Presidential Award to an academically top graduate.)

## "The sun was setting on the battlefield of victory"

The following is the abridged version of the combat activities in Operation Fierce Tiger 5 by Captain Park Dong-won, written in *The Combat Histories of the ROK Army in the Vietnam War.*[1]

Two platoon leaders he loved collapsed, and even his executive officer, Lieutenant Chung Joo-young (Class of 1962), whom he relied on like his own limbs, was lost to the enemy's fire. And yet, even in such a situation, Captain Park and his men defeated the enemy in hand-to-hand combat, killing 75 enemies. This was at a battle in Operation Fierce Tiger 5, the largest operation in the Korean army in Vietnam until March 1966.

On March 23, the first day of Operation Fierce Tiger 5, Captain Park led the entire company at 2:00 a.m., leaving the company base 20 kilometers north of Quy Nhon, on a dark moonless night, killing the footsteps at the sound of frogs and entering the depths of the enemy position. They crossed the knee-high rivers several times and arrived at Phase Line, six kilometers away from the base, by 6 o'clock at dawn, deploying their troops and waiting for an attack signal. Ordered to penetrate deep into the enemy's position and block their escaping route, Captain Park led the company to it with the 1st Platoon to the left, the 2nd Platoon to the right, and the 3rd Platoon from the center in Phung Dong area. And the company successfully occupied OBJs 51, 52, 53, and 54. The operation plan was to form an intercepting position in the Khuong Binh area, advance to the Duong Thanh village, and then occupy Hill 78, the final objective.

When the 2nd Platoon formed the blocking in the Khuong Binh area and reached OBJ 58, the 6th Company, attacking in parallel, was engaged in a fierce fight with the enemy at Phung Dong. Ordered to support the 6th Company by his battalion commander, Captain Park had the 3rd Platoon block the enemy's escape route from the left and tried to help the 6th Company with the 2nd Platoon making a detour and 1st Platoon attacking from the center.

Massive automatic machine gun fire came out from the enemy's fortified position. Captain Park could not maneuver the 1st and 2nd Platoons, so he requested the smokescreen. At the same time, he had the company headquarters members change the enemy's firing direction and, using the moment, tried to advance the 1st and 2nd Platoons. In the process, the company executive officer, Lieutenant Chung Joo-young, was killed. Overwhelmed by the death of this young officer, a KMA four-year junior to him, Captain Park closed his eyes. He wailed in front of his subordinates for the first time, trying to close the platoon leader's eyelids, which kept opening as if he didn't want to die. In this battle, he lost three officers--the 3rd Platoon leader, the company executive officer, and the 1st Platoon leader. Pledging to fight with the Viet Cong until his death, Park stood at the head of the company and ordered the charge toward the enemy position. The company eventually won a great victory thanks to Captain Park's quick, sober judgment and excellent leadership.

At 2 p.m., the heatwave peaked, and the dead body's rotting smell poked his nose everywhere. After lunch in two shifts, the soldiers were busy finding the ditch and filling the empty canteens with water to soothe their thirst.

Although two enemy bodies were floating in the gutter, no soldier thought or said the water was dirty. Leaving the battlefield of victory, watching the sunset, he returned to the company tactical base with the soldiers. It was around 8:30 p.m. on March 23, 1966, the first day of Operation Fierce Tiger 5. His company needed rest again for tomorrow's battle. In the dark silence, everyone went to bed, promising tomorrow's victory again.

## A warm camaraderie that only men could feel on a harsh battlefield

Captain Lee Jong-koo (Class of 1958, later, ROK Minister of National Defense) went to Vietnam as the 7th Company commander, the 2nd Battalion of the Cavalry Regiment, in 1965 as a member of the first combat contingent of Korean forces. Though initially appointed to a battalion staff, he requested to be an infantry company commander to have combat experience in the field. The first battle he fought was Operation Advance 3, conducted by the 2nd Battalion from December 28 to 30, 1965, along the Son Con River in Binh Khe County. In this operation, his company captured tons of medicines and military supplies in a cave, which turned out to be the Viet Cong's 9th Surgical Hospital. The drugs and medical equipment seized were equivalent to two trucks.

Another battle Captain Lee fiercely fought was Operation Lightning Flash 3. Two battalions of Cavalry Regiment were committed to this regimental operation. The operation was conducted three days from February 26 to 28, 1966, in the Tay Son area. In the past several months after Operation Advance 3, Captain Lee's 7th Company had not achieved any remarkable results in some search and destroy missions, so the company members were full of grievances, which highly motivated them to be in combat. At 07:56 on D-day, all company members boarded 16 helicopters and landed safely at Landing Zone 7. While watching the soldiers of the 3rd Platoon, who first landed at the LZ, spot the fleeing enemy and open fire immediately, Captain Lee was convinced that he could win this battle.

Captain Lee gave an attack order to each platoon. While occupying the objectives one after another, the 2nd Platoon on the right came under fire from an enemy squad. The platoon leader, 2nd Lieutenant Kang Eui-yeong (Class of 1964), immediately enveloped the enemy and unleashed a concentrated fire on them, killing four enemies and capturing some rifles and grenades. The 1st platoon leader Kwon Young-dae (Class of 1963) encountered another enemy squad and killed two men. At 13:00, the 7th Company seized its final OBJ, My Duc village, and arrested many VC suspects in a hidden cave. The company also found a very narrow cave curved in an S shape. PFC Shim Mu-nam voluntarily entered alone, arrested six Viet Cong suspects, and came out with one VC soldier who had strongly resisted. When Captain Lee saw PFC Shim, whose face was all sweat and dirt, his heart was filled with indescribable emotion. After completing the first day's battle, he consolidated his company in the village and set up a night ambush.

The next day, a fierce battle took place in the area of the 2nd Platoon. Suddenly, radio communication between the company and the platoon was cut off, along with the sound of an explosion. Captain Lee learned later that the platoon headquarters personnel, including the platoon leader and the RTO, were injured in the enemy's surprise attack. However, the courageous platoon leader Lieutenant Kang immediately responded to it with the company mortar flares and artillery support. After about 30 minutes of firefight, 2Lt. Kang killed 11 enemies. Captain Lee ordered the 3rd platoon leader, 2Lt. Jang Yang-hee (Class of 1964), to support the 2nd Platoon and evacuate the wounded to the battalion aid station. At around 06:00 in the morning, they killed seven more enemies in the subsequent search, and the enemy fled, leaving many pieces of equipment behind. Captain Lee evacuated his injured soldiers by helicopter and consolidated his unit.

During the three-day operation, the soldiers were utterly exhausted from the heat and movements in the jungle. Patting the soldiers on their shoulders, Captain Lee commended them for displaying courage and perseverance. At the encouragement of their company commander, the soldiers happily responded with innocent smiles. At 16:00 on the last day, Capt. Lee received an order from the battalion to withdraw to the company's base at Kien My village. In this operation, his company killed 47 enemies, captured 97 prisoners, and seized some rifles and ammunition. Hearing the sound of the soldiers singing vigorously on the returning vehicles, Captain Lee felt a profound love for his men. He felt a warm camaraderie that only men could feel on a harsh battlefield.[2]

Captain Jang Joon-ik (later, Superintendent of KMA as a three-star general), as the 3rd Company commander of the 26th Regiment, participated in Operation Fierce Tiger 6, the largest and full-fledged mountain operation since the Korean deployment to Vietnam. On September 21, 1966, D-2 day, the 3rd Company split into 22 helicopters for demonstration and maneuvered to the Quy Nhon Peninsula. The open area on the west side of the ridge connecting Hill 361 on the Quy Nhon Peninsula and Nui Sa Den was characterized by thick trees of average height but limited foot movement.

According to the division's intelligence estimate, this area had no enemy situation. As soon as the company's first group landed, though, it encountered the enemy, and the firefight began. While Captain Jang was still looking down mid-air on board the helicopter, he saw the enemy escaping to the north. So, he changed the plan without delay and landed one platoon in the air behind the enemy's rear to block the expected escape route, and at the same time, he led one platoon to land on the enemy's left side. Eventually, his company destroyed all the fleeing enemies. The enemy POWs later confirmed that the platoon leader of the enemy D-555 Company was killed by a surprise attack of the 3rd Company while descending the Quy Nhon Peninsula for a reconnaissance. For his contribution to this operation, Captain Jang was awarded the Hwarang medal.[3]

The role of artillery in the Vietnam War was more significant than in any other war, but there are few battle stories about artillery. Although this is not about the artillery fire support, the first night experience in Vietnam of Captain Kim Jin-gyu (Class of 1958) deserves mention here. He was Charley Battery commander, the 61st Artillery Battalion, directly supporting the Cavalry Regiment. As a battery commander of the first dispatched Tiger Division during his tour, he offered excellent artillery support for infantry operations in numerous battles. In particular, he played a significant role in the Battle of Duc Co by the 9th Company of the 3rd Battalion, which his battery directly supported. For this achievement, he was awarded the Hwarang Medal. Here is the story of the first night he spent in Vietnam.

It was a silent night. The first night in Vietnam was deepening. I leaned against the trench wall and looked at the stars in the sky. But the stars were not romantic. My head was full of thoughts about the Viet Cong. But right now, there was a terrifying enemy, the mosquitoes. We had to admit that our defenses against mosquitoes spreading the deadly malaria were still sloppy. Mosquito nets were rather annoying to the front-line soldiers. I drank coffee from a C Ration can that had already cooled down. After checking the soldiers in the perimeter and the positions of guard duty, I found the darkness had already enveloped them, and everything seemed all right.

I kept instructing the soldiers to be alert for unexpected situations, but my heart was pounding. I believed that the chaos within itself could lead to unnecessary sacrifices. Suddenly the buzzer made a loud noise. The alarm from the battery FDC signaled a surprise attack from the enemy. 'I can see an enemy's light 50 meters from the 6th guard post,' said a soldier on duty. After a while, the situation soldier reported that the soldier at the 6th guard post pressed the buzzer. I thought it couldn't be. The enemy had no reason to show the light when he came. I went to the 6th post.

"Where?" I quietly asked the soldier, who seemed trembling.

"There it is," the soldier pointed towards Route 19.

I saw fireflies. The fireflies in Vietnam were unusually big and bright. The soldier was well acquainted with observation rules and did his job correctly. As a novice on the battlefield, however, he must have imagined something different and was sure of his imagination. I stayed with him for a while. Then all of a sudden, a gunshot, 'Bang,' was heard from the 2nd guard post. And then it got quiet. Strangely, if the enemy had come, either side might have continued to fire. I went out to that post.

"What's the matter?"

"The soldier did not answer quickly.

"Was it a misfire?" I pushed the soldier.

"No, sir! It was a dog."

"So?"

"I fired, sir."

"So, you hit it?

The soldier did not answer. I returned to my trenches thinking I had much to teach them tomorrow."[4]

As we all know, the Vietnam War was guerrilla warfare, and ambush operation was one of the main tactics. Initially, Korean units focused on defending their respective TAORs. And at night, the Viet Cong sporadically attacked the Korean bases as if the night were their main stage. However, Korean military commanders developed night ambush tactics, realizing that victory and defeat in the Vietnam War depended upon who seized the night initiative. Since then, the night became not a stage where only the Viet Cong could dominate. At that time, the ambush tactics became indispensable as an active alert measure within the TAOR of the Korean military.

Captain Kim Dae-heon (Class of 1958) first went to Vietnam as the Reconnaissance Company commander of the Cavalry Regiment. Aware of the war's characteristics as the guerrilla war, he carried out ambush operations independently. And based on his experiences, he spread them to most company commanders of the Tiger Division. His theory of ambush tactics did not come from a field manual but from his own experience. Later it was adopted as a doctrine and used as a demonstration training material for instructing the replacements in the Cavalry Regiment of the division.[5]

By 1968, the Class of 1958 members were promoted to Major. Those who had not joined the Korean deployment in the war's early phase went to Vietnam later when they became Majors. One of them was Major Lee Yong-su. Serving as the operations officer of Korea's 78th Regiment of the 27th Infantry Division, Lee came to Vietnam in August of 1970 after completing Army College in the same year. Lee was assigned to S-3 Operations of the 29th Regiment, the White Horse Division in Vietnam. The regiment commander at the time was Colonel Chun Doo-hwan. After taking office as the regimental commander, Chun said, "Running is better than ginseng or antler." As soon as they woke up, all the troops clapped their hands and sang military songs while running around the base. The regiment commander gave an electric fan as a gift to each soldier who had made an excellent combat achievement when returning home. Such an act boosted the morale of his men and drew loyalty from them.

From January 30 to February 7, 1971, the 29th Regiment conducted Operation Bat 25 to destroy the Viet Cong in the western Hon Ju Mountain area. In this operation, the 29th Regiment found an ammunition depot secretly hidden by the Viet Cong, seized numerous shells and small arms, captured one prisoner, and killed some enemies. Another operation that the 29th Regiment conducted was Operation Eagle 71-1, in the Suoi Cai Mang Mang area from February 23 to March 11, 1971. This operation was a failure on the part of the 29th Regiment. One soldier was missing from the 7th Company, and the adjacent 28th Regiment found his dead body and handed him over to the 29th Regiment.

In addition, the 10th Company was ambushed by the enemy, resulting in two killed and ten wounded. The 9th Company, mobilized to reinforce the 10th Company, was surrounded by the enemy in the dark and took three men killed. In Operation Bat 26, an accident occurred in which three were killed and two wounded in the 5th Company due to a misfire by friendly guns. Major Lee Yong-su learned from all these operations that we should be careful not to seek vainglory and find a way to minimize damage to our own forces.[6]

## Class of 1959 (15th Class)

In 1954, the KMA's Seoul era began. The Class of 1959 took the entrance exam in Seoul, and on July 1, 1955, they entered the Taereung campus. The 1959 Class commemorates this day as the "beginning of the history of 1959 classmates." The first "lion cubs" were 219 and 176 cadets who successfully survived the beast barracks graduated on May 27, 1959. The attrition rate was 21%. Of these, 126 men participated in the Vietnam War, 72% of the graduates. They were mainly assigned as company commanders or battalion staff of the Tiger and White Horse Divisions. To name some, Captains Lee Chun-geun, Bang Seo-

Lt. General Stanley R. Larson, Cmdr. of the U.S. I Field Force Vietnam, visits the Tiger Division's Jae-gu Battalion in 1966 and encourages its company commanders. Right: Capt. Bang Seo-nam (Class of 1959), 12th Co. Cmdr. (Source: Institute for Military History, MND).

nam, Park Chang-sik, Lee Won-yang of the Tiger Division, and Park Jong-sik, Kim Sang-gu, and Cha Jun-hoe of the White Horse Division were company commanders. Among these men was the company commander, Captain Lee Chun-geun, who fought the fiercest battle in the company-level operation of the Vietnam War.

## The Battle of Duc Co: "Hold Back the Night!"

Captain Lee Chun-geun, the 9th Company commander of the Cavalry Regiment with the Tiger Division, won the fiercest battle in the history of the Korean army at Du Co in Vietnam. On August 9, 1966, around 10:00 p.m., the 9th Company, under the command of Captain Lee, defeated a massive enemy attack and demonstrated the bravery of the Korean army throughout the world. With a desperate determination to "hold back the night,"[7] they fought for eight hours from 10 p.m. until 6 a.m. the following day. This remains the battle that created the Vietnam War myth, together with the one that the 11th Company of the Marine Blue Dragon Brigade conducted seven months later at Tra Binh village in central Vietnam on February 14-15, 1967.[8]

Duc Co was an area near the border between central Vietnam and Cambodia and was a crucial place for NVA troops to penetrate South Vietnam through the Ho Chi Min trail. It was located between Route 19 and the Ia Drang River on the east side of Pleiku in South Vietnam and Ratanakiri in Cambodia. In this area, small mountains with an average elevation of 190 to 300 meters formed an ancient forest, and numerous small rivers were intertwined. The area was so wooded with overgrown weeds and trees that walking, observation, visibility, and fields of fire were all limited. The enemy troops stationed in this area were familiar with the terrain and were less threatened by the allies. In particular, they would quickly retreat into Cambodian territory's so-called "sanctuary" if the situation were unfavorable. ROKFV sent the 3rd Battalion of the Tiger's Cavalry Regiment to this area on July 9, 1966, reinforcing the battalion's firepower by attaching some powerful artillery units.

At 10:00 p.m., on August 9, 1966, the enemy launched a night surprise attack, a full-scale attack on the 9th Company base. Thus began the Battle of Duc Co. After four months of service as the intelligence officer of the 3rd Battalion, Captain Lee was appointed the company commander. As a professional soldier, he fully understood the nature of war: "war is hell. You must die to live and fight for peace. Full of contradiction and paradox, war is nothing but the self-infliction of human beings. What is essential is not the matter of noble ideology but victory or death. We have to ask not why we have to fight but how we have to fight and win."[9]

On August 8, the 9th Company went on a combat reconnaissance mission according to unconfirmed information that there was an enemy division command. Captain Lee searched for the enemy headquarters for two days and one night in a 20-kilometer stretch of jungle. He went through an untrodden virgin land with no access road but returned to the base without any result. The next day, at 4 p.m. on August 9, when the soldiers returned to their base and took off their gears, the tension eased, and fatigue poured out like a flood, making it difficult to control themselves. The battalion still required two platoons to go on ambush duty, but the company commander judged that this was too much for the troops who had just come back from a mission, so the company was on two shifts in position and night watch to get as much sleep as possible. It was customary to bathe each platoon in a stream nearby in the evening after returning from an operation. The First Platoon leader, First Lieutenant Koh Kun-young (Class of 1964), requested the company commander to allow his platoon to take a bath in the stream but was not approved. Captain Lee was more afraid of lax discipline than of having to bathe and sleep well.

That evening, Captain Lee was leisurely conversing with Captain Kang Sei-ho, the new company commander who would replace him in a few days. Around 10:40 p.m., a report came that a trip flare exploded in front of the listening post of the Weapons Platoon, followed by another word that they heard the digging sound in the direction of the small stream ahead. Captain Lee ignored both reports, but after receiving the third one, he ordered the searchlight installed in the American tank to be reflected in the sounds direction. As soon as the light shone, the company's machine gun team found something

Captain Lee Chun-geun (right, Class of 1957), the 9th Co. Cmdr. of the Cavalry RGT, guides MG. Chae Myung-shin, CG of the Tiger Division, inspecting the battlefield of Du Co, where his company fought the fiercest battle on August 10, 1966. (Source: Institute for Military History, MND)

moving and fired massively toward it. A group of enemies was seen to penetrate the perimeter like high tide. At the same time, Lee cried out, “Hold back the position,” and tried to ask for artillery fire, but he was thrown into the trench by an unknown force. First Lieutenant Han Kwang-duk (Class of 1964), an FO, was fired upon, by the enemy when he tried to request artillery fire. Still, he regained his composure, calmly asked, and began adjusting the artillery fire. Captain Lee instructed Lieutenant Han to guide artillery fire at all concentration points around the company perimeter as the enemy company was swarming in.

All of the company members were now fully engaged in action. The enemy’s mortar shells were hitting the entire area of the company. The 1st and 2nd Platoon leaders were seriously wounded, and only the 3rd Platoon leader reported completing the combat positioning. At Han Kwang-duk’s quick request, artillery fire began to pound the enemy in and out of the perimeter. Captain Lee had a lot of regrets. “Why did I ignore two reports of enemy discovery? I will never escape this responsibility. I’ll fight and die here.”[10] Captain Lee judged that the enemies would siege the company and launch a decisive assault following fierce attacking preparation fire. Soon after, however, the soldiers’ morale increased with the friendly’s powerful artillery support, conducted under Charley Battery commander Captain Kim Jin-gyu (Class of 1958), a year senior to Captain Lee at the Academy. Then the intensity of the enemy’s mortar fire was remarkably weakened. The only option left for the enemy seemed to be to charge the company positions or flee.

Captain Lee and his men were prepared for hand-to-hand combat. Many soldiers aroused from sleep were fighting without even wearing clothes. The company commander returned them to their bunkers to wear combat gear to prepare for hand-to-hand fighting. Lieutenant Lee Chun-shik (Class of 1964), the second platoon leader, was lying with piercing wounds on his head and abdomen. The company commander couldn’t let go of Lieutenant Lee, a junior to him at the KMA, holding his hand tightly. “Death is the completion of life. But we never die.” He encouraged Lieutenant Lee. Next, Lieutenant Ko Kun-young (Class of 1964), the First Platoon leader, was also seriously injured. Still, even in that state, he did not lose his fighting spirit and encouraged his men to fight. The U.S. artillery also concentrated firepower on the perimeter of the company. All its artillery fire was pounding the incoming enemy, such as 8-inch howitzers, 175-mm guns, 105-mm, and 155-mm howitzers.

Despite the medic's dedicated treatment efforts, Captain Kang Sei-ho, the successor to Captain Lee, died. Around 4 a.m., the enemy in front of the 2nd Platoon suddenly started to charge over the barbed wire entanglements into the company base. The company commander asked Lieutenant Han Kwang-duk to guide artillery fire closest to the company base. Amid the powerful and accurate artillery supporting fire, the men confronted the enemy rushing into the base and held hand-to-hand combat. After that, the enemy's attack slowed down, and only occasional rifle shots flew in. A total of seven hours of fierce and nerve-racking fighting ended at 6 a.m.

The commanders and high-ranking officers of the U.S. and South Vietnamese armies who looked around the battle site uniformly marveled, saying, "It's a miracle. It left a new record and tradition in the Vietnam War history." Later, General Chae Myung-shin recalled the battle as follows: "Captain Lee Chun-geun's calm command and leadership stood out in an extreme situation, and when the enemy invaded the camp with a wave attack, all members of the company were radioed, saying, 'Hold back the night! Fight! Even to the last one! We'll all be killed if this base is broken through.' In addition, 1Lt. Han Kwang-duk temporarily lost consciousness when an enemy mortar fell right next to him, but soon regained consciousness and played a brilliant role in guiding artillery fire." General Walker, the Commander of the adjacent 3rd Brigade of the U.S. Army, said, "It was the first time I felt so embarrassed and anxious about my life tightened through my 20-year military life. I fought numerous battles in World War II. Still, I've never seen so many enemy soldiers killed in front of such a narrow space."[11]

The enemy's casualties were estimated to reach about 500, including 187 dead bodies, hundreds of WIAs, and six prisoners. In addition, the company captured four 60-mm mortars, twelve 40-mm anti-tank rockets, ten RPD machine guns, and 43 AK automatic rifles. Besides this, the company retrieved 500 explosive shells, 563 grenades, firearms, and 140,000 bullets. The company took seven KIAs and 42

WIAs. The enemy that attacked the 9th Company proved to be the 1st and 2nd Battalions of the 88th Regiment of Viet Cong and one Engineer Company, totaling about 700 troops. Sergeant Lee Jong-sei, who had assumed the command of a platoon leader when two platoon leaders were immobilized because of the wound, received the Taegeuk medal, the highest order of military merit, and Captain Lee received Eulji, the second-highest medal, and 10 men received Choongmoo medal, the third-highest one, including the FO 1Lt. Han. And 24 others received Hwarang medal, the fourth-highest one. Captain Lee was also decorated with the U.S. Silver Star.[12]

Another war hero among the Class of 1959 was Captain Park Jong-sik, who went to Vietnam as the 11th Company commander of the 28th Regiment of the White Horse Division in September 1966. As soon as he arrived at Phu Lam base in Tuy Hoa, he gave a speech to his men: "If you want to survive in Vietnam, you must keep these three things in your hearts:

1. Do not kill what is alive: Life was born by the power of God. Those who challenge that power must die under God's retribution. However, enemies and pests are excluded.
2. Don't rape. A jinx necessarily leads to bad luck when raping on the battlefield, such as God's teachings of retributive justice.
3. Dig your foxhole. Constructing a foxhole is the best way to protect your body from the enemy. Dig a strong foxhole wherever you stop during the battle.

My father fought on the Philippine front as a member of the Japanese Imperial Navy in the Pacific War, and my father said, 'Anyone who kills recklessly eventually dies tragically. Also, the raping person dies of a scary disease even if he survives.' In addition, one of my elder brothers, a Korean War veteran, said, 'I had a habit of diligently digging foxholes wherever I went during the battle. So, my nickname was a mole, and thanks to this effort, I came back alive without being injured.' Such experiences remain firm in my head."[13]

One day on a patrol mission, the 1st Platoon of his company delivered a threatening fire toward a house in a village. When Captain

Park went to the scene, he saw a white pig had been shot and killed. The company commander called in Sergeant Cho, who killed the pig, and asked if he knew three combat guidelines he had issued, and he replied, "Yes, Sir." The company commander asked, "Then do you know what you've done?" And then he ordered Staff Sgt. Oh, to bury the pig, Cho's squad leader, and build a tomb. He also apologized to the pig and instructed him to pray for its peace. Sometime later, Sergeant Oh reported holding the pig's funeral well and asked forgiveness for the unintentional killing. Captain Lee once again reminded the soldiers not to shoot people and animals that could not speak.

## "If I die, find my bones and send them home."

Captain Park Jong-sik is one of the company commanders who fought in the fiercer battles in Vietnam than anyone else among the KMA graduates. The dawn of April 10, 1967, is a day he will never forget. The 28th Regiment of the White Horse Division obtained information through the investigation of prisoners captured in Operation Ojakgyo Bridge that about 500 weapons were hidden in an arsenal of the NVA 5th Division. The regiment commander ordered the 11th Company to seize them. Accordingly, Captain Park planned to attack the arsenal on April 10, 1967 and decided to give the soldiers sufficient rest before action. The soldiers were exhausted from the past 45 days because they had been in the field for Operation Ojakgyo Bridge. And also, Park decided to operate only two listening posts that evening, not the six posts usually operated. So, all soldiers fell into a deep sleep.

Then, his deceased mother appeared in his dream, saying, "Jong-sik! Get up, come on, Quick!" Surprised, he got up, immediately called his messenger and RTO and ordered, "Each platoon should check and report whether there's anything wrong." As soon as he said the words, he heard a loud "Bang" from the front of the Weapons Platoon. While Sergeant Chun of the Weapons Platoon on duty was asleep with a radio receiver in his ear at the listening post, the enemy sappers naked lifted the barbed wire fence and infiltrated the base by crawling. The time was 2:30 a.m. on April 10, 1967. An 800 enemies, including two companies of the NVA's 5th Division and the 30th and 85th Battalions of the Viet Cong, attacked by surprise the Tan My base of the 11th Company, which had only 130 men sleeping.

The infiltrating enemy threw grenades and opened fire at the two posts of the Weapons Platoon. From then on, his company conducted hand-to-hand combat for 4 hours and 20 minutes on a pitch-black night. The friendly cannons and aircraft fired flare bombs alternately, but it was difficult to see even one meter ahead. He raised his voice as high as he could and shouted.

"Listen, each platoon, Hold back your positions. The enemies are all around here in our position. Kill all those who move without responding."

The men seemed to have gained courage when they heard their commander's shouts. The sounds of gunshots and grenades exploding here and there, the sound of hand-to-hand fighting, the dull sound of the rifle butts, the moans of the wounded, the sound of shells exploding in mid-air from the friendly artillery fire within the position, the sound of searching for the company commander, in a word, it was simply hell.

He gave the first head-to-head report to the battalion commander by radio.

"The enemy's coming massively."

The battalion commander kept asking him over the radio, not realizing the urgent situation.

"How many were killed?"

At this time, Sgt. Kim shouted, "Grenade, get down!"

Then he floated to the grenade's opposite side and covered Captain Park with his body. Fortunately, the two grenades were blind shells. The battalion commander asked again.

"How many casualties there?"

"The enemy's overrun us. We can't afford to figure out the casualties because the situation's going head-to-head right now. My last wish is, 'If I die, find my bones and send them home.'"[14]

And he immediately turned off the radio. From Park's words, the battalion commander seemed to have sensed the seriousness of the situation. Judging that one of his companies might be eradicated, he requested artillery fire within the base. During the hellish confusion of bombardment and hand-to-hand combat, they could easily recognize the NVA soldiers by touching their heads with their hands because the enemies had their heads shaved. In addition, the men could tell the enemy by smell when they approached. This was because the body of Vietnamese people smelled of soy sauce called nuoc mam,[15] just like the smell of garlic from the body of Koreans. Overhearing his continued shouting and commanding, the enemy seemed convinced that Park was the company commander. So, to capture the company commander, three naked men wearing only black panties rushed at him from the front and back. But they were killed by the orderly escorting him.

On the other hand, the friendly artillery shells falling within the base caused our casualties. Yet the fire within the position contributed to the failure of the enemy's attack. And he thought the B52 bombers would hit the target accurately, even at night, using radar, but one of the napalms dropped close to the company command post, and the gigantic pillar of fire drove them into a crucible of fear. Around six o'clock, the enemy began to recede. Captain Park told his men to take the wounded to his bunker, but there were so many injured soldiers that they couldn't get them in. He put the wounded outside the bunker, but his heart ached since they groaned incessantly.

"Hold on a little bit. The helicopter will arrive soon at dawn," he said with tears. He felt the night was longer than his life's, and dawn came. He ordered them to fire in unison at the fleeing enemy.

As the tide turned, the platoon leaders (Captain Park's KMA juniors), who had not been seen, began to appear before him. They were excited and suggested a counterattack.

"A counterattack? Damn it, Go to hell!"

How can they say so in human clothing? Where the fuck were they? The guys usually seemed to do everything only with their mouths!

"What the hell is KMA? Did I have to fight alone, risking death for fear of damaging that honor?" Covered in gunpowder on his face, he stood alone amid the roaring and moaning soldiers.

At around 10:00 a.m., the first UH-1D flew in and sat at the company base. Deputy Division Commander Brigadier General Baek Moon got off the helicopter. He came to Park, hugged him, and had tears in his eyes. The Chinook helicopter then landed on the outside of the company base. It simultaneously loaded nine men who died on the spot, three seriously injured, and 21 men with light injuries and flew to a field hospital. Later, the number of dead increased to 11. The enemy loaded 17 wagons with the dead and fled inland before dawn. Park's achievement included 43 enemy corpses, two prisoners of war, 18 rifles, and hundreds of hand grenades and live ammunition captured. This tragic disaster gave Captain Park a chance to see an aspect of human beings' nature and how it worked when facing danger.[16]

## The Jae-gu Village in Vietnam

The Jae-gu Village is the first village with a Korean name in Vietnam built for friendly relations with the South Vietnamese people. "Jae-gu" is the name of the 3rd Battalion, the 1st Regiment of the Tiger Division, meaning that it inherits the soul of the late Major Kang Jae-gu (Class of 1960), the 10th Company commander of the 3rd Battalion, who sacrificed himself by embracing a grenade thrown by a subordinate during the training for deployment to Vietnam. Captain Kwon Joon-taik (Class of 1959), the 3rd Battalion Intelligence Officer, contributed significantly to the construction of this village. Captain Kwon completed the irregular warfare course at the U.S. Army Special Forces School in 1963 before coming to Vietnam. The most important thing he learned then was that "victory in irregular warfare depends on the right combination of military and civilian operations while executing psychological warfare." Captain Kwon realized that the Vietnam War was a typical irregular war and focused on civil operations.

The Vietnamese people are from East Asian countries like Koreans, and they were straightforward and highly Confucian and valued etiquette and manners. As the Vietnamese suffered from war for a long time, they hoped to appease the hunger and anxiety of their families rather than ideology. Therefore, when the Korean military first arrived, they were wary of it, but the Koreans treated them like brothers. Later, they provided the Korean units with intelligence and participated in the defection projects of their fellows.

When the 3rd Battalion was initially stationed at Nam Tang, Captain Kwon suggested to his battalion commander, Lieutenant Colonel Park

Kyung-seok, that he organize a civil affairs team and distribute relief rice and medical treatment to the residents. And he also suggested that he put a red cross flag on the road leading to the deep mountains where the Viet Cong were hiding. Three weeks later, a response appeared from the residents, and this first response was that a father and son Viet Cong had defected. This was the first defection case since the Tiger Division came to Vietnam. And from then on, the battalion intelligence officer was officially assigned to the civil psychological warfare service.

After that, the battalion moved from Nam Tang to Phu Cat, further strengthening its civil psychological warfare activities. Many Viet Cong families were living in Hoa Hoi village in this area. Still, before the 3rd Battalion arrived, the South Vietnamese troops conducted a sweeping operation in this village, and the village almost disappeared to the extent that there was no trace of human touch. Captain Kwon suggested to his commander that he build 100 houses in this village to accommodate refugees who had lost their homes in the battle and asked the Chief of Phu Cat County, Captain Loc, to build a small school to teach the children. Accordingly, Captain Loc actively supported for construction of the village, and with the approval of Binh Dinh Province, the village was reborn with the Korean name "Jae-gu Village."

After some months, however, the Viet Cong raided the village, massacred the residents, and burned the school building. The whole village was covered with corpses. Afterward, as the South Vietnamese government again requested that the Korean military take charge of the region's security, the 9th Company took control of the safety and restored peace. Captain Kwon became famous enough to be nicknamed the "Korean Village Chief" in Vietnam, and Commander Chae Myung-shin especially recognized him for the work he had been doing.[17]

Captain Koo Bon-geol (Class of 1959), the 1st Company commander of the 1st Regiment, participated in Operation Fierce Tiger 6. His company's mission was to cut off the Song Dai area and search the 1st Battalion's TAOR. While the 3rd Platoon of Lieutenant Chae Hee-seon (Class of 1963) was in action, six friendly 155-mm shells fell on the platoon, wounding 15 men (1/2 of the platoon) and killing one. This was a close shot, not a stray bullet. As a result of later confirmation, it was revealed that they miscalculated the firing specifications when they were shooting supporting fire. Later, the company took one Viet Cong as a prisoner and found out where they were hiding their weapons through his interrogation. Captain Koo instructed Lieutenant Chung Poong-jeon (Class of 1964), the Weapons Platoon leader, to lead his men to search the Suoi Dai valley, and Lieutenant Chung finally found the arsenal and seized 25 weapons.[18]

On April 28, 1966, a Vietnamese woman visited the 3rd Company base of the Cavalry Regiment. She notified the company that one Viet Cong in her own home wanted to defect. The company commander, Captain Lee Won-yang (Class of 1959), sent several soldiers to bring him to the company. He was a 22-year-old young man named Than Tham. According to him, he left the village of Luc Le in Tuy Phuoc County, the operational area of the Tiger Division, and joined the Viet Cong in September 1964. However, since the Tiger Division's arrival in November 1965, he and his Viet Cong unit had to retreat to the An Truong mountain area, 40 km southwest of their village. This area was a mountainous region with mountains more than 800 meters high, including Nui Tha Duong.

Here the supply problem was a pain for them from the beginning. Crossing Route 19 was their only supply route, but the Korean army's active ambush operation completely blocked their activities, and they could not sustain themselves. Moreover, their main force, far from repulsing the Korean military, continued to be defeated and collapsed, so their morale was unreasonably low. The Korean army was a very fearful existence to them. Three months before he defected, his higher units issued the following instructions: "The Tiger units are potent, so do not attack and run away if you encounter them. If necessary, attack the South Vietnamese army or militia to obtain weapons and supplies."

Everything was hopeless for them. The men's morale was so low, to the extent that even their platoon leader escaped. Than Tham had been thinking about running for a long time and was only looking for an opportunity to put it into action. Still, at the dawn of April 21, he and two of his colleagues plotted to quit the Viet Cong life and defect and escaped through their surveillance network. He was handed over to the South Vietnamese County Office and was under its protection, and he was determined to return to his hometown and become a free citizen.[19]

Chapter 3

# Class of 1960 (16th Class): Company Commanders

The 1960 Class graduates of the KMA, who gave birth to a "military deity," the late Major Kang Jae-gu, have a special relationship with the Vietnam War. They seemed destined to play a crucial role in the Vietnam War for the Korean military. No one will deny the contributions the Class of 1960 has made to its success in the Vietnam War.

The Class of 1960 entered the KMA on June 27, 1956, and graduated on March 23, 1960. The initial entrants were 214, but only 178 had the honor of graduating. When the KMA was first established, it followed the semester system of the American school system. Thus, all the classes entered the Academy in the fall and graduated in the summer four years later. However, as the Korean fall semester system changed to a spring system, students from the Class of 1960 to the Class of 1962 who entered in the fall graduated in early spring. So, the educational period for these three classes was just three and a half years. The KMA applied a transitional period to these three classes in the study cycle. After graduating in March 1960, the 1960 class graduates headed to the frontline units with hope in the first year of a new decade that began the turbulent period of the sixties. At that time, the Cold War was at its peak, and the war that would later develop into the Vietnam War on the Indochina Peninsula was brewing slowly.

At the time, no one thought that the Class of 1960 graduates would become the main characters on the battlefront of the Vietnam War.

Those who had started as platoon leaders for their long professional careers at the forefront of Korea in 1960 began taking company commanders' positions around 1965. However, when the Vietnam War began, and the country finally decided to send the Capital Division to Vietnam as the first deployment unit, the Class of 1960 graduates were thought to be the right men to assume the positions of company commanders in the combat units. Given that the Vietnam War was a guerrilla war and most of the battles were fought by small units, the role of a company commander was the most critical. Among the 1960 graduates, 131 men participated in the war during the entire period, amounting to 74% of the Class. These class members took most of the positions of company commanders in the early period. Some of them later went for a second tour as regimental staff.

In October 1965, the first two regiments of the Tiger Division--the 1st Regiment and the Cavalry Regiment—were designated as the first deployment troops, which consisted of approximately 30 companies, including 24 Infantry, two Reconnaissance, two Combat Support, and two Headquarters companies. All the commanders of these 30 companies were 100% KMA graduates. Except for the four graduates of 1958 and 1959, all 26 company commanders were graduates of the Class of 1960. General Chae Myung-shin publicly announced that almost all company commanders of the two regiments would be KMA graduates of 1960. Most company commanders of the 26th Regiment of the Tiger Division and those of the White Horse Division sent in 1966 were also graduates of 1960 and 1961. These 1960 graduates made a significant contribution to establishing the fame of the Korean military as courageous and fierce in combat in and out of the country in the early period.

## The First Engagement with the Enemy at Phung Son

The First Regiment's TAOR of the Tiger Division was divided into north, southeast, and southwest, with one battalion deployed to each sector. An artillery battery was attached to each infantry battalion. Each battalion deployed its three companies to different places, so the Regi-

Capt. Jang Sei-dong (left, Class of 1960), the 3rd Co. Cmdr., 1st Rgt, Tiger Div, with 2Lt. Kim Kwang-hui, an Artillery FO, at Cu Mong Pass in 1965. (Courtesy: Kim Kwang-hui)

ment's TAOR amounted to about 750 $km^2$. On October 29, just a week after its arrival, there was a surprise attack by the enemy. It might have been an action to test the Korean troops' combat readiness and degree of vigilance. The enemy, disguised as a farmer, observed the Korean position in the daytime or from the top of a hill and may have believed that the vigilance would be poor. And perhaps the enemy intended to scare and dampen the spirits of the newly arrived Korean troops.

But their attempt failed. And the Koreans' powerful counterattack repelled the surprise attack, and the Korean troops showed complete combat readiness in their first engagement with the enemy. The 3rd Company (Commander Captain Jang Sei-dong, Class of 1960, later, Chief of the Presidential Security Service), which was attacked, killed the infiltrator with a surprise fire by pulling him up to three meters from its position. This was the first combat achievement that the Korean army had made since its arrival in Vietnam. Later, military history recorded this first battle of the Korean army as "the Battle of Phung Son," fought on October 29-30, 1965.[1]

## "Blow the Bugles:" The First Night Surprise Attack[2]

"The one who rules the night in the Vietnam War ends up the victor of it."[3] Thus wrote General Chae Myung-shin. In the Vietnam War, victory or defeat depended on who dominated the night. The person who succeeded in seizing the initiative of the night by conducting the night surprise attack by the Korean army for the first time was Captain Yong Young-Il (later 2nd Corps commander as a three-star general), the 9th Company commander of the 1st Regiment, the Tiger Division.

Since the 1st Regiment landed in Quy Nhon in October 1965, the Jae-gu Battalion (the 3rd Battalion) was stationed in the vicinity of the Regiment CP in charge of guarding the southern sector of its TAOR. Upon constructing the company's tactical base, Captain Yong focused on training the soldiers to adapt to the new environment by conducting platoon or squad-size patrol missions. Toward late January 1966, three months after his arrival, Captain Yong received an order from his battalion commander to provide security for a U.S. 101st Airborne Division unit carrying out a large-scale Operation White Wing at Bong Son area north of Quy Nhon.

His company moved to the northern end of the battalion and took a position around the torn-down school building at Khan Poc village. Here he had the baptism of fire for the first time in his life and learned that a good combination of officers' brains and the fighting skill of individual soldiers needed to win over the guerrillas in the terrain of Vietnam. In this first battle, his company completely seized the initiative and laid a foothold to provide security for Route 1. His next target was the enemy's stronghold at Hoa Hip village, where the Viet Cong would shoot at the friendly helicopters recklessly, even in the daytime. Indeed, an ideal place for organizing defensive operations, the village had mountains on the right and Hill 109 on the left. These two sides led to the valley opening to the coastal region. To reach the village, Captain Yong needed to pass a one km-long open area to the foot of Hill 109. Beside the hill flowed a stream eight-meter in width. As the hostile village was strongly fortified, he had to risk severe damage if he wanted to attack in the daytime.

Thus, Captain Yong decided to attempt a night surprise attack nobody had tried since they arrived in Vietnam. And he reported his plan to the battalion commander and got permission. His goal was: 1) the 3rd Platoon was to approach the village along the low ridges of the mountains, infiltrate into the rear of the object, and block the enemy from escaping to the coastal direction; 2) the 2nd Platoon was to advance to the front, and the 1st and Weapons Platoons were to move along the stream on the right of the village. And first, a point squad from each platoon was to close into the village. If all point squads succeeded, the main force of each platoon would move into the village all at once. If any point squad failed, he would give up the operation.

At midnight on February 26, 1966, he ordered three selected point squads to move toward the object along the designated route. The night was perfect for conducting the night attack because there was no moonlight. While they were stealthily approaching the enemy position, he requested a harassing fire delivered on it. In addition, he ordered firing some illumination shells to deceive the enemy while the squads were closing in on them. As he had trained his men thoroughly about how to act in the situation, they quickly hid when they heard the sounds of the illumination shells that left the mortars. If the illumination shells exploded all at once, the bright light of the explosion would blind the enemy. It was then that the men would move into the enemy positions quickly.

Three platoons reported that they had safely made it to the destination two hours after they departed and that, as expected, the enemy was there. So, Captain Yong maneuvered all company men toward the enemy position at about 4 a.m. Positioning with the Weapons Platoon, Captain Yong advanced through the deep water reaching their waists with the rifles raised high over their heads. Fortunately, as their movement was not to go against the water flow, they did not need to use all their strength to cross the water, but what they were afraid of was making noises by splashing the water. When they met with the point squads, it was almost 6 o'clock at dawn. While they approached the village, bullets started to come from behind, and their attack halted. They found the shots coming from Hill 109 in the rear. At the same time, the bullets also flew toward them from the village in front. They were rats in the trap as it was. All men assumed prone positions and waited for his order in a fix. The day was not yet dawning. The only way to break through, Yong thought, was to make a bold attack.

He ordered the bugler to blow on a bugle and all men to make an all-out assault toward the enemy position. Though he knew the assault with the bugle sound was a reckless act, it was the only way to give the enemy a sense of fear and, at the same time, elevate the morale of his soldiers. Embarrassed at first, the soldiers courageously ran toward the enemy position. Instead, the enemy panicked at their assault and did not have any thought to take action. When they entered the village, the enemy had already escaped to the mountain to the right. But as planned, his 3rd Platoon took its position to meet the escaping enemy and killed or captured some alive. Since then, the enemy in this area did not show any sign of attacking them. It seemed they all moved away to other places. The bugle was often used in the final stage of attacking operations among the Korean military from that time on.

From then on, Captain Yong Young-il became well known as an explorer of the night surprise attack in Vietnam. On October 1, 1966, at the desperate request of the South Vietnamese army authorities, the Tiger Division command decided to deploy Captain Yong's 9th Company deep into Nui Phu Cat by an airmobile operation. It was to be put into the heart of the enemy hideouts because the company caught their signs of escaping around Phu Cat Mountain. However, Captain Yong suggested canceling the airmobile operation by helicopter. The reason was that it would reveal the friendly's operational attempt at the enemy. Instead, he suggested that night invasion operations would be better. His proposal was approved. Indeed, Captain Yong was the first person who succeeded in the night invasion of the Korean army in Vietnam through Operation Jae-gu 2.

The company commander started toward the objective simultaneously in the late afternoon of the D-1 at a distance about five kilometers northeast of the target area. Since the 9th Company had to go through a jungle, it took hours to go only one kilometer. When they reached Hill 418, traces of cooked rice were found on the jungle trail. Judging that he had already entered the enemy territory, the company commander led the advance party and searched the jungle. Right at that time, he faced two enemy soldiers in the forest. They were frozen when they saw the Korean soldiers suddenly appearing and were captured.

At this time, all the captured enemies were dressed as NVA soldiers. In other words, this fact confirmed the intelligence that the NVA company was active in the Nui Phu Cat area. The company commander

Capt. Yong Young-il (Class of 1960), the 9th Co. Cmdr. of the 1st Rgt, Tiger Div., receives Eulji Medal, the second highest military distinction in the ROK Armed Forces, from Army Chief of Staff General Kim Kyei-won, for his achievement in Operation Fierce Tiger 6. Behind is Tiger Div. Cmdr. MG. Ryu Byung-hyun. (Source: ROK Army Museum)

found a group of enemies gathered between ridge valleys and concentrated heavy fire on them. In this surprise attack, the company killed 36 enemies and captured 12. In addition, it seized many weapons, including two machine guns and 11 AK rifles. Only eight company members were injured. The success of this operation resulted from Captain Yong's outstanding leadership and tactical knowledge. He voluntarily refused to operate by helicopter movement and chose a deliberate search focusing on night attacks and daytime covert activities. It was the first operation in which the Korean troops destroyed a regular company of NVA since their arrival in Vietnam.[4]

### "Retrieve the remains!"

Until March 1966, the Tiger Division wiped out most of the enemies in its TAOR through the first and second stages of operations. Still, the enemy unit, known as the E2B Battalion, was expanding its power in the area of Nui Phu Cat. The enemy had long dominated Nui Phu Cat and its surrounding area, which had never been occupied, so the South Vietnamese military called it a sanctuary for Viet Cong. Even if the Tiger Division completely wiped out the Go Boi plains, its TAOR still had the problem of being threatened by the enemy unless it occupied the VC's home base, Nui Phu Cat, and the jungle around it. Therefore, the Korean army decided to conduct Operation Fierce Tiger 5 to protect the harvest of residents during the harvest season and deny essential food sources to the enemy in the 100 $km^2$ area ranging from Nui Phu Cat to its southern end.

In this Operation Fierce Tiger 5, the first-ever division-sized operation, 13 companies participated from two battalions, each of the 1st Regiment and the Cavalry Regiment. On March 23, 1966, the operation began. Seven commanders of these companies were graduates of the 1960 class, including Captain Lee In-soo, who would have the harshest experience in this operation. Captain Lee, the 11th Company commander of the Cavalry Regiment, suffered heavy casualties during the operation to the extent that the company could no longer function. His company had to withdraw, leaving the bodies of its dead comrades in enemy territory. But Captain Lee had to go back to retrieve the bodies of his men. The division commander's stern order was not to leave comrades' dead bodies or the injured behind "within the enemy's sphere of influence."[5]

Captain Lee recalls the situation a year after the operation, "A year ago today! The horrors of that time are revived in front of my eyes. I lost six men and injured eight in Operation Fierce Tiger 5. I cannot forget the expression on the faces of my subordinates. It was hot, and I was so thirsty that we even had to drink bloody or rotting water. All the scenes of fatigue, sunstroke, enemy bullets, and scattered bodies are reflected in my eyes like a movie film."[6]

One decisive battle in Operation Fierce Tiger 5 was the Battle of My Thuan, where Captain Lee's 11th Company was trapped by the enemy's luring tactics while pursuing them in retreat. The My Thuan Valley was a vital route for the Viet Cong connecting Nui Phu Cat with the Go Boi plains, and it was the "Valley of Death" where the French army had suffered heavy losses some years ago. The 3rd Battalion of the Cavalry Regiment had little contact with the enemy since it entered the operation on March 23, 1966. As the same situation continued the next day, the tension relaxed, and most soldiers became impatient.

Around that time, the 3rd Platoon of the 11th Company, searching the My Thuan area, encountered the Viet Cong of one squad, pursued them, and entered the My Thuan Valley. However, the Viet Cong, in retreat, suddenly turned around and attacked the platoon. And the platoon suffered heavy casualties in an instant. The 2nd Platoon, reinforcing the 3rd Platoon, was also trapped in the Viet Cong's envelopment. Captain Lee realized he had entered too deep while the Viet Cong offensive grew more potent. Even if he tried to withdraw, evacuating the dead was a problem. It was around 17:00.

The company commander thought it would be even more dangerous if he did not withdraw immediately, so he decided to pull back, leaving the remains of five soldiers for some time. He decided to retrieve the remains during another offensive the next day. After eight hours of fighting with the Viet Cong, he barely escaped the Valley of Death. The problem was the remains he had left behind in the enemy area. The division commander, informed of the situation of the Cavalry Regiment, was outraged. "The Viet Cong will try to force more damage by luring the friendly with the remains as bait," he said firmly. "If you carry out the daylight attack tomorrow, you'll repeat another fierce fight. Therefore, detour the valley into the mountains TONIGHT and retrieve the remains. At any cost, the remains must be recovered immediately."[7]

Upon the division commander's order, the 11th Company started an infiltration maneuver at midnight, breaking through the jungle area west of the My Thuan Valley. As a result of maneuvering over jungles, cliffs, and rocky mountains, it took them six hours to climb just 2 km, and they arrived at the rear slope at 06:00, where they had fought hours before. Five remains were left in the fighting area, and after killing the Viet Cong standing on guard, an artillery battery poured TOT fire on the enemy position. Then, the enemies ran away when the 11th Company attacked the rear. Shortly after that, a helicopter arrived and carried the remains to safety. The company escaped the valley after repeated shooting and maneuvering, and the operation to retrieve the remains was successful.

## "Made in Pyongyang"

In the early days of deploying the Tiger Division to Vietnam, Captain Lee Kwon-bok (Class of 1960), the 5th Company commander of the

Cavalry Regiment, underwent ups and downs in combat activities. At the Battle of Phu Lac from November 30 to December 2, 1965, he lost two soldiers, and eight were injured. Without sufficient time for preparation, his company had to rush to the villages of Kien My and Phu Lac. Moreover, the Viet Cong was a battalion force, but the order was to attack it with one company. Once you got an order, you had no choice but to follow. On the first day, around 11 o'clock, the company crossed the River Con and attacked Kien My, but the Viet Cong had already escaped, so the 5th Company changed its attack toward Phu Lac. When the 2nd Platoon of the company reached 80 meters in front of the village of Phu Lac, the enemy's heavy machinegun fire poured onto them, and the platoon suffered two deaths and another eight wounded.

Captain Lee immediately called for artillery and air support and pounded the enemy position with his available firepower. After nearly two hours of bombardment, they occupied Phu Lac. However, his men could not find Viet Cong, as if they had taken away all their dead and wounded comrades. Captain Lee later responded to the question about why he did not counterattack immediately when they encountered the enemy's machinegun fire. Lee said that he judged that the enemy's strength was a reinforced company, given the intensity of the enemy's massive fire. And he argued that his company would have been annihilated if it had recklessly attacked the enemy with complete defense. But they did not accept his argument. Instead, there was a discussion about whether to punish the company commander.

However, Captain Lee compensated for his losses by winning the next operation. In Operation Advance 3, conducted on December 28 to 30, 1965, Captain Lee Kwon-bok's 5th Company found a cave in the valley and obtained a significant combat result, inflicting 30 casualties on the enemy and seizing many weapons and medical supplies. The company captured some prisoners, including a female nurse. After interrogating her, the cave proved to be the enemy's field medical unit. She said that seven nurses and 20 surgeons were in the cave, but they were all enveloped, unable to escape. The drugs captured were the quantities equivalent to those of two helicopters. Notable was that these drugs were all China- and North Korea-made, and even on the surgical instruments was written in Chinese characters, "Made in Pyongyang."[8] The rumors of North Korea's intervention in the war turned out to be true.

## "Life for those who want to die in battle, and death for those who want to live."

Operation Crazy Horse (aka Operation Lightning Flash 66-7) began at 4:00 p.m. on May 27, 1966. It was a grand enveloping operation jointly carried out by the Korean, American, and South Vietnamese forces. To Captain Park Jong-sung's (Class of 1960) 7th Company of the Cavalry Regiment, this operation was like going out for a field trip after completing training inside the base. The company was divided into three groups and airlifted into the jungle. It was armed with sufficient food and M18 anti-personnel mines to block the enemy's ambush and escape routes. The 7th Company members were still in the same company they had formed in the homeland before coming to Vietnam. However, they had felt bored with daily chores up to that time. This operation was a significant turning point for the 7th Company.

As the center of the battalion, the 7th Company was to move northward along the long and narrow Ha Nha Valley, a tactically unfavorable terrain. Although the path for the 7th Company was tiny and dense in the jungle, Captain Park did not stop advancing. Suddenly, an enemy mortar shell fell on the center of the 7th Company's command post. The enemy's surprise bombardment was not intended as a decisive attack but merely a sporadic action. However, it threw the 7th Company into chaos for a while. The RTO, who was with the company commander, was killed. A combat unit on a bloody battlefield was just a solid mass. If this mass broke, it would be the end. Park had no choice but to take action immediately. "Evacuate the remains immediately. Never let an enemy escape." The 7th Company wandered around the jungle to search for the enemy for six days like a "Crazy Horse."

As the company commander was gripped with disappointment but relieved that he had accomplished the mission, suddenly, a cry rang out, almost screaming, from the radio. "All the chicks here are dead. Send the chicks more quickly. Hurry up..." The sound was a radio message from the 2nd Platoon of the 6th Company, an adjacent company to his. The battalion commander's order came down to the 7th Company to support the 2nd Platoon of the 6th Company. So, Captain Park rushed toward the direction of the sound with his 2nd Platoon at the head. A signal of discovery of the enemy came from the point man, and a firefight erupted. The 7th Company barely made contact with the 6th Company. Thirty meters apart, the two companies attacked the enemy at an angle of 90 degrees. The distance from the enemy was 300-400 meters, and the firefight was fierce and violent. The soldiers were good fighters. "If this scene had been in a movie, it would have been the one worthwhile of getting an Academy Award," recalled Captain Park later.[9]

At this time, the 6th Company commander tried to make up for his platoon's devastating defeat and attempted a single attack to restore his honor. However, without the support of the adjacent 7th Company, the 6th Company's single attack, which was carried out only to revenge and regain the lost honor, was thwarted. Furthermore, because of the setting sun and the necessity to reorganize, the 7th Company withdrew to its position at the valley entrance and prepared for the next day's attack. The fateful morning dawned. The 7th Company advanced along the ridge to the valley's right. Fierce firefights ensued. After a while, the loud gunfire stopped, and silence came.

Finally, at 6 p.m., on the order of the commander's charge, various signal grenades exploded in front of the enemy's position. The men rushed toward the position like angry lions, threw grenades into the enemy's cave, fired, and pressed the enemy. Many enemies died in a scream in the cave. The 7th Company pushed the enemy into the cave until it got dark. Enemy corpses were already starting to smell rotten, and enemy firearms were piled up in front of the company. But a familiar corpse in military uniform made his eyes bloodshot. Lieutenant Chung Moon-ki (Class of 1963), a KMA junior and infinitely faithful and loyal to him, was killed. At the time of his death, Lieutenant Chung led his platoon without knowing an enemy position just 10 meters away from him. In the open ground, Lieutenant Chung's body was lifted by rope while a helicopter hovered. There was no way for Captain Park to suppress his emotion as he bowed his head and mourned for the deceased, who was slowly ascending carried by the ropes.

In the end, Operation Crazy Horse ended with the victory of the friendly forces. After the fierce battle, Captain Park acquired his life's motto: "Life for those who want to die in battle, and death for those who want to live."[10] Initially, these words were uttered by Admiral Yi Sun-shin, a renowned war hero of the Joseon Dynasty in the 16th century, when he confronted the invading Japanese fleet at the southern sea.

## "Do not let others know I'm shot"

No officer among the Class of 1960 had a more dramatic experience than Captain Lee Jae-tae. As the 11th Company commander of the 1st Regiment, he came to Vietnam in 1965 and engaged in many large and small operations. In particular, the night surprise attack he carried out at the battle of Chi Hoa was greatly praised by the division commander Chae Myung-shin. The general said, "His battle success was significant in that the Korean army took the night away from the Viet Cong."[11] Later, he was awarded the Choongmoo medal. Informed of Captain Lee's sacrificial achievement, President Park Chung-hee encouraged him by writing him a personal letter commending his courage and indomitable will. The following is a story about Captain Lee Jae-tae at the battle of Chi Hoa.

The success of the night surprise attack by Captain Yong Young-il's 9th Company for the first time on 17 January 1966 was followed by Captain Lee Jae-tae's 11th Company's attack conducted on the village of Chi Hoa on 27 February 1966. These two operations were carried out as part of Operation Jae-gu 2, which aimed to search and destroy the VC in the strategic hamlets along Route 1. With the information about two enemy companies assembled at the village of Chi Hoa, around 40 km north of Quy Nhon, Captain Lee Jae-tae immediately started his company at 23:00 hours and deployed all men to an intermediate objective at 02:00 hours the next dawn. Then, at 02:30, he ordered an attack on the enemy position. The attached 105-mm artillery battery and 4.2-inch mortars pounded the target and blocked the enemy's likely escape route. On attack order at 06:30, the company made an all-out assault toward the enemy position, with the two platoons ahead on the bypassing movement. The platoons infiltrated deep into the enemy's position with fierce fire.

At that instant, Captain Lee, having commanded at the head of his men, stood still as if he were frozen. Unfortunately, he received a piercing wound to the chest. It was a lethal blow. Seeing his commander wounded, his men were panic-stricken. He said to his men, "Keep your positions!" And he continued to command his men as if nothing had happened. Amid the blurring smoke of heavy fighting, Lee continually cried out in a loud voice. After a while, the battalion commander grasped the situation and promptly urged him to get out of the battle site by radio. Yet Lee responded shortly, "No, I'm OK. Don't worry, Sir." With these words, Lee still immersed himself in commanding troops. After several hours of fighting, the 11th Company killed 41 enemies and captured one prisoner. Also, the company captured 1 LMG, 8 AK rifles, 1 PRG, and many other weapons while taking only four casualties, including Lee.

Lee led all his men to a safe place by walking 4 km with his entire body bloodied. And yet he still let his wounded men evacuate first, and then he boarded the dust-off last. The operation was completed by Captain Rho Young-chul (Class of 1960), Battalion S-1, who replaced Captain Lee. When he was evacuated to the 6th Evac Hospital, the surgeons and medics on standby tried to treat the commander first. But Lee cried out, "I'm OK. Treat my men first." Checking his wounded men individually, he finally lay down on his operating table. The moment the surgeons saw the wound on his body, they could not but be surprised at its condition. The part that the bullet had pierced was the ribs. It was a lethal wound that needed at least 4-5 weeks of treatment. After the operation, the chief surgeon recommended that he be evacuated to Korea and treated there. At this, Capt. Lee refused his offer in a determined voice. "This wound's not one to be treated back home. I have a lot of things to do here. I can't leave my men behind. I'll be with them."

When MG. Chae arrived at the hospital and comforted him with his wound, Lee implored him not to be sent home. The general ordered the chief surgeon not to evacuate him back home. And the general asked him, "Why didn't you get out of the spot when wounded?" Feeling guilty for his incompetence that caused his men to be injured, Lee replied, "Sorry, sir. How could I come out only to save my own life when my men fiercely fought at the risk of their lives? I couldn't discourage them." Later, Chae recalls, "Captain Lee's firm determination, love of his men, sacrificial leadership, and sense of responsibility deeply moved me. He was a typical junior leader and a model of soldiers showing military spirit."[12]

And his immediate superior, Lieutenant Colonel Park Kyung-seok, wrote in his memoir, *The Heritage of Jae-gu* (1984): "With his chest blooded and jacket dyed with red, he didn't get fallen and fought to the last. He led his company for one and a half hours and completed the combat successfully. Indeed, he showed what a leader should be like on a battlefield. When I returned from his hospital bed, I said to myself, 'Living Kang Jae-gu, yes. He certainly is the living Kang Jae-gu!'"[13]

After the battle, the heroic story of Captain Lee decorated the headlines of all the domestic mass media. The following is from the article about Lee's combat performance published in *The Seoul Newspaper* on March 26, 1966: "Jae-gu Battalion of the Tiger Division created another legend in Vietnam. During the night attack operation, Capt. Lee, the 11th Company commander of the battalion, got a piercing wound on the chest. Yet he continued to carry out the search and destroy mission with all his body bloodied. As if nothing had happened, Capt. Lee completed the operation and took care of his other wounded men, and then he was treated. Capt. Lee issued a strict order 'not to tell his men that he got a lethal wound.' Lee's act reminded us of our venerable Admiral Yi Sun-shin, who had said, 'Do not let others know I'm shot,' while he was dying in a sea battle of a lethal wound caused by the bullet of the invading Japanese in the 16th century. Here we seem to see the incarnation of the Admiral."[14]

At that time, President Park Chung-hee read the domestic newspaper reports about Captain Lee and wrote hand-written letters, first to Major General Chae Myung-shin and later to Lee, to praise his spirit of sacrifice.[15]

Dear General Chae Myung-shin,

I read an article in a newspaper dated March 26, 1966, about Captain Lee Jae-tae, the 11th Company commander of Jae-gu Battalion, and I was deeply moved by the performance he had shown in the battle despite his lethal wound and couldn't sleep that night. The noble spirit of Hwarang warriors that our forefathers had initiated one thousand years ago has been bequeathed to the veins of our people through innumerable ordeals and hardships. Seeing it reincarnated and continually working in our people of today, I am overwhelmed with the powerful feelings of pride and deep emotion coming from the bottom of my heart. Please call him in and give my words of sincere compliment and sympathy. I understand that such a courageous and touching story of Captain Lee is not the first one your command has had. I wish you and your men good luck in war and pray for your success.

President Park Chung-hee
March 27, 1966.

Dear Captain Lee Jae-tae,

Your unexpected letter made me pleased too much. Chairman of the Joint Chiefs of Staff General Jang Chang-guk returned from Vietnam and brought me a picture he had taken with you. I treasure it, for you are a role model for our country's national armed forces. Indeed, we have many great soldiers like you now in our military, and many courageous soldiers will come out from those younger generations. Yet you're an exceptional soldier in that you commanded the combat with an undaunted spirit despite the lethal wound you got on your chest. While reading the article, I was deeply moved by your bravery and determined leadership, which reminded me of the time I had fought as a company commander in the Korean War. As long as we have a soldier like you, we will surely be able to build our strong nation with a bright and glorified future... I pray that all of our Vietnam soldiers will make outstanding achievements and return home safely.

President Park Chung-hee
June 2, 1966

Chapter 4
# Class of 1961 to Class of 1963: In the Midst of Politics

## Class of 1961 (17th Class)

If someone is asked to pick the three best classes in KMA history, he will choose the graduates of 1960 to 1962. This means that the most outstanding students throughout the country entered KMA in that period. They began their professional military career in the turmoil of the 5.16 Military Coup in 1961, leading to the Third Republic. Among those three classes, the Class of 1961 graduates later contributed most to establishing the Fifth Republic in the eighties. On July 1, 1957, 203 students of the 1961 class entered KMA; on March 20, 1961, the final 177 cadets received the honor of graduating. The attrition rate of the Class was 16%, the lowest among all KMA classes. One hundred thirty-two of the class graduates went to Vietnam.

The education period for the 1961 graduates was three and a half years. Since the four-year Academy was first established in 1951, the school authorities made it a policy that the regular semester began in September, following the West Point semester system. However, as the semester system changed to the one that started in March in Korea, students who entered September could not complete four years if they graduated in May. Therefore, the Academy authorities shortened the educational cycle of the three classes from 1961 to 1962 to three years and six months for a transitional period. From the Class of 1963, the Academy regularly tracked today's Korean school system.

In 1965, when the Tiger Division began to be deployed to the Vietnam War, the Class of 1961 graduates were yet to be captains, so they had to wait another year or two to become company commanders. Around the year 1965 to 1966, they served as company executive officers or battalion staff after completing time as platoon leaders in Korea. So, there were few positions for them in the early dispatching units. However, the graduates in 1961 became company commanders or the battalion's staff in late 1966. In fact, on April 28, 1966, First Lieutenant Kim Jin-young, who served as a personnel officer for the 1st Regiment of the Tiger Division, returned to Korea for a TDY to recruit the company commanders to replace those officers, mainly the class of 1960, who would soon be rotated.

Along with their promotion to captain in July, it was the turn of the 1961 class graduates to take the positions of company commanders in Vietnam. To most junior officers, the position of company commander in the Vietnam War was the most coveted position they wanted to assume. Lieutenant Kim Jin-young returned to Vietnam again with a bulging heart, confident that the Class of 1961 graduates would eventually show their competence as company commanders in the war.

When the Korean government began sending its non-combat unit, the Dove force, which comprised an Engineer battalion and a Security battalion, some graduates of the 1961 class were selected for these units and set foot as platoon leaders in Vietnam six months before the Tiger Division arrived in October 1965. First Lieutenants Park Kyung-won, Kim Tae-seop (the 8th Corps commander), Ahn Hyun-tae (Chief of Presidential Security Service), Lee Sang-ho (the 28th Division commander) went with the Dove Force as the platoon leaders of the Engineer and Security battalions.

With a few exceptions, many 1961 graduates went to Vietnam as company commanders of the Tiger and White Horse Divisions from 1967 to 1969. Some Tiger company commanders include Captains Kim Jin-young (Army Chief of Staff), Park Jae-hwa (the 9th Corps commander), Lee Moon-seok (the First ROK Army commander), Min Kyung-gu, Lim In-jo (KMA Superintendent), Lee Hyun-woo

(Chief of Presidential Security Service), Lee Byung-tae (ROK Defense Minister), Kang Myung-oh (Air Defense Artillery commander), Lee Jung-lin (ROK Deputy Defense Minister), among others. The company commanders of the White Horse Division include Kim Seong-ki, Kim Jin-wook, Yang Seung-bok, Chung Yu-hee, Choi Bong-ryong, etc. In addition, many officers, including Captains Kim Yeon-gak (the Second ROK Army commander), Heo Sam-soo (Senior Secretary for Inspection to ROK President), and Kim Sung-cho (ROK Army HQs Asst Chief of Staff for Strategic & Planning), served as company commander-level officers in the ROKFV, 100th Logistics Command, and MIG, respectively.

Captain Kim Jae-gyu, the 7th Company commander, the 29th Regiment with the White Horse Division, had a painful experience. When his company was put into operation in the Nui Hon Ba area during Operation White Horse 9, drinking water ran out. He had the company members drink their urine mixed with coffee powder in C-ration, and they survived. The 1961 class graduates are proud that, as KMA graduates, they were called upon to serve the nation during difficult times, went to war, and contributed to making it what it is today.

Among the 1961 graduates, one of the first to go to Vietnam as a company commander was Captain Park Jae-hwa. He went to Vietnam in July 1966 and participated in Operation Fierce Tiger 6 as the 3rd Company commander of the 1st Regiment. After the operation, he moved to the 1st Battalion's S-3 Operations. On January 23, 1968, however, Captain Park returned to his former position, the 3rd Company commander, replacing his classmate Captain Sohn Tae-ik who had been killed in the battle of Operation Fierce Tiger 9. As the 3rd Company commander, Park participated in the final stage of the operation, and then, a few weeks later, on February 12, 1968, he boarded the ship to return home. The following describes Park Jae-hwa's combat experience in Operation Fierce Tiger 6, emphasizing the importance of small units in Vietnam.

## “A squad is the smallest military tactical unit”

In Operation Fierce Tiger 6, Captain Park Jae-hwa’s 3rd Company moved north of Dai An River to Nui Phu Cat. It was to search the plains south of Nui Phu Cat. At dawn, on September 23, 1966, they crossed the Dai An River and participated in the attack on the D day, but, as expected, they did not contact the enemy. The 1st Battalion commander judged that the enemy had fled south of the river and ordered the 3rd Company to search the area, but they did not find the enemy there, either. On September 25, the company was assigned a new mission after being assembled in Hoa Dong. To prevent the enemy from escaping to the north of Nui Phu Cat, the 3rd Company was to block the Toh Mo Pass by 08:00 hours on September 28. So, the company landed on Hill 403 by helicopter and began secretly penetrating northward along the rugged ridge of Nui Phu Cat for three days. On September 26, while encamped at Hill 612, the company captured two Viet Cong.

On the morning of September 27, the 3rd Company departed Hill 612 and advanced to Nui Hon Heo around 16:00 under the cover and direction guidance of the reconnaissance aircraft O-1. While camping there, it caught sight of one Viet Cong moving, and Captain Park told the 2nd and 3rd platoons to pursue him. They found a cave where the Viet Cong was and surrounded it. They captured 16 Viet Cong and learned that the cave was the place for making the Viet Cong garments. While interrogating prisoners on the scene, the O-1 plane, which guided the 3rd Company, flew too low, collided with the eastern slope of Nui Hon Heo, and crashed. Captain Park immediately sent the 1st Platoon to the scene to find the wreckage of the plane and the remains of the two pilots and evacuate them.

Until then, the 3rd Company had mainly fought on the plains. They did not have an idea of breaking through a shrub jungle and lighting the equipment because they had no trekking experience in the mountains. The company commander learned a valuable lesson that they could only advance less than 500 meters per hour in the jungle. Captain Park ambushed the trail at the top of the Toh Mo Pass. Then he spotted a Viet Cong approaching along the path and arrested him. He frisked him and found that the enemy had an operational order issued to its subordinate units from the Viet Cong E2B Battalion headquarters. The translation by the interpreter was about the friendly Operation Fierce Tiger 6, published on September 22. The content was this:

*Operational Order: 1) The enemy is expected to operate in the Phu Cat area for a considerable period. They will make their way to the Di Gi region. 2) All units in the area should avoid the decisive battle when the enemy advances, retreat into the cave, and strike from behind when the enemy withdraws.*

Captain Park was surprised to see this document. Indeed, the division maintained the operation confidential so that even the battalion commanders who took part were unaware of its duration or progress. How come the enemy knew about the operation beforehand? One reasonable judgment was that a senior South Vietnamese army staff member informed the Viet Cong of the operation. It was an open secret that spies had communicated with the enemy among the high-ranking officers of the South Vietnamese army.

In the second half of Operation Fierce Tiger 6, the 3rd Company split into 27 helicopters and maneuvered in the air to Hill 462, repeatedly exploring the Suoi Da Valley area. There, the company discovered a cave and captured two enemies. In a detailed search of this area, traces of the Viet Cong E2B Battalion were found, for there were traces of installing eight cauldrons and a latrine, a rare field toilet in the Viet Cong camp. Probably somebody like the enemy E2B Battalion commander had been staying at the location. Here, the company captured one enemy and questioned him. He confessed that their main force had already retreated to the north.

After Operation Fierce Tiger 6, the 3rd Company carried out an operation, focusing on Route 503 and Lu Siem Xiang River banks and securing regional stability. Captain Park emphasized the importance of the operation of squad units in commanding his company, saying that "a squad is the smallest military tactical unit. The company begins with a squad. So, we need to have a strong squad." As the squads conducted most of the ambush and patrol activities, it was necessary to unite the squad members with their leader as the center. Combat training focused on infiltration, firing, and immediate action training. Because, every 23 days, about ten soldiers were rotated, the training had to be intensive, even in the war zone. During the attacks, the soldiers were ordered to use musical instruments like kkwaenggwari (gong), drums, trumpets, and whistles to scare the enemy and boost friendly morale. Captain Yong Young-il (Class of 1960) had used this method while conducting a night surprise attack a year before. As it was very effective, the higher unit authorities widely recommended using it in combat.

## "If the Korean army attacks, you must hide for a few days."

On September 27, 1966, Operation Fierce Tiger 6 started. The 10th Company, the Jae-gu Battalion of the 1st Regiment, searched through a large-scale rocky area where the enemy was believed to be concealed in the Ro Khan Valley of Nui Phu Cat. This combat activity was conducted under the command of the 10th Company commander Captain Lee Jung-lin (Class of 1961). First Lieutenant Kim Gil-bu (Class of 1964), the 3rd Platoon leader of the company, found a hole between the rocks in the valley that would barely allow people to enter. Finding human traces near the entrance, he decided to go into the cave himself. Due to the narrow opening, he threw off the carbine rifle and cartridge belt and entered with Corporal Han Sang-guk, with only three grenades and a flashlight.

As Lieutenant Kim descended the hole about 10 meters below, there was ample space of about two meters in height and five meters in length. Kim, who had been lighting the flash, sensed the presence of someone. Something fell in front of him and the corporal. At that moment, he felt it was a grenade and fell flat. Subsequently, a grenade exploded, but he was not hurt. At this time, platoon members rushed into the cave and fired toward the inside. They also threw grenades. After a while, the inside of the cave became quiet. When the men searched, numerous bodies of the enemy were scattered. The number of guns captured in this cave alone was 18, including a machine gun and a cabin rifle. The number of enemy bodies reached 23.

The division captured various munitions and supplies up to the fifth day of operation, including 69 individual and five crew-served weapons. Many enemies hiding in multiple places in the cave were dead, and it was difficult even to count their bodies. The Viet Cong command perhaps misjudged, "If the Korean army attacks, you have to hide for a few days. If they do not find us, they will immediately end the operation and withdraw."[1] Korean army did not act as they expected.

## "God is not Here"

Captain Kim Sung-cho (Class of 1961) went to Vietnam as a POW Interrogation Officer of the Military Intelligence Group (MIG) for its detachment to the White Horse Division at Ninh Hoa, 20 km north of Nha Trang. He participated in many large and small unit operations, including Operations White Horse 9 and 10, analyzed and translated the enemy documents, and gathered information about the enemy and his actions by interrogating the POWs and Viet Cong suspects.

After arriving in Vietnam at the end of March 1968, Captain Kim Sung-cho spent months without any particular enemy situations.

Capt. Kim Sung-cho (standing, Class of 1961), an MIG member of the White Horse Div, interrogating a VC prisoner.

Then one day, a report came in that a battalion of the 30th Regiment, the White Horse Division, had captured dozens of prisoners while conducting a search operation in the area. Though Captain Kim was skeptical, he went to the scene and found they were aged men and women and little boys and girls. They all proved to be family members from the mountain village. He was sorry for messing up an opportunity for the operating company to receive piles of medals, but he arranged to release them all. They were the Montagnard people, an ethnic minority in the Central Highlands of Vietnam who lived a primitive life deep in the mountains.

The operation that Captain Kim remembers the most during his tenure was Operation White Horse 9. The enemy hiding in Dong Bo Mountain, 970 meters above sea level, 10 kilometers southwest of Nha Trang, was commanding the city of Nha Trang and occasionally fired rockets at the Nha Trang Airport, which was always a threat. The operation commenced on 11 October 1968. The mountainous area where the enemy hid was complex terrain with countless natural caves and crevices among the gigantic rocks. It was difficult for large units to penetrate the caves in such a place.

A chemical operation was attempted by pouring over 100 drums of diesel into the caves but failed, and no progress was made for several days after the operation started. Instead, reports of damage began to come in. Captain Lee Jae-eon (Class of 1959), the 11th Company commander of the 30th Regiment, was seriously wounded. It was difficult to launch the dust-off helicopter due to the strong resistance of the enemy within only 100 meters. Captain Lee Jae-eon was a strict senior who was the object of fear to his juniors during his cadet days.

In the meantime, Captain Kim was preparing to interrogate the captured prisoners with his team at the 11th Company base. Right then, a helicopter landed, and an officer disembarked. Upon closer inspection, it was his classmate, Captain Kang Deok-hwang. "Hi! Kang, what's the matter? Oh, I know. You're the replacement for Captain Lee Jae-eon, aren't you?" Kim said to Captain Kang, "Listen! If you go to the field, don't wander around needlessly, and keep your position as a company commander!" Captain Kang probably had no idea what he was saying at that time.

Soldiers at a prayer session just before leaving for the battles in October 1970. (Source: Institute for Military History, MND)

The 11th Company built a tent church to accommodate 24 Christians on the base and hung a sign saying "Joshua Company" on a plank under a small wooden cross above the tent door. On the morning of the operation, the company members took off their heavy backpacks containing the 3-day MRE and ammunition. They went into the church to pray, "Merciful Lord, we'll go out for the operation soon. Please protect us from getting hurt until we finish the operation and safely return to the base." The soldiers' prayer in chaplaincy was earnest, but Captain Kim was only anxious inside. A few days after they were put into the operation, news of the company commander's serious injury came. A helicopter flew over the yard before the company base, dropping a dozen rubber bags containing corpses and flying away. The cause of their deaths was to defend the country and freedom, but his heart broke when he thought that these youths died because of the decisions of old politicians.

Captain Kim once read a book called *Confessions of a Humanist* when he was a cadet. It was a collection of letters sent by soldiers on the battlefield to their parents, wives, and lovers during the fierce battle of

Stalingrad in World War II. A letter from a German soldier to his pastor father remains in his memory, "Dad! I'm writing this letter in a trench where shells keep falling. No matter how much I cry out to God, God is not here. Next to me lay the corpses of my fellow soldiers, bleeding and dead, and the smell of rotting corpses." [Italics mine] The soldier who wrote this letter was killed a few days later. And Kim said to himself, "People say there are no unbelievers in the trenches, but God seems to be indifferent to human affairs."

Captain Kim obtained some critical information from some prisoners. While the soldiers were eating MRE, sometimes cooked rice was air-dropped for some field-grade officers. The point was that the rice dropped in the wrong place by mistake, and the enemy picked it up and ate it. Captain Kim reported this incident only to the regiment. According to what he heard later, the angered regiment commander warned that he would not keep him alive if any man were found to commit such wrongdoing.

Another vital piece of information that Captain Kim obtained was that some North Korean agents had participated in the Vietnam War. In June 1969, during Operation White Horse 10, a friendly unit searched the place where a regiment commander of the NVA had just stayed and obtained many documents. Among them were papers written in Korean and recorded tapes of conversations in Korean. Captain Kim analyzed these documents and confirmed the participation of the North Korean army in Vietnam.[2]

According to Captain Kim Sung-cho, Captain Chung Pyoung-yang was the most heroic officer among the graduates 1961. Captain Chung, who came to Vietnam as the 9th Company commander of the 30th Regiment of the White Horse Division in early 1968, made the first combat achievement in Operation White Horse 9, becoming a catalyst that affected the entire operation. Captain Chung's company established a company base in Dien Kanh district, west of Nha Trang. This place was a granary only with paddy fields, and Captain Chung took a drastic measure by setting up a barbed-wire fence in the middle of the paddy fields. As the Viet Cong were active every night to obtain food and medicine, the 9th Company set up an ambush every night. The company displayed excellent fighting power in these skirmishes and was evaluated as the most outstanding company in the regiment. Captain Chung received the Eulji medal, the second-highest distinction, and had the privilege of taking special leave to Korea. When he came to Seoul, he appeared in various media and TV shows to inform the public about the activities of the Korean military in Vietnam.

## Class of 1962 (18th Class)

The Class of 1962 graduates, the last Class of the transition period, entered the school on July 1, 1958, and graduated on February 26, 1962. Like the Class of 1961, they graduated after three and a half years of education. At first, 233 students enrolled, but only 181 graduated. In their senior year, they faced the May 16 Military Coup led by Major General Park Chung-hee (KMA 2nd Class) on May 16, 1961. The military government wanted a support demonstration from the KMA cadets to secure the legitimacy of its action. Despite controversies, all the KMA cadets went to Seoul City Hall and held a street march, supporting the coup. With this act, the KMA was involved in Korean politics and raised its voice at every critical moment. Later, once again, when another general, Chun Doo-hwan (Class of 1955), a member of the first regular KMA class, led a military coup on December 12, 1979, and established the Fifth Republic, some Class of 1962 graduates took part and played vital roles in its government.

Of 181 graduates, 153 officers went to Vietnam, amounting to 85%. This number was the second-highest among the entire KMA classes who had been to Vietnam (The 1966 graduates, 88% of them, went to Vietnam, the most among all KMA graduating classes). At first, when the Korean units were deployed in 1965, there were few seats for the 1962 graduates. The positions for them to be assigned were company executive officers because they still needed to be promoted to captain. However, when rotation began after a one-year tour, many graduates in 1962 who had become captains by mid-1967 could go as company commanders or battalion staff members for Tiger and White Horse Divisions.

The last company commander in Vietnam among the Class of 1962 was Captain Hwang Won-tak, who went to the war in October 1969 as the 10th Company commander of the 1st Regiment and returned in June 1971 after serving as a briefing officer at the Command of the ROKFV for the last six months. He was the "Model Hwarang," the most honorable prize given to the best graduate among his classmates, who had shown outstanding and exemplary leadership over four years. Hwang was promoted to a major general and served for President Kim Dae-jung's government as his National Security Advisor after retirement from the military. His final public office was as the Korean ambassador to Germany.

In an interview, Captain Hwang sharply criticized the overall behavior of the Korean military in the Vietnam War rather than narrating his own combat experience. And he expressed his opinion on how to use the Korean experience of the Vietnam War for the future development of the ROK military. He said that his dispatch to Vietnam in 1969 was one year after the Tet Offensive had swept the country, and there were no significant battles because it was when the U.S. withdrawal and truce talks began. In Vietnam, the overall mood among the Korean soldiers was to show off their courage by inflating their combat results. The problem was that such sentiment among the troops caused false reports (about the number of enemy kills and captured guns) to be made. In particular, when the U.S. military and the Korean media pointed out this problem, Lieutenant General Lee Sei-ho, Commander of the ROKFV, successor to General Chae, conducted a large-scale investigation. In case such a false report was found, he said, "he will never tolerate it, cancel the medal, and take measures to send him back to Korea immediately." He issued a stern order throughout the units under his command.

Captain Hwang also thought conducting the Vietnam War in the Korean army meant gaining combat experience. Still, he felt that a commander should not sacrifice his subordinates simply because of his desire for fame or publicity. Hwang's opinion was that the Vietnam War did not significantly contribute to the development of the combat power of the Korean military as expected. According to Hwang, the Korean troops mainly carried out operations such as "search and destroy" and ambushes by small units like platoons and squads. They didn't have the excellent combat experiences their elders had in the Korean War.[3]

Unlike the latecomer Captain Hwang, many other officers of the Class of 1962 experienced the Vietnam War early in their roles as junior-grade officers. First Lieutenant Lim Chung-soo (Class of 1962) was a pilot of an O-1 in the Tiger Division's Aviation Corps. There are few stories of aviation officers and their activities in the Vietnam War. However, Lieutenant Lim's experience as an aviation officer who guided the ground troops to victory in the battle of Operation Typhoon 7-4 draws our attention. After being commissioned, the KMA graduates rarely transfer to other service support branches from the initial one they choose, mainly combat arms. Lieutenant Lim, though, was the only one among the 1962 graduates who transferred from infantry to the aviation branch. His story unfolds in Operation Typhoon 7-4, which started on April 19, 1966:

At 7 a.m. sharp, First Lieutenant Lim took off from the base airfield with First Lieutenant Oh Hyun-guk (Class of 1963), Field Artillery Air Observer (FAAO), for a flight mission. High in the air, the pilots were sometimes fascinated by the view of the ground below. Each mountain ridge was covered with a round white ring of thin clouds. The rising eastern sun crossed the calm sea and was reflected in a multicolored rainbow, making the quiet morning fresh and more beautiful. Often, Viet Cong used this valley passage to move troops and transport supplies at night. Lim found enemies' footsteps when they reached the Phu Cat Mountain Valley. When he flew low to check the suspicious spot, he found the Viet Cong hiding after hearing an airplane from afar. They passed through the low air and identified about ten more enemies in the foxholes built on the left and right sides of the mountain path. He told Lieutenant Oh to ask the Alpha Battery of the 60th Artillery Battalion for supporting fire.

Because of geographical conditions, however, artillery fire was not effective. A radio message came from FO of the 3rd Company of the 1st Regiment, asking for air surveillance of the village of Thin Hoa. Since the village was a little south of its current location, the plane flew there with its altitude lowered. He noticed two soldiers caring for one wounded soldier when he reached the destination. Lieutenant Lim kept the height at 100 feet and flew over the designated village. Lieutenant Oh discovered about 30 suspicious men whose actions looked like the Viet Cong. Lim immediately informed the 3rd Company FO and the 60th Artillery Battalion of their movement. A little later, the number of men increased to a small company of about 70. Lieutenant Oh tried to request air bombing or artillery fire, but the distance between the suspicious group of men and the friendly troops was too close. Just at that time, the 1st Company members appeared. The point was that they were exposed to an open area as they were advancing from the east. Furthermore, the enemies were already completing their perimeter defensive deployment for action, which was unfavorable to the friendly soldiers.

Therefore, he tried hard to inform them of the enemy's ambush on the radio. Despite his desperate effort to inform them, they were unaware of his intention. He even wanted to signal the enemy's presence by crying out loud and waving his hand while flying his recon plane over their heads as low as possible. But the soldiers on the ground misunderstood his gestures for waving say-hello and only waved back at him, continuing to close toward the enemy. In the meantime, the VC waited for their advance to be within the effective firing range. In such a split second, what was passing by in his head? Could he save those soldiers at the risk of his life? With firm determination, he nosed his plane toward the VC and fired. As expected, the VC violently fired back at the plane and inflicted penetration on both wings. And then he looked at the approaching soldiers.

At last, the soldiers were aware of the presence of the enemy at the sound of firefights. Suddenly, the enemies' shooting, headed at the ground forces, was directed into the air. He thought about 200 shots might have come up. Suddenly Lieutenant Lim heard the explosive sound and felt some part of the plane might have fallen away. Later, he discovered that the explosion was the sound of a bullet piercing the fuselage. And he checked the plane's instrument panel but found everything was OK. For fear of something wrong happening, Lim returned to the base. Sometime later, Lim and Oh were told that the troops had destroyed the enemy unit, killed 45 enemies, and captured many firearms. Lim couldn't forget the Thin Hoa village. Because of this heroic action, he was awarded the Hwarang medal.[4]

## "Yes, Sir, I will!"

It was the last word uttered by First Lieutenant Chung Joo-young (Class of 1962) when he started an assault to carry out his superior's order. It was a sign of unconditional obedience to his company commander's order on the battlefield. Among the 1962 class graduates, he was the first to be killed in Vietnam. Chung went to Vietnam as the company executive officer of the 5th Company of the 1st Regiment. On March 23, 1966, he died of a gunshot to the abdomen while rushing toward the enemy position during Operation Fierce Tiger 5. Captain Park Dong-won (Class of 1958) led his company to attack a village. However, the three platoons under his command were met with powerful resistance from the enemy. A platoon leader was killed in the move. Two others reported to Captain Park that it was impossible to move further because of the enemy's unrelenting fire.

Then, Captain Park ordered Company XO, Lieutenant Chung, to lead all the headquarters personnel except for the RTO and attack the enemy position at the front of the 3rd Platoon, whose leader was killed. Upon receiving this order, Lieutenant Chung only said, "Yes, Sir, I will," and immediately advanced with some company headquarters personnel. Unfortunately, he received a penetrating wound to the abdomen while moving. Although he knew the order would risk his life, it was the mandate issued by his immediate superior, the one that had to be followed in any situation. He was a KMA man.

In February 1978, about twelve years after his death, a member of the Institute for Military History under the Ministry of National Defense interviewed his father. When asked about Lieutenant Chung's upbringing, his father gave him the answer: "Joo-young's personality has been mild and delicate since childhood. When he was in middle and high school, his house was poor, so he cooked himself and even delivered newspapers, but he didn't want his parents to worry and always comforted them. In his letters from Vietnam, he also showed that he had always expressed concern for his parents in every word. Joo-young was the filial son. He even asked his parents to marry his younger sister and brothers first because he would come back home victorious, no matter how many years it might take."[5]

## Class of 1963 (19th Class)

Today's KMA education cycle began with the Class of 1963. Before this time, a transitional academic calendar was applied to three classes from 1960 to 1962 to match the school system in Korea. Then, in the spring of 1959, KMA adopted a new semester system, and after four years, the Class of 1963 graduated in February. The competitive rate for the KMA was higher than any other university in Seoul, and this Class of 1963 passed the entrance exam with a 20 to 1 competition. On March 2, 1959, 229 entrants enrolled at KMA, and 205 cadets graduated on February 26, 1963. All these graduates served as platoon leaders in the frontline divisions.

When the Capital Division was designated as a combat unit to be dispatched to Vietnam in 1965, many 1963 graduates volunteered. They had already finished the positions of platoon leaders on the front. Almost all the graduates of the 1963 class were selected as platoon leaders. The policy was that the officers chosen for the war must have already completed the corresponding positions in the homeland. In other words, the 1963 class was the most senior officer among four platoon leaders in a company because they had already finished their tours as platoon leaders. So, in the units to go to Vietnam, their position was the Weapons Platoon leader, usually assigned to the most senior officer among the four platoon leaders in a company. Most 1st, 2nd, and 3rd platoon leader positions were allotted to the Class of 1964 members, one-year junior to them. And almost all 1963 class graduates were appointed the Weapons platoon leaders or company executive officers.

In September 1966, when the White Horse Division went to Vietnam, many 1963 graduates were appointed as assistants to the regiment staff, battalion staff, regiment combat support company's operations officer, firing battery commander, etc. During the entire war period, 148 graduates among them participated in the war, amounting to 72% of them. Of these, four officers were killed. In particular, First Lieutenant Paik Cha-rang, an aviation officer whose kind was few among the KMA graduates, was killed by enemy fire during the reconnaissance mission onboard the L-19 observation plane. While executing his task as an aerial observer for the 628 Artillery Battalion, his aircraft was shot down by the VC fire, and he was killed together with the pilot. At 5:25 p.m. on September 27, his plane guided the artillery fire to support the combating troops at the area about 25 km north of Quy Nhon. But his plane was downed by the VC's concentrated, intense fire from small arms.

## Coup de Grâce[6]

In warfare, a friendly soldier or captured enemy, fatally wounded, becomes a critical impediment to the unit's activities and is often killed to alleviate the suffering. First Lieutenant Chae Hee-seon (Class of 1963), the Weapons Platoon leader of the 1st Company of the Tiger Division's 1st Regiment in Operation Fierce Tiger 6, tells the story of how he decided to kill the enemy he had captured. The following is Chae's action in operation: During Operation Tiger 6, Chae's platoon found several enemies hiding in the water near a small lake while searching for a village. They looked clear under the water with the bamboo tube in their mouths. The platoon captured three men and one woman. When asked, "Where's a rifle?" An enemy replied, "I don't know." A soldier insisted that they would kill the prisoner as an example.

And further, he said, "Even though you caught or killed an enemy, the higher authorities would not recognize you. The number of captured weapons is the only thing they can recognize as your combat achievement. And then you will be able to win a medal." The point was that you could aim at an enemy running away in the front sight of your rifle, but you couldn't kill the man point-blank or right next to you. Even if the platoon leader threatened the prisoner, he only replied, "I don't know." So, he decided to turn the prisoner over to the battalion headquarters for further interrogation. He saw expressions of disappointment appearing on his men's faces.

During the search, Lieutenant Chae's platoon found a cave, and two soldiers entered to find the enemy. In it, they found an enemy who strongly resisted and waged hand-to-hand combat with him. The two soldiers almost killed him but dragged him outside alive. The enemy soldier was seriously injured. When asked about the rifle, the enemy said, "I don't know." The problem was that it was too far away to lead him to the bivouac area. Maybe he'd die even before he got to the camp. The platoon leader thought about what to do, but the RTO said next to him, "Sir, we'll kill him." The RTO had an order from the company commander to get back quickly.

"Yes," the platoon leader said curtly. But he was reminded for a second when taking an English exam in his Academy days. Filling the blank with the words "coup de grâce," he had left the classroom. On a mountain path in Vietnam, far from home, the words suddenly came to him, and he told his men to shoot the prisoner who looked young like them. Chae silently walked down the hill and returned to the camp.[7]

## "Why are you ahead, sir?"

First Lieutenant Kim Joo-cheon (Class of 1963) participated in Operation Fierce Tiger 5, in the Go Boi plains on March 23, 1966, as the Weapons platoon leader of the 5th Company of the 1st Regiment. Lieutenant Kim's Weapons platoon assigned one 57-mm RR squad and one 60-mm mortar to each of the 1st and 3rd platoons. When the 3rd platoon leader was killed during the operation, the company commander appointed Lieutenant Kim as the 3rd Platoon leader. Upon receiving the order to charge, he and his men ran over the river bank and rushed toward the objective. When he occupied Tinh Binh village, he was again ordered to continue his mission as the 1st Platoon leader, leaving the command of the 3rd platoon to the platoon sergeant. From then on, Lieutenant Kim served as the 1st platoon leader until he returned to Korea.

Lieutenant Kim has never forgotten Sgt. Kim Gwang-ma in his platoon. As the 1st squad leader, he was the hero of the company and a symbol of courage. He was the bravest and most loyal soldier of all. Even if he was reticent at the usual time, he always went ahead of the platoon. Suppose Lieutenant Kim was in front, Sgt. Kim grabbed his back and said, "Why are you ahead, sir?" He would push Lieutenant Kim aside and come forward. By the time replacement troops arrived, Lieutenant Kim had given him the privilege of growing a mustache and serving as a training assistant for the recruits to continue the shining tradition of the 5th company.[8]

Lieutenant Park Seong-hwan (Class of 1963), who came to the war as the 2nd Platoon leader of the 5th Company of the Cavalry Regiment, participated in Operation Lightning Flash 66-5 in the Tay Son area from April 2 to 12, 1966. In this regiment operation, the 5th Company was tasked with clearing the Viet Cong in the Nhon Thuan area. At that time, the Viet Cong in Nhon Thuan was not inferior to the friendly forces in firepower. There were also some automatic rifles, which blocked Park's platoon's advance. His platoon suffered three men killed and seven wounded. You didn't dare to fight if you lost ten troops in a platoon. The 2nd Platoon was stuck with its target 100 meters ahead.

At this time, the company commander Captain Lee Kwon-bok (Class of 1960), shouted on the radio, "Charge! You guys, platoon leaders, Charge at the head of your men." The bugler blew the charge trumpet after a while, but the soldiers hardly moved, and the enemy strengthened their firepower. SSgt. Han Sang-hee, the bravest 1st squad leader in the 2nd Platoon, attempted to charge, but it was difficult to subdue the enemy's firepower from behind the tangled thorn bushes and fences. The sergeant, risking his death, bravely ran into the enemy position and threw three grenades at their automatic machinegun mount—all the members of the squad charged in unison and seized the target. However, Park's platoon could not wholly annihilate the enemy as they resisted while fleeing. Eventually, they withdrew on the battalion commander's order and returned to My Dien to conduct a night ambush. Lieutenant Park had to pull back with ten casualties.[9]

Promoted to Captain in 1968, the 1963 class graduates could go to Vietnam as company commanders. These officers were those who had missed joining the first Korean deployment in the earlier period. Captain Kim Jin-seon (later, Second ROK Army Commander as a four-star general) and Lee Jun (later, ROK Defense Minister) were among them.

## "Viet Cong, Viet Cong, Mau Le, Mau Le"

Captain Kim Jin-seon was an instructor of Physical Education at KMA before he went to Vietnam as the 11th Company commander of the 1st Regiment in December 1969. His immediate superior was battalion commander Lieutenant Colonel Park Hee-do (Class of 1956, later Army Chief of Staff), who entrusted small unit battles entirely to company commanders and only provided the necessary support for the operation from behind. During his one-year tenure, Captain Kim Jin-seon participated in many ambush battles and operations on the company, battalion, and regiment levels. Among them, Kim remembers the most memorable incident, when his company repelled the enemy that attacked the base of the 222nd Company of the South Vietnamese militia. On April 27, 1970, after the "Operation Jaegu 70-1," Platoon Sgt. Jee Hae-wook, set up an ambush at Chanh An and captured an enemy document requesting a large amount of ammunition. Captain Kim went over the documents closely and predicted that there would be a night attack on them.

Accordingly, Captain Kim strengthened the readiness posture of the company base and alerted the 222nd Company of the South Vietnamese local militia. He also maintained close cooperation by establishing a hotline between the 11th and 222nd companies. Meanwhile, at midnight on June 7, 1970, a flash of light was seen, followed by the explosions at the South Vietnamese military base located 400 meters east of the 11th Company base, and, at the same time, five to six 82-mm mortar shells flew into the 11th Company base. As soon as Captain Kim dialed the hotline without delay, he heard the urgent and trembling voices shout, "Viet Cong, Viet Cong, Mau Le,

Mau Le (hurry up)." Seven 81-mm and 60-mm mortars focused on the South Vietnamese base and poured heavy fire in unison while intense gunfights broke out between them simultaneously.

After some time, when the eastern sky began to brighten, Kim led the troops to the South Vietnamese base and found it was in complete chaos. Dead bodies were scattered here and there, and there was still a lot of smoke. The South Vietnamese soldiers from the bunker hung around Captain Kim's neck to express their gratitude. Shortly after that, when they inspected the base, they found 17 corpses of the Viet Cong commandos who wore only underpants and painted their entire bodies black, and 120 grenades improvised with B-40s and empty cans filled with TNT. In this battle, the commandos had no personal weapons and were naked, carrying only ten grenades. Captain Kim admired their courage, though they were enemies. The friendly loss that day was only one dead and one wounded. After this battle, the Chief of Binh Dinh Province appeared on the scene and presented Captain Kim with the Bronze Star Medal on the spot.[10]

Captain Lee Jun's experience in the Vietnam War is unique. He went to Vietnam in December 1969 and served as the commander of the 6th Company with the 1st Regiment for one year. During his service period, he had no contact with the enemy. At that time, the 6th Company of the 2nd Battalion was stationed at the foot of Nui Phu Cat overlooking the Phu My Peninsula. In 1970, the Viet Cong or North Vietnamese regular troops significantly reduced their activities in Phu Cat County, occupied by the 1st Regiment. Thus, the residents could lead a peaceful life in that area. However, the company had some contacts with the Viet Cong appearing on the route to provide goods and supplies for the North Vietnamese troops based in Hon Noc Mountain.

Since the Viet Cong had given up using Nui Phu Cat as a refuge, the stationing of Korean troops in the area meant nothing more than maintaining stability. Therefore, Captain Lee focused on education and training as the company base became a permanent position to maintain a stable life system. Particular emphasis was placed on firing practice. Lee used the M16 cartridge cases as targets at 15 meters away and had the sharpshooters carry the hit cases. By doing so, Lee tried to make them excellent combatants.

Captain Lee learned from the war that it took a lot work to clear the enemy, who had known the operational methods of the Korean army for the past several years. For example, if they saw Korean troops landing by helicopter to search a cave, they immediately left it and found a new hiding place to evade the Korean search. In addition, they knew that Korean operations would usually cease after 17:00 and took advantage of the fact effectively. When the enemy contacted the friendly forces after 16:30, they preemptively attacked from their prepared position, thus forcing the friendly units to bleed heavily. Lee knew it was essential to quickly develop appropriate countermeasures in response to such changes in enemy tactics to gain a comparative advantage.[11]

Chapter 5

# Class of 1964 and Class of 1965: Platoon Leaders

## Class of 1964 (20th Class)

In the history of dispatching the Korean military to Vietnam, no graduate of KMA fought as fiercely as the Class of 1964 did. At the forefront of the battlefield, they spent most of their tours leading their men as platoon leaders through densely forested jungles. Thanks to the class of 1964, the Vietnam War may have earned the nickname of the war of platoon leaders. They seemed destined to fight in the Vietnam War since birth. The Class of 1964 was the class with the fewest graduates in the history of KMA. On February 28, 1960, 218 students entered the school, like any other class, but when they graduated on March 11, 1964, only 156 received the honor of graduation. The attrition rate of this class was 29%, and in their first year, many cadets failed in Math and English and had to leave school. The minimum score required to pass each subject was 67 points out of 100 on average per semester. The cadets at the time called 67 points "Hill 67" to overcome. This phrase has been used as such among the cadets from that time on. As mentioned earlier, another hill at the KMA campus is "Hill 92," called Devil's Height by cadets.

When the 1964 class graduates were about to finish the platoon-leader positions in the homeland, the Korean government decided to dispatch its troops to Vietnam. When the Tiger Division was first designated as the dispatching unit in the summer of 1965, its policy of organizing the unit was that almost all the platoon leaders of the division would be KMA graduates of 1964. Major General Chae Myung-shin, then commander of the Tiger Division, also confirmed that he would appoint the KMA graduates of 1964 as the entire division's platoon leaders. Accordingly, all the 1964 graduates in the frontline divisions were transferred to the Tiger Division as platoon leaders. During the Vietnam War, 137 graduates out of 156 participated, accounting for 88%.

Although some had two tours to Vietnam, most 1964 graduates served as platoon leaders and assistants to regimental staff for a year in the dispatching units. Of all the platoon leaders of the 1st and the Cavalry Regiments initially deployed, 61 platoon leaders were graduates of 1964. They all took great pride in fighting most of their battles as the leaders at the forefront of the Vietnam War. Second Lieutenant Shin Jung participated in the war as the 3rd Platoon leader of the 11th Company with the 1st Regiment. Surviving the difficult period in the early days, he returned home after a one-year tour and compiled his war experiences as a platoon leader into the book *The Wall of Jungles* (1967). After being promoted to captain in March 1969, twenty-four graduates were still in Vietnam for the second tour as company commanders or its corresponding-level staff, including Myung San-ok, Lee Gyu-wan, Sohn Yong-nam, Han Kuang-duk, etc. In particular, the 1964 class men wrote many stories and memoirs based on their war experiences.

## "Sir, I killed a man, our comrade"

Second Lieutenant Lee Yoon-gu (Class of 1964), the 3rd Platoon leader of the 3rd Company (Commander Capt. Jang Sei-dong, Class of 1960), the 1st Regiment of the Tiger Division, was exposed to firefight in Vietnam for the first time in his life. On 28 October 1965, the seventh day after the main forces of the Tiger Division landed in Vietnam, the 1st Battalion of the 1st Regiment to which he belonged arrived at the battalion assembly area north of Quy Nhon and started to construct the tactical base. The position of the 3rd Platoon was at the end of a series of low hills. Very few trees were in the area, a large portion of which was mainly paddy fields. At the platoon's frontage was a road running parallel to the line of platoon position. Forward of the platoon were some paddy fields where a few Vietnamese villagers were farming, and on the road were quite a few villagers coming and going. As they acted too indifferently, the platoon members deluded themselves into thinking as if they were out on the field training.

They had to dig trenches against the enemy's night attack in the daytime. Bordered by the paddies, the trenches around the base area were often filled with rainwater, which caused lots of trouble for the men. Still, they kept on digging and finally completed the construction of the trenches along the perimeter of the base. While working, they had to keep an eye out for the farmers. The same was true of these farmers. They also seemed to look to the platoon positions stealthily under their conical leaf Non la hats. To communicate with the squad leaders at night, Lieutenant Lee connected his foxhole with their holes by a rope, at the end of which was hung an empty can so that they could give signals. Now that they had completed construction work for the platoon positions, it was time to commit the men to the trenches for night duty. The Tiger Division had had no combat up to that time except for capturing some weapons by a unit of the Cavalry Regiment on patrol. So, all of the men in the division were nervous about what actual combat would be like.

Lieutenant Lee thought it was time to commit his men to their positions when the darkness was near. But what made him worried was that if the farmers who had worked on the paddies in the afternoon had seen their positions, they might have figured out the position of each trench accurately. So, he decided not to commit his men to each trench but to position them between the trenches. The squad leaders did what they were told to do. They placed their men on the bare ground between the trenches and were prepared for the enemy's coming. As the night went on, the voices of the people at the front village were heard. After a while, the voices became quiet; the only sound was the loud barking of dogs.

Toward eleven o'clock, a report came from the second squad leader that the enemy had appeared. He quickly went out of his tent and positioned himself at the spot where he could see all his men. About 30 minutes later, the M2 Carbin rifle sounds broke the silence of the night from the second squad. At that instant, he heard the sound of a man's scream, followed by the violent sounds of the rifles. And then came the horrible silence. The final death throe of life was heard. Corporal Koh Guang-bae approached him and said, trembling, "Sir, I killed a man, our comrade." Recovering from his dizzy senses, he checked each squad to know whether it was OK. Every squad said, "OK." That meant that the man killed an enemy. With confidence, he reported to the company commander about what had happened. He thought they would collect the body when it was bright the following day. But the commander told him to immediately bring it here, saying that the VC would always take their fallen comrade away with them.

Dragging a dead body from the no man's land on such a pitch-dark night was extremely dangerous. If an enemy was somewhere around the body, they might take unexpected damage. But the order had to be obeyed. In the dark, his patrol team went out. The other members of the platoon were all alert. Thirty minutes later, the team brought one body back to the platoon. The body received lots of bullets in its abdomen. His platoon won the first encounter with the enemy. Later, he recalled, "That night, if I had positioned my men into their trenches, nobody would remain alive now."[1]

## "Sir, get me up to look over to the homeland."

First Lieutenant Sohn Yong-nam (Class of 1964), the 2nd Platoon leader of the 6th Company with the 1st Regiment, participated in Operation Fierce Tiger 5 conducted in the Phu Cat Mountain area on March 23-26, 1966. This operation was carried out to expand the division's TAOR and secure the Go Boi plains area to protect the residents' harvest and cut off the Viet Cong's food supply. The 6th Company was the left unit of the 2nd Battalion and was to attack in the direction of An Hoa. Lieutenant Sohn's 2nd Platoon was to be separated from the company and carry out an independent mission alone. Lieutenant Sohn advanced in column formation in the order of the 1st, 3rd, and 2nd squads. He crossed the river of Tun Giang and climbed the trail up to take Tu Cung village. After searching it, he left the village, but the platoon received the enemy's heavy fire. The LMG leader had a piercing wound on his right hand. The platoon leader saw that the enemy entrenched themselves north of the village. He ran along the river's west bank to find out what the enemy was doing.

However, the enemy did not find Lieutenant Sohn and fired at the 2nd squad. Lieutenant Son killed the enemy LMG gunner and rushed to his LMG position, but PFC Lee Sam-rang, following Sohn, shouted, "Sir! It's dangerous" from behind. And when Sohn fell, a bullet came straight from his front. At the same time, the 1st squad advanced from the right, saw the scene, charged, and destroyed the enemy position. Following the fire support team, the Platoon Sgt. and SSgt. Jin Seok-gon arrived at the single wooden bridge 500 meters behind the platoon leader. And while crossing the bridge, the enemy's 60-mm mortar shell flew in, killing Sgt. Jin and wounding the Platoon Sergeant. Eventually, the platoon seized Tinh Hoa village, and the enemy fled.

The platoon leader evacuated the casualties and ran into the village to chase the fleeing enemy, but no one was there. At the order of the company commander, he deployed his men to secure a 400-meter-diameter LZ near the village of Tinh Hoa. At this time, Corporal Kim Hyung-seok found that the reeds moved even though the wind was not blowing, so he searched through the reeds to find an enemy hidden there and jumped into the water to capture him. The enemy detonated a grenade while fighting him in the water, killing Corporal Kim and himself. After that, the enemy pumped intensive fire into the platoon while moving to block the adjacent village. Corporal Sohn Seon-bae received 5-6 bullets and suffered a penetrating wound in the abdomen.

The medic treated him because the helicopter could not land, but he lost consciousness due to too much bleeding. He pleaded, "Sir, get me up to look over to the homeland and lay me in the shade under the tree." These were his last words. Even as he died, the Corporal thought of his hometown and beloved family far away. Lieutenant Sohn did what he said, and soon the Corporal closed his eyes and went unconscious. The platoon members brought him about 2 km on a stretcher and then sent him by helicopter, but he died on the way.[2]

### "On the battlefield, only the brave can survive."

On January 9, 1966, Operation Flying Tiger 6 began. Second Lieutenant Hahm Duk-sun (Class of 1964, later, Corps commander as a three-star general), the 3rd Platoon leader of the 2nd Company, the 1st Regiment, hid in the fence of an independent house and communicated with his company commander. His men were firing violently back at the Viet Cong. When Hahm came out of cover to advance the 3rd Squad into the enemy line, the 1st Squad's rifleman PFC. Park Dong-soo grabbed his stomach and ran to Hahm. "Sir, take this out. Hurry up. The guy shot me and is running away. I have to go after him and catch him," he said, asking him to pull out the bullet stuck in his stomach. When Lieutenant Hahm approached Private Park, he found that he had been shot in the stomach and had a bullet lodged in him. Of course, the platoon leader evacuated him immediately. Motivated by Park's actions, the men rushed into the enemy lines and could complete the operation. Lieutenant Hahm thought he won the battle because of the courage and devotion shown by Private Park.[3]

A month later, Lieutenant Hahm experienced another crisis at Nam Tang, Binh Dinh Province. On February 17, 1966, he set out on a search-and-destroy mission. The mission this time was to intercept the fleeing enemy while the 1st Company was attacking target A. This area was an ancient forest that denied humans access. However, it was the soldier's mission to carry out orders. He instructed his men to "enter" and advance toward the valley step by step. At that moment, an enemy machine gun fire poured out from the valley. It was indeed a critical moment.

At this time, he remembered the words of a letter from a girl in his homeland. "On the battlefield, only the brave can survive, and he is the one who acts calmly in the face of adversity." The words that were so natural and so often heard in his Academy days were brought in vividly through the medium of the girl's writing. He immediately called the platoon sergeant and said, "I think we should prepare ourselves for death today. But those who wish to die will survive, and those who wish to live and act cowardly will die. As a soldier, let's fight courageously and die," he said. And he divided the platoon into two groups and ordered one team to advance with him and the other with his platoon sergeant.

After that, while moving for about 20 minutes, the squad leader ahead suddenly opened fire with a Calvin rifle, shouting, "Viet Cong!" The enemy fire poured out from the front and both sides. Bullets rained from 30 meters ahead so that he couldn't lift his head. The enemy shot the squad leader, who screamed 10 meters away, and his helmet fell to the ground. In an instant, anger and hostility boiled up all over his body, and he couldn't lie down and shoot. He told the soldiers to get up, screaming, "Charge forward!" When the enemy saw the platoon's massive fire and assault, they overestimated its strength and ran away. He reported the current situation to his company commander, asking him to send a chopper, and withdrew to the company base.[4]

## "The girl they helped turns out to be a spy."

The soldiers of the Tiger Division always tried to help the villagers by being careful not to get them hurt. So, the Vietnamese people's trust in the Korean soldiers was increasing. On March 24, 1966, the 11th Company of the 1st Regiment received an order to search again for Than Danh village, which the company had seized a couple of days ago. Thus, the company searched for the area over the next two hours. During the search, the 1st Squad leader brought a girl who appeared to be about 12 years old. First Lieutenant Shin Jung, the 3rd Platoon leader of the company, asked with a puzzled expression.

"Squad leader! What happened?"

"Yessir, I got one on patrol."

"You should've sent her back, shouldn't you?"

"She's so pitiful, so I got her here."

According to the squad leader, he found a crying girl in the yard of an empty house during a village search. More pitiful than suspicious, he asked a few words. She answered that her parents had died and begged for help, saying she was left alone.

Lieutenant Shin ordered his men to send her back immediately, but the girl noticed the platoon leader's expression and pleaded for pity, clinging to the squad leader. However, the men returned her to the village with cookies and chocolates at the platoon leader's order. Again, after completing its assigned search mission, the platoon attempted to camp that night in Thanh Danh village. Around 18:00 hours, the squad leader passed a house in the village on the way back from his mission. And the girl he had met in the morning stood in the yard of the house with a very startled expression and stared at the squad leader with strange eyes. Having a suspicious hunch, the squad leader had his men take a prone position by the fence and found three men openly reading a map. As the squad leader entered the house, the three men began to run away. The squad opened fire on them, killing all three. The squad members searched the house but found no guns. Eventually, they caught the girl. They were too disappointed by the girl's betrayal.

Later, after interrogation by a military intelligence agent, the girl proved to be a Viet Cong spy. The maps they obtained showed the company's campsites and search routes. After this incident, Lieutenant Shin regarded the South Vietnamese residents he met on the battlefield as sympathizers of the Viet Cong, regardless of age or gender. It was an opportunity to know the truth of the guerrilla war, as he had only heard about it.[5]

## A Female VC

First Lieutenant Jeon Sang-woon (Class of 1964) was the 1st Platoon leader of the 6th Company with the Cavalry Regiment. His company participated in Operation Lightning Flash 66-5 conducted in Binh Khe village on April 2, 1966. The target area of the three-day operation was 40 km northwest of Quy Nhon. As bamboo forests across the Con River surrounded the village, it was difficult to determine the exact location of the enemy. His platoon lost Staff Sergeant Lee, the 2nd Squad leader everyone loved. He had been the first medal recipient in the Tiger Division and was the best sergeant in his platoon who had graduated from the 1st Army NCO School. He grew up as an orphan in the Korean War. Sergeant Lee's death filled the hearts of all men with thoughts of revenge, which led his platoon to ravage the enemy position

A female VC in action (Source: ROK Army Museum)

and take the victory.

Among the countless corpses of the Viet Cong, the one that surprised them was the corpse of a female Viet Cong. A maiden dressed in typical Vietnamese peasant attire, black pants, and a white blouse was found dead, holding a pistol like a cowboy from a Hollywood Western. There was no way to suppress his complicated feelings as he watched the dead body, a female VC he had only heard of in words. All the battles were over at dawn the next day. Although he felt joyous after the operation's victory, he would recall this three-day battle as a painful memory for the rest of his life. The greatest victor in the Vietnam War would be the one who would suffer the fewest casualties and come back home in triumph. But he had already lost a subordinate. The more he thought about it, the sadder he was. He was a man before being a soldier.[6]

## The Significance of Artillery in modern war

First Lieutenant Han Kwang-duk (Class of 1964, later the 11th Division commander as a two-star general), a Forward Observer, the 9th Company of the Cavalry Regiment, had the most unforgettable night of his life on August 9, 1966, at Duc Co. It was the night he fought the fiercest battle in the history of Korean troops' deployment to Vietnam. After returning to the base from an overnight patrol mission with his company, he and his fellow officers had a farewell party for the outgoing commander. The 2nd Platoon leader, First Lieutenant Lee Chun-sik, Han's classmate, said, "The change of command party in the Vietnam jungle is rare and historic, isn't it?"

After the party, he fell asleep while lying on his bed, reading *Endless Journey* by the world traveler Kim Chan-sam. It was around 22:30. At around 22:50, the Weapons Platoon leader, First Lieutenant Lim, who was patrolling the base, received a report from a soldier that he seemed to hear people talking and digging in front of him and relayed this report to his company commander. The commander, Captain Lee Chun-geun (Class of 1959) contacted the tank platoon and instructed them to turn on searchlights for the area where the sound was coming. At this time, a trip flare buried 200 meters ahead of the Weapons platoon exploded and illuminated the entire region. Then, many black figures were seen swarming into the company base. At the same time, the tank and light machine guns started to spit fire at the enemy. Lieutenant Han ran to the observation bunker reflexively at the sound of gunfire, hearing the shouting voice of the company commander from the bunker. "Lt. Han, Lt. Han!" His shouting was an order to request artillery fire.

Crawling into the observation bunker, he instinctively grabbed the radio. He requested the artillery battalion to fire within the position, and in less than three minutes, the first shots flew in at around 23:00 hours. He kept on demanding firing at all concentration points around the perimeter. The enemy mortar fires paralyzed the company's command function. When the situation started, the 1st and 2nd Platoon leaders immediately left the sleeping bunker, but both were seriously injured. In the first 20 minutes, 170 enemy mortar shells fell on the base. These mortar shells caused nearly 90% of seven friendly deaths and 42 wounded. Lieutenant Han started the correction after 4-5 shells fell on concentration point #311 and requested the firepower of all available batteries. First, Lieutenant Han mobilized all the 105-mm and 155-mm howitzers of the 61st Artillery Battalion and the 105-mm, 8-inch howitzers, and 175-mm guns of the U.S. Artillery units to strike all around the company base. The FO neared some of these fires 200 meters from the perimeter and continued Fire for Effective (FFE).

After almost two hours, at 24:30, the enemy fire began to quiet down. At dawn, the enemy completely retreated. However, the new company commander died around 4 a.m. on August 10. Darkness was breaking in the evening sky. First Lieutenant Kang Bo-gil (Class of 1964), an FO of the 11th Company, who was coming as reinforcements, called Lieutenant Han and said, "Countless enemy corpses are lying in front of your company base." Except for the six prisoners, all those who resisted were shot and killed. His soldiers brought a prisoner before Lieutenant Han. Tears were flowing from the eyes of the prisoner, who fell helplessly. Han asked his age through a Vietnamese interpreter and why he came to the battlefield. He said he was 19 years old and was dragged out like a cow by the Viet Cong. Han said through an interpreter, "Our enemies are communists, not Vietnamese. We are your friends."

As a result of the battle on this day, the company had 187 enemies killed, 150 estimated killed, six prisoners, five 60-mm mortars, and many guns and ammunition captured. Most of the enemy casualties were caused by artillery shells. In this respect, the Battle of Duc Co was an excellent example of how significantly the artillery could play a role in winning modern war. The fiercest battle in the history of Korean deployment to Vietnam ended. Lieutenant Han was awarded the Choongmoo Medal for his achievement in this battle.[7]

Since the Korean combat troops were deployed to Vietnam in October 1965, the largest and fiercest operation was Operation Fierce Tiger 6. In this operation, two war heroes were the members of the 1964 class: Kim Gil-boo (Class of 1964) and Kwon Seung-joon (Class of 1964).

First Lieutenant Kim Gil-bu, the 3rd Platoon leader of the 10th Company, Jae-gu Battalion, the 1st Regiment, made a surprise attack on the VC caves on Nui Phu Cat. At the head of his platoon, he rushed headstrong into a cave. Afraid of being separated from his men, they connected their bodies with a string and crept into it. On their way, the

VC threw a hand grenade at them, but it hit the rock at the corner of the cave and exploded. As if it were a signal to fight, massive fire-fights erupted. After some minutes, the fighting ended, and his platoon killed 23 VC and captured 18 weapons.

To First Lieutenant Kwon Seung-joon, a platoon leader of the 11th Company from the same battalion, the operation was the first since he came to Vietnam. He wondered what he would be like in a fight. However, when he was in action, his mind was calm and full of confidence. He was courageously engaged with the VC, and his platoon killed 24 and captured 30 weapons. In this fierce battle, both men again showed the indomitable spirit of KMA. Especially noteworthy was that no one in Kwon's platoon was killed, only some soldiers were wounded. That was the most significant success ever up to the time since the deployment of the Korean troops to Vietnam.[8]

## Class of 1965 (21st Class)

Another class of platoon leaders in the early period of Korean deployment to Vietnam was the Class of 1965. Most of the Tiger Division platoon leaders were graduates of 1964, and almost all the platoon leaders of the White Horse Division that went to Vietnam in September 1966 as reinforcements were graduates of the Class of 1965. In other words, the positions of the White Horse Division's platoon leaders were mainly reserved for the 1965 class graduates. Of course, many of these class members were also assigned to the Tiger Division as rotating personnel for their seniors returning after completing a one-year tour. Some of the 1965 class graduates had managed to join the first deployment troops. Jung Chang-soo, Yang Soo-il, and Kim Gwang-hong were among those who had boarded the first troop ship to Vietnam. It was October 1965, less than a year after their commission.

The history of 1965 graduates began on February 20, 1961, when 216 entrants came to KMA. After four years of study, only 178 cadets graduated on February 23, 1965, six months before the first combat troops left for Vietnam. In 1961, when they entered the Academy, the country was in chaos due to the May 16 Military Coup. This military coup broke out less than three months after they entered the Academy. When all cadets went to Seoul City Hall to support the coup, they followed orders as freshmen without any particular thought on their outings. They graduated after four years of unchanging, arduous cadet life. Upon completing the OBC, they were assigned to each frontline division and served as platoon leaders. When the government decided to send the First and the Cavalry Regiments of the Tiger Division to Vietnam that summer, the 1965 class graduates on the frontline for less than two months had no place in the dispatching units soon to be deployed.

Even if there were some room for platoon leaders, assigning those fresh second lieutenants just out of school as the leaders of the combat units in a foreign war was impossible. Further, the policy was that all positions in Vietnam would be filled only by those who had already completed corresponding ones in the homeland. Thus, they had to wait a year or so to go to Vietnam, and in September 1966, a large number of 1965 class graduates were selected as platoon leaders for the White Horse Division. By the end of September 1966, one and a half years after graduation, some sixty classmates went to Vietnam, about one-third of all the graduates, mainly as the platoon leaders of the White Horse Division. Others went as replacements for those of the Tiger Division or the 100th Logistics Command. Some volunteered for a second tour as company commanders or its corresponding-level staff after being promoted to captain in 1969. During the entire war period, 145 graduates of the class went to Vietnam, accounting for 82% of its total graduates. The Class of 1965 was the class that suffered the highest number of deaths in Vietnam among all the graduating classes of KMA, with seven deaths. Six of them were killed in battle while fighting as platoon leaders, and one died of a disease he contracted during a battle after being evacuated to his home country.

### "If you focus on thinking about women and money, you will be in trouble."

Second Lieutenant Lee Jae-kwan (later, First ROK Army Commander as a four-star general) was appointed 1st Platoon leader of the 10th Company of the 29th Regiment of the White Horse Division to be deployed to Vietnam in August 1966. And his unit arrived in Cam Ranh on September 25, a month later. During his tenure, he participated in Operations Bulldozer 5, Stagecoach 2, and White Horse 1 and 2, including many ambushes, searches, and reconnaissance actions. In particular, his company yielded a considerable achievement in Operation White Horse 1, conducted for a month from January 29, 1967. It captured 150 weapons, including 57-mm recoilless rifles and crew-served weapons such as machine guns.

Lieutenant Lee's platoon played a crucial role in detecting and seizing the weapons his company captured in that operation. For this achievement, he received a Choongmoo medal. He attributed the honor he earned to the sacrifice of his subordinates. In particular, Private Oh man-ok was a brave soldier, short stature, and cheerful soldier. He was an exemplary who always took the lead in the platoon's dirty work and often volunteered for the position of the point man, whom everyone hated. In the case of a cave search operation, it was necessary first to check whether there was an exit (secret passage) other than the entrance and to set up strict all-around security. One of the mottos that Lieutenant Lee had was, "You must be a selfless soldier on the battlefield" and "If you focus on thinking about women and money, you will be in trouble."[9]

## The Baptism of Fire

The story of a KMA graduate's participation in the Vietnam War is best described in the memoir of Kim Hyung-seok (Class of 1965). In his book *Fighting Spirit: A Platoon Leader in the Vietnam War* (2016), he vividly describes his experience as a platoon leader in Vietnam. It is not an exaggeration to say that his experience represents the experiences of his contemporaries in Vietnam. Second Lieutenant Kim Hyung-seok went to Vietnam on October 2, 1966, as the 1st Platoon leader of the 6th Company, the 30th Regiment, the White Horse Division, and returned to Korea on September 28, 1967. During his service, he participated in many large and small operations, including Operation Hong Gil-dong, a Corps-level operation conducted from July 9 to August 26, 1967.

November 6, 1966, was the 31st day of Lieutenant Kim's arrival in Vietnam. His second battalion was the reserve unit of the ROKFV which should be ready to move at any time. For the past month, the soldiers were nervous and curious, wondering what the battle would be like. The 30th Regiment was ordered to provide security for the South Vietnamese 23rd Infantry Division, whose 566 supply trucks were going from Duc My, some 50 km from Nha Trang, to their base at Ban Me Thuot near the Cambodian border. The regiment was asked to pay special attention to some VC companies in both Eachuchay Valley and Eakrong Valley on Route 21, running from Ninh Hoa on the eastern coast to the intersection of Route 9B at Ban Me Thuot to the west. The terrain of both sides of Route 21 comprised a jumble of high mountains, valleys, ravines, and ridges that runs westward. This operation, Stagecoach 1, was the first combat duty assigned to the 30th Regiment since its arrival in Vietnam. Its combat performance was of significant concern for the ROKFV in Saigon.

At 06:00 on November 7, Kim's battalion left its base camp near Cam Ranh by motors with his first platoon of the 6th Company as the advance party. After a five-hour ride, the platoon arrived at Duc My, where his regiment commander and principal staff members had already arrived in advance and met with Lieutenant Kim and his company commander. Major Paik, an Intelligence officer of the Regiment, said to the two officers, "Since we arrived in Vietnam, we've had no chance of engaging with the enemy, so your men may have slacked off a bit. Now you need to alert them to watch out." And he continued, "Lt. Kim, your mission as an advance party is critical. I'm sure you can do it well, can't you?" Given Major Paik's repeated warnings, Kim thought he would surely encounter the enemy this time.

Some minutes later, Kim spread the map and looked into Route 21 and its surrounding mountains, envisioning what the natural terrains would be like. And he looked 10 km ahead to the winding road going over the hills as if it were spiraling up to the sky. He called his squad leaders to his side, explained the enemy's situation, and told them not to slack off a bit. He said, "We'll come in the enemy position in a minute. When we reach the target area, 40 km from here, around 17:00, you must occupy some part of it and set up a security guard. On the way, you need to give special attention to both sides of the road. And you have to find the enemy first and suppress him. Got it?" All the men on hand answered "Yessir!" in chorus.

After a while, the company's motor column reached the entrance to Eachuchay Valley. The road's right and left sides soared high with Hill 833 and Hill 993 like back-dropped screens, so they felt like their trucks were climbing toward the sky. While riding up the road slowly at 10 km per hour, his company was attacked by the enemy three times. The first enemy shot came from the ridge of the high precipice on the right overlooking the uphill road. At 200-300 meters from the company with a valley in-between, the enemy positioned themselves on the long ridge from where they could secure good observation and fields of fire and neutralize any counterattack. All troops got off the trucks in no time, but they were at a loss for what to do. Some lay on their belly on the asphalt road, others jumped off the road to the ditch, and still, others wandered on the road with bewilderment. The enemy directed their machinegun fire toward the troops, and lots of bullets flew 2-3 meters over the head of Kim, now lying prone on the road. But he tumbled three meters off the road at the thought of being killed there.

Aware that he was separated from his men, Kim wanted to go to them but couldn't move his body because it became stiff. For the first time, he experienced the physical phenomenon that his body and mind seemed to work as separate entities. The sounds of the bullets hitting the ground beside him were like those of a death messenger's footsteps. The friendly artillery shells began to burst on the left ridge at this moment. Encouraged by this fire support, Kim raised his body and climbed the road. When the company commander saw Kim, he told him to launch an attack toward the VC on the ridge to the right, which, he believed, had fired upon them moments ago.

Kim spread his three squads in the line formation at the clearing in the front and went on a search. However, the clearing was so thick with tall bamboo trees that the movement of the open ranks was almost impossible. It took thirty minutes to break through a 100-meter distance.

1Lt Kim Hyung-seok (left, Class of 1965), 1st Pt Leader of the 6th Co, the 30th Rgt, the White Horse Div. (Courtesy: Kim Hyung-seok)

Kim's platoon conducted a precision search but couldn't find any signs of the enemy. The company commander ordered to stop the search and withdraw because the company had to arrive at the designated place by 17:00. In this skirmish, the company suffered one dead and several wounded. Their baptism of fire ended up a terrible disaster. Kim was deeply disappointed at his poor combat performance as a platoon leader. Returning to the initial place with empty hands made Kim miserable and heartbroken.[10]

## "An army without the will to fight is useless."

First Lieutenant Jang Chang-gyu's (later, Superintendent of KMA as a three-star general) experience in the Vietnam War is unique. After graduating in 1965, he served as a platoon leader in the 2nd Infantry Division on the Korean front, then as a battalion operations officer. Later he went to Vietnam as replacement personnel for the Tiger Division in December 1966, where he was appointed platoon leader of the 10th Company of the 1st Regiment. While a platoon leader, Lieutenant Jang was assigned as the chief of the training center established in the regiment to educate the South Vietnamese militias. The militia in South Vietnam was similar to the reserve forces of the Korean army. A class of militia consisted of about 60-70 members at a time, and they entered the training center and received a week's training. The training center trained people nearly 2,000 a year. In the Vietnam War, where there was no clear front, solidarity with the residents was essential. Thanks to the militia training, the units of the Tiger Division could maintain good relations with them in their TAOR.

The South Vietnamese militia was an organization that was founded not long before. However, the problem with the militia was that they did not have the will to fight. There was no hostility toward the VC because they believed that "Ho Chi Minh is their nation's leader." Jang was critical of such an attitude of the South Vietnamese people. He stated, "An army without the will to fight is useless."

Lieutenant Jang was promoted to a temporary captain in February 1969 and appointed the 3rd Company commander of the 1st Regiment. He was lucky, given that many other officers waited several months to be assigned to a company commander in Vietnam. Recognized for his contribution to the militia training, Jang was privileged to be a company commander. While Jang served as a company commander for about six months, the Commanding General of ROKFV, Chae Myung-shin, was replaced by General Lee Sei-ho. General Lee's policy was that officers who had served for a long time in Vietnam should be replaced. So, Jang returned to Korea in August 1969, just six months after his appointment as the company commander.[11]

## "Helicopter Transport of 100 Pigs"

First Lieutenant Kim Jeong-jo (Class of 1965) had two tours to Vietnam. From June 1967 to June 1968, Kim served as an operations officer in the 30th Regiment's Combat Support Company of the White Horse Division and as an assistant to the Operations officer at the regiment headquarters. After his one-year tour, he returned to KMA, serving as an operations officer at the Corps of Cadets. After that, he went to Vietnam again in April 1971 and served as the 7th Company commander of the 29th Regiment. During his first tour to Vietnam, he were an assistant for Major Paik Un-taek (Class of 1955), the 30th Regiment Operations officer. He was proud of what he had done when he served in that position. He drew up an operational plan of "Arc Light Bombing" for the area around Nui Hon Ba as a preparation fire. His fire plan was for the B-52 strategic bombers to sortie from Guam first, followed by naval gunfire, close air support, and ground-based 8-inch and 155-mm howitzers.

Once, he went on a reconnaissance patrol with ten men from the Regiment Reconnaissance Company. On the way, his team was attacked by the Viet Cong and took some casualties. And he, too, was seriously injured and evacuated. Later, Lieutenant Kim reflected that the failure of the patrol was due to his misjudgment. He had taken the jungle trail used by the enemy waiting in an ambush. Another reason was that he had only been in the regimental situation room without an experience in jungle operations. He did not repeat the same mistake when he went to Vietnam again as a company commander.

The second time he went to Vietnam, Captain Kim (promoted as of April 1, 1969) served as the 7th Company commander of the 29th Regiment of the White Horse Division. His regiment commander was Colonel Chun Doo-hwan (Class of 1955), the 2nd Battalion commander was Lt. Col. Park Jong-nam (Class of 1957), and his regimental Intelligence officer was Major Jang Sei-dong (Class of 1960). The 7th Company had suffered heavy losses during operations before he arrived, with dozens of casualties. It had to be supplemented with as many as 30-40 recruits. The company's discipline was very slack, and morale fell to the lowest level. Therefore, Captain Kim, first of all, tried to boost the confidence and health of the men while tightening discipline through the fortifying base defense and intensifying training.

During Operation Bat 26 in June 1971, Captain Kim's company discovered an enemy farm and 19 huts on a mountain slope, and the enemy had already disappeared. However, only a large number of pigs and chickens remained. Upon the order of the battalion commander to send the pigs and chickens to his base by helicopter, the company launched an operation to capture them. After placing guards, scores of 2-3 men teams were formed to catch the animals, but the chickens were so fast that they could not catch them, and only 100 pigs were caught and transported by helicopter. An incident unprecedented in the history of the world war called "helicopter transport of 100 pigs" was executed in Vietnam by his company. It was also reported in the Korean Army's *Comrade Newspaper* dated July 8, 1971. Through Operation Bat 26, Captain Kim learned several lessons: 1) The brave lives, the coward dies; 2) A person who has usually been brave was not always brave on the battlefield; 3) A leader can get things wrong when he becomes eager for fame.[12]

Captain Park Young-il (Class of 1965) served in Vietnam for 24 months, from October 1968 to November 1970. He performed three missions during this period. He initially served as a platoon leader guarding the Tiger Division headquarters for six months. After finishing the Vietnamese Language Training Course, he served as the Reconnaissance Company commander of the 26th Regiment. He participated in the operation to commemorate the Founding ROK Armed Forces and Operation Tiger during his tenure. What is especially unforgettable was that he discovered the enemy arsenal near Hill 624 during Operation Tiger and captured many rifles. During Operation Tiger at Nui Voc Cat, the company conducted interception missions starting on July 18, 1970. The 2nd Platoon reported that they discovered a natural cave in the valley on the left of Hill 624.

Captain Park immediately ran to the 2nd Platoon and scouted the cave with its leader, then selected five soldiers to enter. Inside the cave, a soldier shouted that he had found three Soviet-made LMGs. Captain Park sent in three additional soldiers to explore the caves connected in all directions, and as a result, 14 more weapons were captured, including three M-1s and five CKSs. This cave was the command post of the NVA's 3rd Division, and they abandoned their weapons while fleeing from a surprise attack by the Korean troops. A characteristic of the Vietnam War was the meeting engagement (ME). Hid in favorable terrain, the enemy surprised Korean troops when they appeared and hid again. Therefore, confronting such an enemy and quickly subduing it in meeting engagement was the key to success in the Vietnam War.[13]

## A Rendezvous at the Ojakgyo Bridge

Captain Byun Hong had been to Vietnam twice. His first tour was from March 1966 to April 1967 as the 2nd Platoon leader of the 11th Company, the 26th Regiment, the Tiger Division, while his second tour was as the 10th Company commander of the 1st Regiment from August 1970 to September 1971. This company was that of the late Major Kang Jae-gu, who sacrificed himself for his men during the grenade-throwing training of the Tiger Division in Korea. During his tenure, he participated in Operation Fierce Tiger 8, Operation Ojakgyo Bridge, and Operations Fierce Tiger 16 and 17. The most memorable event he had in Vietnam was when he, as the platoon leader of the Tiger Division, met with his counterpart of the White Horse Division during the first rendezvous between the two divisions in Operation Ojakgyo Bridge. This historic operation was conducted from March to May 1967 to open Route 1 from Song Cau to Tuy Hoa. Symbolically, the meeting of two divisions was named a "Rendezvous at Ojakgyo Bridge"(A legendary meeting of a male star and a female star on the milky way called Ojakgyo Bridge once a year on July 7).

When he was appointed the 10th Company commander in the second Vietnam tour, the company had already become a well-trained unit by the former company commander, Captain Hwang Won-tak (Class of 1962), a Model Hwarang in his cadet days. Byun inherited the commanding policy of the former company commander with my emphasis on combat training and unity among the men. According to intelligence, the enemy had fled to the Xa Rang Mountain area during Operation Fierce Tiger 16, which started on April 22, 1971. The company sent three platoons to Hill 719 by helicopter to search for the enemy. They found a hut on the upper part of the ridge left of the hill, and eventually, the men found four corpses of the enemy who had died a few days ago. Also, the 2nd Platoon leader crawled around the hut, killed four more enemies, and captured two LMGs and 500 grenades. The company continued to search in the direction of Xa Rang Mountain and found the 12 enemy bodies killed by friendly shelling. The company buried them well. Then, on the way back to the base, they found fleeing enemies from Thanh Ban village and killed three of them.

On June 26, 1971, his company was fiercely fighting with the VC at Nui Hon Sung near the 1st Regiment Headquarters. The combat was so violent and intense that Captain Byun was immersed in it. The division commander MG. Lee Hee-sung (later, ROK Army Chief of Staff), who grasped the battle situation of Byun's company by overhearing the radio communications, was impressed and immediately sent a message to all soldiers of the Tiger Division. He commended the 10th Company for its courageous fighting and requested that all soldiers use the 10th Company as their model and attain outstanding achievements. He still keeps the note of the message in his album. Further, he feels proud that his 10th Company was once designated as the best among all the Korean forces companies during his tour in Vietnam.

Captain Byun was aware of the importance of small unit leaders in Vietnam. He was convinced that he could win any battle if a platoon leader had the firm determination to win and the obedience of the soldiers following him. In addition, he learned from this battle that the Vietnam War was a war fought by small units, so it was essential to acquire small unit combat skills (platoon leader, squad leader). Byun also pointed out the problem of enforcing combat for fame by commanders, false reporting of combat results, and the unfairness in awarding medals.[14]

Capt. Byun Hong (second from right, Class of 1965), the 10th Co. Cmdr, 1st Rgt, Tiger Div. with Artillery FO, RTO, and U.S. Army officer (far rght), at Nui Hon Sung in June, 1971.

Chapter 6

# Class of 1966 to Class of 1968: In the Footsteps of their Elders

## Class of 1966 (22nd Class)

Second Lieutenant Kim Jun-sik, a member of the Class of 1966 who served in the 1st Infantry Division Reconnaissance Company as a platoon leader in Korea, went to Vietnam as the first among his classmates on December 2, 1966. It was just eight months after he graduated from the Academy. The Class of 1966 entered the Academy on February 1, 1962, and graduated on March 2, 1966. The first entrants were 234, but only 181 cadets graduated. The attrition rate was 23%. One hundred sixty officers out of these graduates participated in the Vietnam War, accounting for 88.4%. The Class of 1966 recorded the highest number of Vietnam War veterans among the KMA graduates.

The 1966 class graduates knew they would soon be called upon to fight in the Vietnam War. They had already seen the Tiger Division go to the Vietnam War in their senior year at the Academy. They thought their road to Vietnam was not far away, even though they knew they had to serve their first tour of duty as platoon leaders in the homeland. They were holding back their thoughts of going to Vietnam for the time being. Moreover, in September 1966, when they started their service as platoon leaders in the front divisions, they saw another combat infantry division, the White Horse Division, leaving for Vietnam. It was only a matter of time before they would board the ship to Vietnam. A few months after the 1966 graduates started service as platoon leaders in Korea, their minds were already flying over the sky of the Vietnam War front.

In 1967, the KMA graduates who had already completed their one-year tour as platoon leaders in Vietnam began to return home. Just as the Classes of 1961 and 1962 graduates were dispatched as company commanders and their corresponding-level staff to replace their seniors of 1960, the 1966 graduates began to alternate duties of platoon leaders with their one or two-year seniors. According to *Asadal*, the KMA Alumni Association Newsletter, three-quarters of 1966-year graduates had already gone to Vietnam by the end of 1967. The Section of the Class of 1966 in the Newsletter proudly announced that they were holding one-third of the positions of the whole platoon leaders of the Korean forces in Vietnam.[1] In other words, 130 graduates of the 1966 class were already fighting in the Vietnam jungles. Then, the remaining 30 officers went to Vietnam after 1967, some of whom did as company commanders after being promoted to captain. To name some, Captains Byun Gil-nam (1st Company, the 1st Regiment), Kim Jong-hwan (5th Company, 1st Rgt.), Lee Eui-ho (11th Company, 1st Rgt), Jee Man-won (Artillery Battery Commander), among others.

Second Lieutenant Oh Hyung-geun succeeded in a night ambush operation, killing 10 Viet Cong and capturing 14 weapons, including two ARs, three Carbines, two Claymores, and two hand grenades. Greatly satisfied with what 2Lt. Oh had done, General Chae Myung-shin encouraged him by saying, “Was that in the course they taught at KMA?” In addition, Second Lieutenant Choi Moon-gil, a platoon leader of the 9th Company of the 30th Regiment, killed 28 enemies and captured one in the Battle of Dien Khan. In addition, Lieutenant Chung Young-moo, a platoon leader of the 11th Company of the 29th Regiment, White Horse Division, received the Inheon medal directly from General Chae Myung-shin for his outstanding achievement in an ambush during Operation Ninh Hoa 1 and 2. Unfortunately, Kim Jun-sik, Park Guk-woong, and Shin Yang-ho suffered penetrating wounds during operations and were transported to Korea or a local army hospital for treatment.

## The Longest night at Chai Mountain village

Among the Class of 1966, one who sustained the most severe trials in combat would be Second Lieutenant Choi Young-ha. On June 8, 1967, a year after graduating from KMA, Second Lieutenant Choi Young-ha went to Vietnam as the 1st Platoon leader, the 2nd Company, the 28th Regiment, the White Horse Division, and returned to Korea in June 1969, two years later. When he got to his battalion, it was being committed to Operation Horse Head 2, and its base was empty. He also had to be brought into combat immediately. Boarding a Chinook helicopter leaving for the operational area with a whole load of ammunition and supplies, Lieutenant Choi landed in the middle of the jungle where his company command post was located and reported to his commander. His life as a platoon leader that began in this way was a series of tears, resentment, and regret.

One day in late June, Lieutenant Choi took the first casualty while searching a cave in the jungle. He was Corporal Jei Jeong-do. While the platoon was exploring the cave, Jei was on duty as a guard. While sitting on a rock, keeping watch, he was shot and killed by the Viet Cong inside the cave. Having one of his men killed in the first encounter, he was so disappointed by his performance as a leader that he couldn't contain his tears. The company commander scolded Choi, who was crying over the man's dead body. He greatly rebuked Lieutenant Choi, saying, "How the hell will you fight in the future if you

2Lt. Choi Young-ha (Class of 1966), 1st Pt leader, 2nd Co. of the 28th Rgt, White Horse Div. during Operation Horse Head 2 in June 1967.

are so discouraged by such an incident?"

Lieutenant Choi lost several members at the very beginning of his appointment. He had to evacuate four dead and nine wounded. Lieutenant Choi will never forget how his platoon was hit by the enemy at Chai Mountain, where he lost two squad leaders. On September 3, 1967, his battalion went into action for Operation Hoa Mai Dep Lam, which was carried out on the information that there would be an offensive by the NVA forces ahead of the presidential election in South Vietnam. On August 31, Lieutenant Choi's 1st Platoon went into an ambush under the direction of his company commander.

Two squads of platoons were waiting in an empty house near the village, and as darkness fell, they encountered an enemy entering the village while moving along the bank of a small stream to an ambush point at the bottom of the hill. The enemy saw the 1st Platoon first and fired. The guards said the enemy was about two squads equal to Choi's platoon. His platoon immediately poured out massive fire. Tracer bullets rained down on both positions. After about an hour of fighting, two squad leaders were killed. The RTO also suffered shrapnel wounds. A dust-off flew through the darkness in the night, carrying the two dead bodies and the wounded.

The following morning, Choi found that the enemy had all escaped. His platoon had been exposed to an open area, and his loss was inevitable. Yet he couldn't believe what had happened. Looking up at the sky far, far away, he was stunned for a while at the tragedy he had been facing. The battalion commander flew to the scene by helicopter. He must have been very heartbroken, but he calmly comforted Lieutenant Choi. None the less, Choi felt miserable, thinking it was all his fault. It was *the longest night of his life*. He could not forget the memory of having picked up the bodies of two subordinates with his own hands and loaded them onto a chopper. What was more painful was the moment when the wounded men cried out for help all night.

The next day, on September 1, as the day dawned, all his superiors arrived at the scene. Amid the heavy silence, his battalion commander comforted Lieutenant Choi, who fought courageously through the night. Later, the battalion commander would often recall that he could not forget Choi's pitiful appearance of that day. The battalion commander, who had assumed the position a few weeks before, kept informed of the situation all night. Watching Lieutenant Choi and his platoon members fight to the death, he was deeply moved by their desperate struggles in such a horrific situation. Later he told Choi he felt an indescribable sadness and heartbreaking sympathy toward him and his men.

Lieutenant Choi lost his orderly to an enemy grenade while searching a cave on September 26, 1967. The orderly was a recruit who had just arrived in Vietnam. When he was searching for the entrance to the cave right in front of the 3rd Squad leader, Sergeant Lim Byung-ok, he was struck by a grenade thrown by the Viet Cong from inside the cave. It instantly killed the orderly and seriously wounded Lim. All five enemies fled while the platoon was treating the casualties, but a thorough search through the woods found them all. Choi then shot and killed an enemy. He killed a human being for the first time in his life.

What would you expect from an angered man who had lost his temper at the death of his subordinates? He ran into an enemy right in front of him. The two were dumbfounded, and they looked at each other blankly. Choi emptied the M-16 rifle magazine instantly. The enemy's body became a beehive, but no bullets were in his rifle. What would have happened if his rifle had had live ammunition in it? And Lieutenant Choi is still traumatized whenever he thinks about it. Sgt. Lim had to have his leg amputated later at the Evac hospital. Afterward, when he went to the hospital, Choi saw him crying and vowed to look after him for the rest of his life.

Lieutenant Choi's tour as a platoon leader in the jungle for four months stopped when he was dispatched to the Vietnamese Language Education Center of the White Horse Division. After completing the language course, he was ordered to extend his tour one more year to work with the battalion civil psychological warfare staff (S-5). He returned home in June 1969 after two years of service in Vietnam. Back home, he often would be immersed in mixed emotions. He lost some of his men and could not return with them together. What was the war? What did they fight for, and why did they have to die? What were the consequences of their deaths? Should those young men on the other side have been killed? Sometimes he would return to the jungles he had wandered decades ago and be immersed in deep thought about the war. In later years, he became the Military Attaché, the first ever to the Korean Embassy in Russia as a one-star general, and Korean Ambassador to Uzbekistan after retirement from the army.[2]

### "Sir, it's dangerous. I'll lead the way."

Second Lieutenant Lee Jong-geun arrived in Vietnam on September 21, 1967, and served three years until September 1970. Upon arrival in Vietnam, he was appointed the 3rd Platoon leader of the 11th Company of the Cavalry Regiment. The life of an infantry platoon leader in Vietnam began with an ambush. He had to fight a swarm of vicious mosquitoes in tension and nervousness, not knowing when a situation would develop into a firefight. During his first three months, he conducted day and night ambushes and search operations to clear the Viet Cong, operating within a 5-10 km range around the battalion base. About this time, news of the deaths of three classmates came to him.

His first combat experience was in Operation Fierce Tiger 9, a divisional operation conducted for two months. It was to clear out the enemies hiding in various places after the Tet Offensive in January 1968 by the North Vietnamese forces. Lieutenant Lee lost two men in this operation but killed 19 enemies and captured two crew-served weapons and six small arms. During the operation, he had no time to shave his thick beard. He let it grow to 12 cm. Sometime later, this beard even grew to as long as 15 cm. At that time, his division commander Major General Chung Soon-min, officially recognized his beard in consideration of his role as an officer cooperating with the minority people of the Montagnard in the mountains. Since then, he earned a nickname, the "Hairy Chief," the rarest case in the division and the entire Korean army.

On December 24, 1967, during Operation Fierce Tiger 9, around 14:00, a violent gunshot was heard from about 400-500 meters away. He forced his men to strengthen their security and listened to the ra-

2Lt. Lee Jong-geun (Class of 1966), 3rd Pt Leader of the 11th Co. Cavalry Rgt. Tiger Div, picture-taken during Operation Fierce Tiger 9 in Dec. 1967. He was well-known as a "Hairy Chief" throughout the division.

dio. After a while, the company commander issued an operation order. "Seems that the enemy ambushed one platoon of the 6th Company, and the platoon leader and his RTO were injured. Organize a rescue team to open the path to the beleaguered platoon." Accompanied by 1/2 of his platoon, Lieutenant Lee flew to the battalion TACP by helicopter and barely advanced 300 meters to the target point in two hours.

As the sun went down, Lieutenant Lee was examining the terrain to select an ambush site near the target, and suddenly, with two gunshots, "Ta Tang!" the point man standing right in front collapsed, and a battle ensued. The number of the enemy was around 15-20. But the fierce firefights for some minutes ended without any further damage. And the situation was buried in the silence of the night. Around 21:00, Lee heard the moaning sounds of the injured soldiers here and there. "Sir, give me some water. Please, Mr. Platoon leader..." but everyone couldn't move. The moans subsided at dawn. They were the soldiers of the 6th Company who got sniped during the day. Even in that desperate situation, the soldiers called out "platoon leader" for help rather than "mother." Soldiers know who they are fighting for and who will save them on the battlefield. Do they believe in such high-sounding words as "fighting for a nation and the people"? What makes them fight is the platoon leader and comrades fighting side by side. It was a scene that made Lieutenant Lee realize how vital the job of a platoon leader was on the battlefield.

As the day dawned on the 25th, Lieutenant Lee and Captain Park

Pan-gil (Class of 1960), the 6th Company commander, took control of the site and evacuated the fallen Lieutenant Suh Soo-geun (Class of 1965) and the soldiers by helicopter. Lieutenant Lee also lost two platoon members in this operation. His heart was overwhelmed by the extreme hostility and revenge. There were several things Lieutenant Lee had realized through the operation. Before the first operation, he strictly instructed all his men not to "harshly harm the lives of petty livestock, animals, or innocent civilians captured." Even though you were in a war situation, it would be best not to take the lives of living things recklessly. Later, Lee found that the deceased squad leader and his squad members had shot and killed a roe deer that had appeared around the platoon base before they moved out for an operation. He thought that this tragic incident might be some causal retribution for what they had done.

On one occasion, his platoon captured a Viet Cong female nurse suffering from a piercing wound on the thigh while searching a cave. The soldiers had other thoughts, and Lieutenant Lee vaguely guessed what they were. He firmly stopped them from committing any wrongdoings against her. Lee also saw his messenger quenching his thirst by drinking his urine with a water purifier. When Lee inadvertently asked the messenger for water, his messenger handed the precious cup of water from his canteen to his platoon leader. He had reserved the clean water for his platoon leader, just in case. Gulping it down, Lee was deeply moved by his man's loyalty and warm heart toward him and shed tears.

And Lee cannot forget the squad leader, who was shot and killed by enemy fire. He was the man who had volunteered to be a point man on the jungle trail, saying to the platoon leader, "Sir, it's dangerous. I'll lead the way." Indeed, a battlefield is where you can see men's most heartfelt comradeship and spirit of sacrifice. Perhaps you will be able to see such sublime human love nowhere else but on the battlefield. During his ten-month tenure as a platoon leader, Lieutenant Lee always feared death and yet felt comforted by the presence of his men together. Maybe that's why men love war. After completing his service as a platoon leader, Lieutenant Lee was transferred to the division's Vietnamese language center, received training for six months, and was then assigned to the security unit of the

ROKFV in Saigon, serving another 20 months before returning home.[3]

## "Hard to win a unilateral victory"

Captain Byun Gil-nam (later a division commander as a two-star general) was one of his classmates of 1966 who went to Vietnam directly as a company commander, not a platoon leader. Many of his classmates had already returned to Korea. In April 1969, he was appointed the 1st Company commander of the Tiger Division's 1st Regiment and participated in Operation Eagle 70-1 and Operation Tiger.

During Operation Eagle 70-1, the 1st Company had a lot of contact with the enemy in An Hoi Valley. The Valley area was steep, and a small stream flowed through it. In addition, as it contained many natural caves, it was the best place for the defenders to hide, but hard for the attackers to detect the opponent. On the first day of the operation, while the 1st Platoon ambushed at night, they found a Vietcong moving upstream of the valley at around 03:30, and Corporal Yu exploded Claymore to kill him. The next day, Byun found traces of the enemy in a natural cave and instructed the 3rd Platoon leader to conduct a deliberate search. After a while, the 3rd Platoon leader reported that the enemy fire shot one of his men above the valley and requested a medic to be sent. A company medic rushed in to rescue him, but he was dead.

The next day, Captain Byun ordered the 3rd Platoon to conduct another search of the spot where his company had taken two casualties. The reason was that, although the enemy had fled, Byun judged they would be hiding nearby. Around 11:00, the 3rd Platoon found the Viet Cong moving north, and while the 2nd Platoon detoured and blocked the ridge, the 3rd Platoon killed three Viet Cong and captured two CARs and an M79. This action was retaliation for the casualties the company took the day before. His company killed 11 enemies and seized nine weapons during this operation, but it suffered four dead and four wounded. Captain Byun recalled later, "It's not easy for either side to win a unilateral victory without damage in a war where each side kills and is killed."

In the Vietnam War, more than 60% of the damage was caused by buried objects such as booby traps. It was why the enemy, inferior to the allies regarding human resources and materials, had to wage guerrilla war. And in Vietnam, one rifle was considered the same as three enemy kills. This became a factor that influenced the front-line leaders and commanders to succumb to the practice of false and exaggerated reports. It was not desirable when looking into the distant future of the Korean military. Captain Byun learned some lessons from the Vietnam War that the military's contradictions and corruption, such as greed for material and unreasonable medal awards, should be eradicated.[4]

Another member of the 1966 class who served as a company commander was Captain Lee Eui-ho (later transferred to Korean National Police Agency and promoted to Senior Superintendent General). He went to Vietnam twice. On the first tour, from August 1967 to April 1969, he was the Weapons Platoon leader in the 8th Company of the 26th Regiment of the Tiger Division. On his second tour, he served as the 11th Company commander of the 1st Regiment from March 1971 to March 1972. During this second tour, he participated in Operations Fierce Tiger 16 and 17. The former commander of the 11th Company was Captain Kim Jin-seon (Class of 1963), who made outstanding achievements in every operation. The North Vietnamese Army and the Viet Cong were so afraid of the 11th Company that they even offered a bounty for successfully raiding the 11th Company's base.

However, the successor to Captain Kim Jin-seon was poor in leading the company. Due to his lack of leadership competence, many accidents occurred, including frequent self-inflicted incidents and enemy rocket attacks on the company base. He should have united the company members and engaged in transparent administration. The company searched the surrounding area one day to eliminate the enemy threat. But the enemy enveloped it at the village of Vinh Long, about one km from the base. Here, the company suffered enormous casualties, amounting to the number of a platoon. Then, the higher command immediately dismissed him and appointed Captain Lee Eui-ho as the new commander. The change of command took place in the middle of the fighting. After Captain Lee took office, the unit gradually regained stability.

In Operation Fierce Tiger 16, which started on April 23, 1971, the 11th Company secretly set up an ambush around Da Giang Mountain. The enemy judged the friendly troops had utterly withdrawn and came down to infiltrate Hoi Son village from the mountain where he had been hiding. At 09:50 on May 8, when an unknown enemy group appeared in front of the friendly ambush team, the 2nd Platoon leader First Lieutenant Kim Tae-woong (Class of 1970), triggered Claymore, killing four men and capturing two AKs and a radio.

The problem was Operation Fierce Tiger 17. It started on June 25, just one month after Operation Fierce Tiger 16 ended. It was conducted in Huu Giang Mountain and the An Hoi Valley area. The 11th Company was scheduled to land on Hon Xong Mountain on D-day. The 9th Company, to the left of the 11th Company, was hit hard by enemy fire while landing at the target around 09:00. A U.S. helicopter crashed, a pilot was killed, and three others were injured. The 11th Company attempted to land at the target, Mount Hon Xong, around 13:30 but could not land due to fierce enemy fire and heavy rain. Eventually, the company returned to the base and attempted to land again at 09:20 the next day. However, the enemy's fire still was so fierce that it failed to land. After being hit by enemy fire, a helicopter rose again and crash-landed on the sandy beach of the Song Con River in front of Huu Giang Village, 5 km away. The U.S. pilot died while being evacuated to a field hospital.

About 20 minutes later, five soldiers from another helicopter succeeded in landing, but the enemy's fire was so fierce that all the helicopters could not land in time. However, the 2nd Platoon leader, Kim Tae-woong, led 14 men and succeeded in landing on the unknown hill. Then, he joined the soldiers who had gotten off earlier and occupied the target hill. However, two company soldiers were killed in this battle, and two others were wounded. On the other hand, the company killed four enemies and captured one LMG. Through this battle, Captain Lee once again confirmed that all Vietnam War battles were conducted by squad leaders, platoon leaders, and company commanders.[5]

## “If only I could return to my homeland alive!

Second Lieutenant Jee Man-won, an Artillery officer who fought in Vietnam for 44 months, spent the most extended period in Vietnam among his classmates. At first, he served for 22 months, from July 1967 to May 1969, as a FO and Assistant to S-3 Operations of the 30th Artillery Battalion, the White Horse Division. During his second tour, from November 1969 to May 1971, he served seven months as an aide-de-camp to the Chief of Staff of the ROKFV and 12 months as the Bravo Battery commander of the 30th Artillery Battalion.

When he came to Vietnam for his first tour, he was a FO of the 3rd Company of the 28th Regiment, the White Horse Division. The 3rd Company commander was Captain Min Chan-ki (Class of 1960), a rugby player when he was a cadet. The day before Operation Hong Gil-dong, everyone prepared the things necessary for the operation with hardened faces. An orderly and RTO came, greeted Lieutenant Jee for the first time, and brought military gear for him. He also carried four canteens full of water. They said to the new lieutenant: “Sir, save your water. Do not give water to or ask for water from others. Rules! There is no drinking water in the jungle. Water in the mountain is toxic, so drinking it may cause serious problems. Rumor has it that the Viet Cong poisoned the water. You must only drink canned water. Even if you buy a watermelon at the market, it will be a disaster. They say they put a poison sting in the watermelon.”[6]

On July 9, 1967, at the dawn of D-Day, Operation Hong Gil-dong,[7] Lieutenant Jee went to the Pickup Zone (PZ) to board the helicopter. The number of passengers on board was five. It was the first time boarding an aircraft in his life. His helicopter flew over the field, and he looked down without closing the side door. He fastened his seat belt but felt like he would fall out, so he grabbed the chair tightly with both arms. A vast plain was visible, and a winding river came into sight occasionally. The jungle-covered mountains unfolded endlessly. It was a strange foreign landscape, unique and beautiful.

“What if they fire guns in the jungle at the helicopter flying low?” It made his heart flutter. The helicopter descended and stopped above a reed field taller than a person. The soldiers jumped from a height of about two meters. They quickly assumed prone positions and took all-around security. What a movement of the body! Each company had a different landing place. Two divisions were laid out like a checkerboard in the vast jungle mountain. Operation Hong Gil-dong, the longest one ever up to that time, was launched in this way.

The 3rd Company discovered a large number of weapons during the search of a labyrinthian cave. It was equivalent to two truckloads of weapons, including mortars, machine guns, and rifles. Captain Min's face turned pale blue. It was too much of a combat achievement at an operation. At the radio report of the company commander, the happy voices of the battalion commander and the regiment commander in the rear rang through the receiver. Helicopters were sent to start carrying the captured weapons. It was late at night. However, the commanders only cared about securing the weapons without regard for the soldiers' safety. The soldiers were painfully thirsty. They would drink coffee with urine in C-ration cans. The thirst was unbearable. Everywhere they heard the sound of a water-flowing ditch, they felt like running and drinking, even at the risk of death.

When they spent the night in the jungle, they pitched tents. They sprayed poisonous mosquito repellent on the ground to keep scorpions and snakes away. They also applied it to their faces and hands. Lieutenant Jee's orderly blew air into a rubber bed, put a poncho raincoat on it, and put a blanket on top of that. However, the orderly and RTO slept only on raincoats and blankets on the damp ground. The tedious operation continued for over a month. A month later, groups of helicopters flew in. Finally, the operation was over, and the company returned to its base. The company commander called his platoon leaders to his tent at the base. He told them to drink a cold can of beer that was cooled to just before freezing. Lieutenant Jee had dreamed of being able to have a moment like this. He pinched his flesh several times to know whether it was a dream or real life. Somebody played Korean pop songs on the record player. They were the ones that he had been tired of at home, but at this moment, every word and note penetrated his heart deeply. He prayed, "If only I can return to my homeland alive! I'll do anything, and I'll be able to have anything."[8]

## Class of 1967 (23rd Class)

Two hundred and thirty-five students enrolled as the KMA Class of 1967 on February 1, 1963, and 177 graduated on February 23, 1967. The attrition rate reached 24%. The idea of Vietnam gradually began to creep into their minds in 1964 when the Korean military deployed its troops to the war. By the time they graduated, the Vietnam War had peaked, and the word "Vietnam" was lodged in their minds. After three months of OBC, all 1967 graduates assigned to the front divisions as platoon leaders were already determined to go to the Vietnam War.

Eight early birds left for Vietnam at the end of 1967, the year they graduated. To be qualified for being selected as Vietnam volunteers, they had to learn and be used to the practical skills of leading troops in the field. The following year, 84 officers, almost half of the graduates, went to Vietnam, 26 in 1969, 10 in 1970, 6 in 1971, and 10 in 1972. A total of 144 graduates participated in the Vietnam War for its entire period. This number accounted for 81% of the class's graduates, almost all of whom were platoon leaders of combat units, with some appointed as assistants to the support units' staff.

Since they had yet to wait to be promoted to captain, except for a few, they could not become company commanders. Therefore, almost all of the 1967 graduates and their juniors who would later come to the Vietnam War remained only as platoon leaders. The war was slowly coming to an end. The year 1968 was the peak of the Vietnam War, and the 1967 graduates engaged in many battles and made remarkable achievements. A platoon leader of the Tiger Division's Engineer Battalion, Second Lieutenant Kim Gu-woong went out on an ambush operation with his fifteen members, killed four Viet Cong, and captured two AK rifles, a rare case for the non-infantry unit.

Second Lieutenants Kim Young-won and Chun Gwan received the Choongmoo medals. Besides these two officers, forty men of this class were awarded the Hwarang and Inheon medals, the fourth-and fifth-highest distinction, respectively. And some twenty other members received the Combat Commendation Medal and Presidential Award. The Class of 1967 was the most decorated class among the KMA classes. Unfortunately, however, four officers died in action.

Also noteworthy was Second Lieutenant Yang Gil-yong, who went to Vietnam in July 1968 and returned in late August 1969 after a tour of 13 months. During his tenure as the 3rd Platoon leader, the 2nd Company with the 29th Regiment, he was engaged in Operation White Horse 10, conducted on June 4 to 20, 1969. He made an outstanding achievement, killing ten enemies and capturing one AK rifle, three B40 Rocket Launchers, two hand grenades, and some documents. Yang was awarded the Hwarang medal. In special recognition of his heroic actions in operation on June 4, 1969, USMACV conferred the Army Commendation Medal on August 26, 1969, just before he left Vietnam.

*The Certificate of Award*

*For heroism in connection with military operations against a hostile force: First Lieutenant Yang Gil-yong distinguished himself by heroic action on June 4, 1969, while serving as Platoon Leader, 3d Platoon, 2d Company, 29th Regiment, 9th Division, Republic of Korea Army. On that date, the lead elements of Lieutenant Yang's platoon encountered an enemy position on the far side of an open area. Pinned down because of the advantageous fields of fire and elevation afforded the enemy by their position, Lieutenant Yang consolidated his platoon and, with complete disregard for his personal safety, ran from his covered position across the fire-swept open area toward the enemy position. His platoon, inspired by his fearless act, charged behind him and overran the enemy position killing ten and capturing their weapons, including three rocket launchers and stores of TNT. First Lieutenant Yang's heroic actions reflect great credit upon himself and the Republic of Korea Army.*[9]

Commissioned in 1967, Second Lieutenant Jang Pan-yong was one of the earliest members among the Class of 1967 who went to Vietnam with the idea that soldiers should have combat experience. As a member of the 28th Regiment of the White Horse Division, he arrived in Vietnam toward the end of 1967. Although they recommended that he go to the Regiment Combat Support Company as a platoon leader, he wanted to be sent to an infantry company as a boonierat to experience actual combat in the jungle. Eventually, his first assignment was as the 1st Platoon leader of the 7th Company of the 28th Regiment.

After arriving at the company, he was engaged in usual actions like ambushes, reconnaissance patrols, and small-scale operations for several months. On April 5, 1968, around 13:00, however, an operational meeting was suddenly held at the battalion headquarters with all the officers above the platoon leaders. That was the beginning of Operation Dokkabi 6, a large operation conducted on a regimental size. It was the operation requested by the governor of the Phu Yen Province, which would be carried out in the TAOR of the South Vietnamese army's 47th Regiment. According to the intelligence that the governor revealed, two battalions of the regular Viet Cong entered the villages of Phu Lok and My Thanh in the southwest of Tuy Hoa.

The first thing that the regiment did was to evacuate the area's residents. After they were safely moved out of the villages, the preparation attack fire started with bombing from US fighter jets and gunships in the area. Along with this action, the regiment began moving into the area through air maneuvers, with the 2nd Battalion to the northern part and the 1st Battalion to the southern part and establishing an envelopment net. Under aircraft illumination, bombing continued through the night, with the support of artillery fire in full brightness. On the night of April 5, the first encounter was made in front of his 1st Platoon. The enemy tried to break through his platoon position but failed. After fierce fighting, the enemy tried to retreat to the village but only took tremendous damage. The 1st Platoon killed 18 enemies and captured many weapons, including one 82-mm mortar, one 60-mm mortar, 7 AK rifles, two pistols, and others.

2Lt. Jang Pan-yong (sitting, Class of 1967), 1st Pt leader, 3rd Co, White Horse Div's 28th Rgt, with his RTO and orderly during a lull at My Thanh village in Operation Dokkabi 6 in April 1968. (Courtesy: Jang Pan-yong)

Even now, Lt. Jang cannot understand why the enemy had to suffer so many casualties. His platoon took a line formation with little depth. Thus, if they had concentrated on any spot of his formation, they might have succeeded in breaking through his position with slight casualties. The enemy were looking for a way out here and there like fish caught in a fishnet, and they had nothing to do until the day dawned. Two days later, his platoon was given the task of clearing the battlefield and went back to My Thanh village to bury the enemy's dead bodies. The 28th Regiment's Operation Dokkabi 6 was completed in three days. It was fortunate his platoon had no casualties. For this achievement in the operation, Jang was awarded the Hwarang Medal.[10]

### "Make them scream!"

Second Lieutenant Lee Jong-gyu (later, 15th Infantry Division Commander as a two-star general) went to Vietnam in June 1968 and returned in December 1969 after a one-year tour. In Vietnam, he was appointed 2nd Platoon leader of the 2nd Company, the 29th Regiment. Lieutenant Lee participated in Operation Eagle 1, carried out from October 5 to 20, 1968, and underwent some painful experiences for the first time in his life. His 2nd Platoon found two caves in the jungle, searched them, and captured many weapons and food. His platoon also took the enemy's families as prisoners, 25 women, 15 children, and some sick people. These civilians lived together with the enemy troops inside the caves. In the meantime, however, his platoon was enveloped by enemy forces. So he immediately set up a defensive line against the enemy's attack. On October 13, 1968, around 2:20 a.m., a group of Viet Cong attacked the defenses of the 2nd Platoon. It was a night, and torrential rain was pouring down.

Lieutenant Lee ordered the soldiers to enter the cave while guarding the entrance. Then, he requested fire from an artillery battalion about 10 kilometers away. Meanwhile, the number of enemies increased to about 300, and they were attacking the defense line of the 2nd Platoon. The friendly artillery fire was exploding over the head of the attacking enemy inside the platoon's defense line. According to the report later, the friendly artillery battalion fired more than 1,500 rounds, and many enemy soldiers were killed by this fire. But the enemy did not stop attacking. They continued to attack the positions of the 2nd Platoon.

Lieutenant Lee thought his platoon would be annihilated without taking extraordinary measures. He told the soldiers, "We will break through the enemy's envelopment at 4:30 a.m." Lee planned to exploit the enemies' families. Lee had each soldier tie a woman and a child to his belt and come out holding their hands. The only way for a platoon to succeed was to have the enemies know that their families were in the custody of the Korean troops. But the women didn't want to reveal that they were in the hands of the Korean soldiers. So, Lieutenant Lee told his men, "Strike them hard on the head and the face. Thus, make them scream so the enemies in front of us may know their dependents are with the Koreans in the cave. Then they will not fire at us." The soldiers carried out the platoon leader's orders perfectly. This way, the 2nd Platoon could break through the enemy's siege. The 2nd Platoon suffered heavy losses, though. Five men were killed, eight were seriously injured, and Lieutenant Lee was shot three times. After this battle, the platoon leader and several other soldiers were awarded medals. However, Lieutenant Lee always feels heartbroken at the thought of his men killed and wounded in this battle.[11]

## A Descendant of Cain

Second Lieutenant Kim Myung-sei (later, 22nd Infantry Division Commander as a two-star general) fought in Vietnam for precisely one year, from September 1968 to August 1969. He was appointed the 2nd Platoon leader of the 12th Company with the 30th Regiment, the White Horse Division. In Vietnam, he got shot by a bullet that was meant to kill him for the first time. Yet he was proud of himself for overcoming the fear of death. Kim wrote a letter about the battle to which he had been first exposed as the platoon leader in Vietnam to Lieutenant Colonel Lee Dong-hee, a professor of Political Science at KMA. LTC. Lee was the instructor who taught Lieutenant Kim and greatly influenced his academic and personal growth during his cadet days. Kim confessed such anguish in his first battle to Professor Lee, who published his letter in *The KMA Press*, a monthly paper of the

Academy, in May 1969. The following is an edited version of his letter published in it.

Dear Professor Lee:

I was engaged in combat for the first time since I became an officer and killed a fellow human being. Astonished at my action, I doubted whether it could be justified in the name of war or whether the word was fitting that a human being is a descendant of Cain with his black blood flowing through his veins. However, I recognized that the battlefield was not where we would be immersed in such sentimental thoughts because we would die if we did not kill others.

I am writing to tell you how I fought in the jungle of Vietnam and killed the Viet Cong. You were so considerate and enthusiastic in teaching cadets. Most of us were grateful for your valuable advice in pursuing a professional military career during our cadet days. After fighting the VC commando in a jungle operation, my men killed eight people. Keeping in mind your words not to get swept up in emotions when entering the battle, I led my platoon to the enemy positions as my rational mind told me. My platoon was moving forward cautiously by conducting a search mission in the jungle that made us tired. It was a time when the forest, where the morning sun was shining faintly, gradually began to come to life. And the raindrops held in the leaves were dripping onto our helmets.

2Lt. Kim Myung-sei (Class of 1967), 2nd Pt leader, 12th Co., 30th Rgt, White Horse Div. at Dai Dien Dong village, in 1968.

As I looked up at the thick tree with fearful eyes, a signal came from a point man that he had found a trace of something on the creek. When I saw it, it was undoubtedly a trace of a human, the distinct footprint of man. I thought someone had been watching us. And when I lifted my head and looked up, I could not see anything except the shadow of a thick tree. At that moment, the tension I felt only in 007 movies gave me a headache. And I told the soldiers to go forward with strict vigilance. After 15 minutes or so, when we struggled through the jungle, we noticed a group of troops passing by from about 30 meters away. I immediately dispersed my men and gave them the order to start firing, which I emphasized daily to "squeeze the trigger in the aiming position." Then, gunshots began to sound like an angry lion, and the enemy in front began to fall one by one without being able to respond in panic. I began to drink water from the canteen, feeling a burning thirst in my mouth as I shouted like a half-madman.

The moisture seeping through my throat gave me more strength and courage, and I started screaming again. At the sound of "Sir, there's an enemy on the right," I boldly detoured the platoon and had them charge toward the enemy in front and on the right. Seeing the enemies start running away in confusion, we pursued them without giving any leeway. After the battle, I sat down on the paddy field, took a deep breath, and did not have time to enjoy the thrill of surviving the first battle. I just grabbed the radio and reported to the company commander. He responded, "Congratulations. On behalf of the company, I congratulate you for making such a great achievement through your outstanding leadership." A sense of release enveloped my entire body, and I closed my eyes. "Good job! Congratulations!" said the fellow officers who patted my shoulders. Walking down the mountain, I had never felt before how beautiful the sun shining was.

I've heard that the pleasure a man feels after a victory in war is comparable to the one a woman feels after the pain of her childbirth. My first battle in the war was a great success. Please stay healthy, and goodbye until I see you again, sir.[12]

Two 1967 class members served as team leaders in the long-range RECONDO (Reconnaissance Commando) units. One is Second Lieutenant Kim Sang-deok, and the other is Second Lieutenant Yang Sang-jin. Though they were platoon leaders, they were tasked with special missions of reconnoitering and collecting information about the enemy situation on a long-range basis.

Second Lieutenant Kim Sang-deok participated in Operation White Horse 6, a division-level operation, as the team leader of the long-range RECONDO of the 29th Regiment of the White Horse Division. This operation took place three months after he arrived in

Vietnam. The operation lasted 33 days, from May 5 to June 7, 1968. In particular, during this operation, the White Horse Division aimed to attack and destroy the NVA's 18th Regiment threatening three military bases on the central coast of South Vietnam—Cam Ranh, Nha Trang, and Ninh Hoa—and the local Viet Cong guerrillas on Nui Hon Ba, the stronghold of VC People's Commission in Khan Hoa Province and Ninh Hoa.

On the morning of May 10, Sgt. Hyun Sang-il, a point man, guided the teams to discover the cave. After receiving the enemy discovery report, Team A's leader, Second Lieutenant Kim Sang-deok, in consultation with Team B's leader, concluded that the cave must have an exit to the rear. So, Team B deployed its men to block the rear exit, 100 meters opposite the cave entrance, and, in the meantime, Team A fired a volley, killing three enemy guards and charging into the front of the cave. As expected, the enemy fired back and began to run away through the escape route at the back of the cave. When Team B on standby fired a volley, the enemy fled to the front again.

As a fierce fire came from the front again, the enemy became rats in the trap, moving back and forth between Teams A and B three times. Ultimately, the teams annihilated the enemy in 30 minutes but were unharmed. The most significant achievement was the killing of Mai Shao, the Commissar of the local Viet Cong Commission. The South Vietnamese government had put an enormous bounty of 10,000 piastres on him and tried hard to catch him. This man served as the Commissar of the Viet Cong Committee in Khan Hoa and Ninh Hoa Provinces and the commander-in-chief of the local guerrillas. He was the man who wielded a pistol and wandered here and there in the central coastal region of South Vietnam and was notorious for his brutality. However, a ghost-like figure, Mai Shao, was no match for the RECONDO teams. In addition, they killed 25 Viet Cong commandos and captured six small arms, one pistol, 45 hand grenades, 3,700 live ammunition, and about 100 valuable documents, including Mai Shao's camp diary.[13]

## "Gunfight at the O.K. Corral"

Second Lieutenant Yang Sang-jin's experience in the Vietnam War is unique. He arrived in Quy Nhon in August 1968, a year and a half after graduating from KMA. He was appointed 2nd Platoon leader of the 2nd Company with the 1st Regiment. At the same time, he was ordered to be a platoon leader leading a RECONDO team rather than a general infantry platoon leader. The 2nd Company commander was Captain Koh Young-bok (Class of 1962) and Yang's senior on the KMA soccer team. Yang was also a soccer player, so they had something in common. With the 41 platoon members, Yang was dispatched to the Tiger Division's special training center to receive four weeks of RECONDO training necessary for survival, infiltration, and escape in the jungle.

The task given by the battalion commander, Lieutenant Colonel Lee Bum-cheon (Class of 1955), was to search the Viet Cong bases in the battalion's AOR, report on their activities, and strike when necessary. And his mission was to reconnoiter their secret bases around Nui Phu Cat and Nui Jung Tanh, which commanded the Go Boi plains and controlled the coastal road, the main supply route for the Viet Cong. Lieutenant Yang's reconnaissance team infiltrated the enemy area in Viet Cong's black outfits. Usually, they operated once a week for three nights and four days. After returning to the company base, they had two nights and three days of maintenance. Yang always divided his platoon into two teams, one with the platoon leader, code-named "King Spider" and the other with the platoon sergeant, code-named "Fire Ant" when moving in the enemy area.

In October 1968, just two months after being appointed platoon leader, Lieutenant Yang arrived at Nui Jung Tanh with nervousness and anticipation to carry out his first infiltration mission. Lieutenant Yang's team found a group of 15 enemies and ambushed them on their path. In this encounter, his platoon captured an unarmed female Viet Cong, about 19 years old. He reported the female prisoner to the company commander. But he could not send her to the company base because the helicopter could not fly in due to the darkness. So, Yang had to spend the night with the female prisoner. There were many gruesome stories about women's rights on the battlefield, especially the women of the enemy. Lieutenant Yang met with his squad leaders and reconciled their conflicting opinions regarding her treatment. He made a sober judgment. It was a moment of overcoming the temptation of war crimes. Yang spent the night protecting her directly by tying himself and her with the straps. He returned to the company base safely and handed her over to the battalion headquarters the next day.

One day Yang's platoon conducted a surprise attack against an enemy base in the Nui Jung Tanh area. The operation was successful, but one soldier got a penetrating wound in the abdomen. On his request, the dust-off helicopter arrived in the air of the operation area, but for some reason, it only circled the sky a few laps and disappeared. Lieutenant Yang immediately protested and learned the reason by radio from the company. The American helicopter pilot said that he only found the enemies when he tried to land. His remarks made sense because his men wore enemy clothes to disguise themselves. Eventually, Yang evacuated the wounded man safely by signaling the color of the star cluster to the pilot.

Lieutenant Yang has special memories of the operation carried out in May 1969, about three months before returning home. His mission was to ambush the enemy's main supply line, the coastline road, in the Go Boi plains. This operation was conducted in a deserted corral in the region, and Yang named it "Gunfight at the O.K. Corral." After waiting patiently for two nights and three days, he lured the stealthily approaching enemy to close range. Then, with his signal to start firing, they fired a volley, killing about 15 enemies. This long-term ambush was made possible due to thorough camouflage and management of the platoon's position, especially food and feces. Lieutenant Yang killed about 40 enemies, captured one person during his one-year tour as RECONDO platoon leader, and seized 23 weapons. Three men in his platoon suffered severe injuries. He still feels sorry for those living in maimed bodies and only wants them to have God's blessings.

Lieutenant Yang had heard unbelievable stories about the atrocities committed in the Vietnam War. The level ranged from the summary execution of a captured enemy in retaliation for the friendly killed to the story of the enemy who killed a friendly soldier and mutilated his body. He had also heard horrific stories of detonating enemy corpses with grenades under their bodies to instill a killer instinct into recruits. All are unconfirmed stories. Lieutenant Yang wondered if the rules of engagement to make an inhuman war human would make any sense at all in combat.[14]

## "Go by boat and return by boat, not by plane"

A guerrilla war is a war that emphasizes psychological operation more than any other type of conflict. Several officers of the 1967 class participated in the war as members of the Civil Affairs & Psychological Warfare Unit. And one of them was First Lieutenant Cha Ki-moon (Later, the Senior Member of the United Nations Command Military Armistice Commission as a three-star general). First Lieutenant Cha Ki-moon completed his tour as a platoon leader in the 17th Regiment of the 2nd Infantry Division in the DMZ, Korea. While serving as the training officer of the same regiment in December 1968, he received an order from the higher command that he should go to Vietnam. It was nothing to do with his intention. As 1968 was the most intense year of fighting in the Vietnam War, he was slightly nervous and afraid of being in the war. But he gladly accepted the order because he took it for granted that a soldier should have a combat experience on the battlefield. In fact, at that time, many casualties from Vietnam were being transported to Korea by plane. So, it was a widespread rumor among the officers that if one went to Vietnam as a junior officer, he would die. Indeed, there was a common saying like this, "Go by boat and return by boat, not by plane." If a man returns by plane, it means that he is dead.

After a while, he was ordered to learn Vietnamese before going to Vietnam. So, he entered the Vietnamese language training center in Seoul. After months of language learning, he arrived in Nha Trang and was appointed platoon leader in the Civil Affairs & Psychological Warfare Unit under the ROK Field Command. His mission was to produce flyers, distribute them to the areas where the Viet Cong were hiding, and use a loudspeaker to make a broadcast inviting the enemy to defect in mountainous areas. After conducting a flamethrower operation

A PSYOP member of the White Horse Division and a South Vietnamese lady broadcasting to encourage the enemy to defect in the Song Da Rang area in Operation Sacred Horse 72-1 near Tuy Hoa in July 1972. (Source: Institute for Military History, MND)

on the Vietcong hiding in the cave, he persuaded them to come out using a portable loudspeaker. Because the Viet Cong conducted guerrilla warfare, the operation was impossible without the help of the locals.

Therefore, the civil affairs operation that separated the Viet Cong from the civilians was the most important. In the village, the Viet Cong were hiding in the basements of their houses, so persuading the residents was an operation to risk one's life. As the Korean military's civil operation was successful, the villagers were very friendly in the area where the Korean army was stationed. After one year of platoon leader's tour, Lieutenant Cha worked for the Chief of Staff of the Field Command of the ROKFV as his aide-de-camp for one more year and returned to Korea. Lieutenant Cha's service in Vietnam for two years gave him valuable experience in pursuing his professional military career.[15]

## "Do the White Horse's Artillery shells have eyes?"

An artillery officer, First Lieutenant Noh Nam-seop (Later, the 72nd Infantry Division Commander as a one-star general) landed in Nha Trang via Da Nang on November 24, 1968. The soldiers of the White Horse Division returning to Korea after completing their one-year tour in Vietnam waved their hands to say goodbye to them. It indicated that they had survived the hell of war and were now returning alive. But it also stated pity and sympathy for those poor fuckers who would spend a year in the terrible jungles. He arrived at the White Horse Division headquarters in Ninh Hoa and was ordered to go to the C Battery of the 52nd Artillery Battalion. The C Battery was on a low hill near the Cai River in Dien Kahn. When he got to the Battery, Captain Lee Yoo-shin, C Battery commander, told him to "prepare to leave for the 11th Company as its FO." The company was building a tactical base on Hill 42, about 6 km northwest of the C Battery.

The 11th Company's tactical base was significantly lower than Hill 114 in the west and Hill 216 in the north. It was an unfavorable terrain. He was puzzled why the company placed itself on such a low hill that the enemy could command. After listening to the company commander's explanation, he understood that Hill 42 was low but a tactically crucial strategic point as a hill that could control the crossroads leading to the Viet Cong's hideouts in Nui Hon Ba and Nui Hon Juh. On the 3rd day of his appointment as an FO to the 11th Company, the Viet Cong attacked the company with mortars. "Bang! Bang!" Debris splattered along with the mortar explosion sound, and the soldiers shouted all at once as they rushed to the temporary bunkers or mortar positions. The gunners of the 60-mm mortar squad of the Weapons Platoon immediately fired back. The company fired mortars toward Hill 216, but the range was too short. However, it was a great help in boosting the morale of the company members. Lieutenant Noh immediately analyzed the mortar shells that fell within the base and urgently requested artillery fire by radio to his battery.

"Fire around Hill 216, Hill 114, and the expected escape route." At his request, the shells flew in no time and hit Hills 216 and 114, as well as the expected escape route. The Viet Cong fired only six or seven mortar shells and escaped. Only one shell fell into the 11th Company base, and the rest fell into the jungle outside the barbed wire fences. Fortunately, the 11th Company took only one casualty, a soldier who received a minor abrasion on his shoulder. It was so dark that they could not go out to check that evening. The following day, the 11th Company thoroughly searched the entire area of Hill 216. As expected, they confirmed the traces of the Viet Cong's mortar plates in the open space on the top of Hill 216 and the footprints of countless Viet Cong in the retreat area. Lieutenant Choi, the Weapons Platoon leader with extensive combat experience, spoke quietly to him. "It gives me goosebumps when I see the footprints of the Viet Cong shoes. This area is a Viet Cong stronghold, so it's perilous."

On March 4, he returned to the C Battery of the 52nd Artillery Battalion after completing the 11th Company's FO service for about three months. He was assigned a new Fire Direction Officer (FDO) position in the battery. Around 9:00 p.m. on April 8, Lt. Koh, FO of the 9th Company, sent an emergency report, saying the 9th Company's ambush team saw about 40 Viet Cong entering the village of Dai Dientai to find food. Around 01:00 a.m. the following day, Lt. Koh requested an illumination fire, saying, "The Viet Cong is coming to the Claymore killing zone soon, so please continue to fire flares at the sound of an explosion."

A Forward Observer (FO) directing the artillery fire in Operation White Horse 2 in March 1967. (Source: Institute for Military History, MND)

He immediately gave the order to the Firing Battery to prepare for firing, and the moment he looked at the 9th Company area, the sound of claymore explosions and mortar flares lit up the night sky. He heard the loud sounds of the friendly fire of M60 machine guns and M16 rifles. He fired flares with the two main guns and fired the remaining four tubes to block the Viet Cong's escape route for one hour and ten minutes. After 02:00 a.m., the illumination was conducted by aircraft, and it lit the area of the 9th Company brightly as if it were broad daylight. As the morning sun rose, the 9th Company's ambush team checked the site, found six Viet Cong killed, and collected two AK rifles, two long guns, two CAR rifles, and 140 kg of rice. A Viet Cong POW later confessed to the Korean interrogating team: "Do the White Horse's Artillery shells have eyes?"[16]

## Class of 1968 (24th Class)

The *Asadal*, the KMA Alumni Association Newsletter, dated December 1969, writes, "At this point, the initiative in the Vietnam War has been transferred to the Class of 1968. We advise that many of our classmates participate in the actual war."[17] These words were an expression of the determination of the Class of 1968, "We're here. Now is the time to replace you, Seniors. It's our turn to take care of the war in Vietnam." In 1969, many graduates of the Class of 1968 headed for the war front in Vietnam. A total of 111 graduates went to Vietnam, almost all as platoon leaders.

The history of the Class of 1968 began on February 1, 1964, when 252 students entered KMA. After four years of harsh life, only 180 cadets graduated on March 3, 1968. The attrition rate was 29%, the highest one ever among whole classes up to that time. In fact, after the NVA's Tet Offensive in early 1968, the United States was thinking of ending the Vietnam War, and peace talks started to be held in Paris. In 1969, thus, large-scale combat operations were relatively rare, and almost all combat was conducted as small-unit fighting. Therefore, there was less enthusiasm for going to Vietnam than before. Regardless of these political and military changes, however, all graduates of 1968 thought they should have combat experience as a soldier as long as the war was going on. There was also impatience that they had to go and experience the war before its end.

Starting in early 1969, the time to complete their platoon leaders' tour in Korea, they volunteered to go to Vietnam. Until the following year, 1970, most graduates of the 1968 Class went to Vietnam. The Class of 1968 Association of Graduates encouraged its members to participate in the war by emphasizing their role in the Vietnam War as part of the Blue White Line. Although fewer in number than their predecessors, 64% of 1968-year graduates went to Vietnam. After months of completing their tour as platoon leaders in Vietnam, some officers moved to higher positions like company executive officers or assistants to battalion staff. Few were even assigned to the positions of company commanders with the rank of provisional captain, including Kang Ho-tae, Park Min-soon, and Hwang Eui-seok.

### "Our Platoon leader closed his eyes"

The following is the description of Second Lieutenant Na In-hwan's (Class of 1968) combat experience, published in *The Asadal*, No. 59, in October 1969. He explains the aim of his writing in this way: "At present, a great many KMA graduates are fighting as platoon leaders in the jungles of Vietnam, and more graduates will come to serve as platoon leaders in the time to come. I will briefly introduce what I have experienced in a recent operation to the potential platoon leaders. In this operation, the company commander was Captain Oh Kil-hwan (Class of 1962), and the executive officer was 1Lt Lee Sang-hong (Class of 1966). I was the 3rd platoon leader, the 5th Company, the 28th Regiment, the White Horse Division."[18] This is the abridged version edited by this author.

Operation Dokkabi 2-7 started at 08:30 on May 4, 1969, when six UH-1H helicopters and three C-47 Chinooks loaded the company and took off one by one from an airfield in Tuy Hoa. The enemy in the area was two battalions under NVA's 95th Regiment and local guerrillas in Phu Yen Province. Though there was an engagement with the enemy at an ambush the previous night, the company only killed two enemies, even without rifles. The company felt a little disappointed. Yet it was comforting to know that the enemy was at least somewhere. Until May 9, the company killed nine enemies and captured three rifles, an 81-mm mortar, and 47 rounds of ammunition. On May 9, the 6th Company unexpectedly encountered the enemy's main force. It fell into an enemy trap and suffered considerable damage. Therefore, the company had to change the initial plan of moving from Mang Mang Valley to Suoi Cai Valley.

First Lieutenant Na's company made a two-km movement to contact across the jungles of Suoi Cai Valley and started to advance in a line formation toward Nui Hon Ba in the center of the area. On May 11, his platoon moved to a place 300 meters south of Nui Hon Ba from its ambush site the previous night in search of a clearing in the jungle to get a resupply. There, all the company's platoons got resupplied. At 13:00, after lunch, they started to move toward a position along the ridge. Suddenly, the violent sounds of AK47s erupted in front of the 1st Platoon in the lead. Immediately Lieutenant Na spread his 3rd Platoon to the left and moved the squads by bounds until they reached a small creek. There he rearranged his squads and waited for some minutes. Violent firefights between the two sides continued in front of the 1st Platoon.

Amid this appalling confusion, repeated voices came on the radio to call the 1st Platoon leader. Lieutenant Na's platoon made a detour to block the enemy's escape and began crawling to approach the top of the hill. After a while, the 1st Platoon Sergeant appeared on the radio, reporting, "Our platoon leader closed his eyes. We're now treating four injured chickens. We captured one AK rifle. The enemies are regular forces, over." When the report ended, the enemy in front of the 1st Platoon poured their fires upon Na's platoon. Their fire was so intense that his men couldn't raise their heads. The bullets hit the C-ration cans on their backs, and the contents flew out. The LMG gunner of his platoon got a piercing wound on his thigh.

Immediately Lieutenant Na gave the first-aid treatment and sent him to the company headquarters. Then he tried to advance his men toward the front little by little. But he had no idea where the enemy was entrenched. The bullets whizzed in out of nowhere. The hand grenades thrown forward got caught in tree branches. And the M79 grenade launchers were ineffective because the distance to the enemy was so short that many shells remained unexploded. They fired the M72 rockets on the enemy position, making them shrink a moment, but again the enemy concentrated their massive fire upon Na's platoon. Amid the increase of casualties in the 1st Platoon, the 2nd Platoon in reserve was ordered to replace the 1st Platoon, but it was not easy because the enemy's fire was so intense. Further, their firepower was not likely to be weakened.

The gunships that the company commander requested were hovering in the air to find the target on which to strike. The dust-off evacuated some casualties, and the gunships strafed the enemy position. Seeing the 1st Platoon leader carried on one of his men's backs with his head bleeding to get aboard the dust-off, Lieutenant Na was overwhelmed by extreme anger. At that moment, an order came down from his battalion to step back for about 200 meters because artillery fire would start as soon as the gunship strike ended. Lieutenant Na pulled his platoon back about 250 meters to the rear while keeping it in the line formation.

At last, the artillery barrage began. The fragments of the artillery shells rained down around, but he did not move back anymore and waited for the artillery fire to be lifted. The artillery bombardment continued for as long as two hours. Lieutenant Na chain-smoked two packs of cigarettes instantly out of fear and tension. At 19:00, an order of night attack came down. The 6th Company started to advance to the creek. The company killed two enemies and captured one LMG and one AK rifle in this move. The enemy began to escape when the friendly artillery and small arms fire stopped. Entirely in the open formation, the platoons crawled forward to climb the hill. Though the illumination shells by the artillery battery continued, densely forested jungles prevented the surrounding area from becoming as bright as expected.

While moving forward, his men found an unidentified dead soldier with a piercing wound on his stomach. That made them full of revenge. All around were the sounds of exploding hand grenades. At last, the enemy came out of their hideouts while responding nominally. Lieutenant Na's platoon occupied the hill and reported to the company commander after piling the parapet and reorganizing his platoon. The time was 21:30 sharp. The following morning, the company checked what had happened the night before. Feeling the bloody and strange smell of the battlefield, the men scoured the enemy stronghold, foxhole by foxhole, and pulled out the dead bodies. The enemies killed were 21 men, all the regular NVA soldiers. The company also captured two LMGs, four AK-47 rifles, two CKS rifles, and many other documents.

From this battle, Lieutenant Na learned a lesson that a soldier's combat skills and techniques should be improved. As a regiment com-mander said, "A soldier's wound on the head is a commander's respon-sibility, and the one below the neck is that of a leader." The newly arrived soldiers usually got wounded below the neck, which meant the platoon leader should be held responsible. The casualties of Na's company were four killed, including a platoon leader, and six wounded.

## "The fear of death has nothing to do with rank and position."

Captain Kang Ho-tae was among the few in the Class of 1968 who had served as a company commander in Vietnam. He went to Vietnam in March 1971 as platoon leader but was assigned to S-2 Intelligence of the 2nd Battalion with the 1st Regiment. Four months later, he had a chance to take the position of 5th Company commander of the regiment whose predecessor, Captain Park Ho-jang (Class of 1965), was to return home after completing his one-year tour of duty. Thus. First Lieutenant Kang was appointed as the 5th Company commander with the provisional rank of captain. During the twelve months of his tenure there, he was exposed to numerous life-and-death moments while engaging with the enemies in some forty large and small operations and ambushes. Whenever Captain Kang Ho-tae led his unit to fight the enemy, he wondered if he could return alive by walking back on this road or by helicopter. Perhaps Captain Kang was one of the men who had had the harshest combat experience among KMA graduates. He survived the hazard of death four times during the countless battles he fought.

The first moment of death occurred in June 1971, just a few days after being appointed as a company commander. One day, when the rainy season began, Captain Kang was working to build trenches with his men at the top of Nui Hon Ju (about 640 meters above sea level). The top of the mountain was covered with thick clouds. Captain Kang, who had had some experiences with the sudden change of weather in Vietnam, was anxious as he heard the low thunder roaring from the clouds. Thus, he ordered his platoon leaders to take back the buried claymores. However, before the platoon leaders carried out the order, hail suddenly began to pour down with lightning strikes. In the clouds they were working in, the top of the mountain became a hell where the lightning of one million volts struck, and 30 Claymores exploded instantly.

The blue flame bolt of lightning spread with a roar and bounced off rocks and the ground 40-50cm high. It fixed them in the trenches they had dug. Smoke rose from the rocks, branches were broken, and everything hit by lightning exploded, shattered, and bounced. Claymores, whose primers had not been eliminated by the lightning, lay inert for several minutes, then exploded. The aftermath of the 30 Claymore explosions was as terrifying as lightning. Two soldiers died instantly in this abyss, and two others were seriously wounded. Thunderbolt, hail, and the claymore's explosions were over, and the bare chest of the dead soldier, still gushing with blood, was warm. A prayer letter and a family photo, including his mother, were found in the soldier's upper pocket. He was a Catholic.

Another incident happened shortly after that tragic one. The early mornings of the dry season in Vietnam were refreshing, as usual. Captain Kang's 5th Company was fortifying its base near his battalion's TAC-CP. One day, his company base was heavily mortared by the

Capt. Kang Hoe-tae (left, Class of 1968), S-2 Intelligence officer of the 2nd Bn, and a few month later 5th Co. Cmdr., 1st Rgt, Tiger Div. Right: Capt. Chung Yeon-soo (Class of 1966), S-3 Operations.

enemy from a place nobody knew. Amid a relentless pouring of shells and a hail of bullets during the fierce fight, Captain Kang was struck by blazing shots along with his men. He was blown away by a loud noise and tossed around. A fleeting moment passed, and small stones, dirt, and strange things fell from the air in the black smoke that smelled of gunpowder. They were torn limbs of the human body and pieces of flesh and hair that had blown into the air. They fell like hail on the worn-out face and body of the company commander.

After a while, when the shells stopped, the two soldiers' torn bodies were recovered and wrapped in ponchos. The body fragments of Sgt. Lim and Sgt. Kim, who had been working on radios a while before, were as small as old pumpkins. The remnants of the bodies were caught on high branches of trees in the jungle or flew farther away. Captain Kang should also have been torn and blown away with his men.

The third incident occurred when Captain Kang's company moved on several helicopters to Mt. Bak Mah, the company's operational area. The UH-1H helicopter the company commander was on board lowered its altitude to land and approached the airspace above LZ. The enemy fired from the ground, and the pilot and co-pilot were shot in the face and head. For a moment, Captain Kang saw two Viet Cong, dressed in black, jump out of the trenches and plunge into the jungle under the helicopter. The death of the two pilots dumbfounded him. While looking at the map, Captain Kang had talked with them about the landing site a minute ago. Kang alone survived by a hair's breadth.

In a moment, the helicopter shook violently and crashed as the two dead pilots let go of the controls. After 5-6 seconds, the aircraft bounced off with both legs broken due to impact on the ground and fell again, making a crash landing by the fuselage. The soldiers riding with the company commander desperately crawled out and ran in all directions, fearing the crashed helicopter might explode. The aircraft did not explode, though; the long rotors spun slowly for several minutes without the engine being turned off. The company commander realized that he was not dead. But he had survived. What made him live? He had to ask God.

The fourth incident occurred during Operation Fierce Tiger 17, conducted from June 25 to July 11, 1971. On the first day of the operation, Captain Kang's 5th Company was deployed to Dung Xuan village in the Go Boi plains. The village's vast swampy area was used for rice farming during peacetime. The battle started 30 meters above a swamp, even before the UH-1H helicopter landed. The enemy concentrated their fire from machine guns and B-40 rocket launchers on the 2nd Platoon's No. 1 helicopter, descending into the thick bamboo forest in the swamp. However, the pilot refused to land on the ground and suspended his aircraft about two meters above the swampy area. Captain Kang told everyone to jump.

Following the command of the company commander, everyone jumped into the swamp, where the water rose to their waists. Enemy fire was coming from three sides when everyone jumped. The 2nd Platoon leader, Lieutenant Kim Jeong-su, was seriously wounded, and six soldiers were killed and seven others wounded. One of the dead was a senior Sergeant Chun, who approached the company commander shortly before death, explained the situation, and returned with ammunition replenished. He served as a substitute for the squad leader, who had been evacuated. Captain Kang couldn't see him returning because his back was obscured by cactus, and then there was a loud sound of gunshots from somewhere, and 2-3 minutes later, Sergeant Chun was killed by the enemy bullets.

When he had come to Kang to explain the situation, he had said with a smile that he would return home after this battle ended. Sergeant Chun was killed with a grenade in his hand, just 15-16 meters ahead of Captain Kang. At that time, Corporal Kim Sam-jung, the company commander's operation soldier, was exchanging messages with a P-77 Radio with a long antenna. The enemy targeted this long antenna and concentrated B-40 anti-tank guns, grenades, and rifle rounds on Kang. Captain Kang managed to escape death in a foxhole dug by the Viet Cong, but all the equipment around him was smashed, including the radio and backpack. C-Ration broths dripped through the bullet holes of the bee-hived gear.

Captain Kang thought a lot about war later as he went through four critical death moments: "What are you most afraid of in human life? It would be the fear of death. Life and death are always with us, but we don't think about them in our ordinary lives; however, the fear of death is amplified on a desperate battlefield where survival matters. The fear of death caused on the battlefield has nothing to do with rank and position. It is common for all human beings to feel it. Yet the sense of duty and responsibility with rank and position hides the fear."[19]

## "You should not have a love of gain in Vietnam."

First Lieutenant Hwang Eui-seok's (later, 60th Infantry Division commander as a one-star general) Vietnam War experience began in October 1971 as an assistant to G-2 Intelligence of the White Horse Division. After three months of service there, when the 12th Company commander of the 30th Regiment went back home early, he was immediately replaced as his successor. Although he had more than two years left to become a captain, the rank of a company commander, he was appointed to the position by the division commander Major General Kim Young-sun. With the status of temporary captain, he was immediately sent to the location where the company was in action and started working.

At that time, his immediate superior, the 3rd Battalion commander, was Lieutenant Colonel Lee Byung-gan (Class of 1956), a professor of Electrical Engineering at KMA before coming to Vietnam. He was a bit clumsy in managing combat operations. At that time, the 12th Company was deployed to an independent base at the southernmost tip of Phan Rang (the White Horse Division's TAOR). Later, as the Battle of An Khe Pass became more intense, his battalion moved its garrison to Binh Khe, where the 3rd Battalion of the Cavalry Regiment of the Tiger Division was stationed before being committed to the An Khe Pass battle. Two platoon leaders were KMA graduates of the Class of 1970, Lieutenants Koh Jeong-sang and Song Pil-won.

As a company commander, his major operation was Operation Sacred Horse 72-1, conducted from July 16 to August 3, 1972. The primary purpose of this operation was to open Route 7 and clear the enemy on Hill 236. Captain Hwang designated the 2nd Platoon as an AP on July 21 and infiltrated the operational area on foot before the BMNT. The company carried out several ambushes, searches, and engagements, but there were no casualties. On July 28, there was an

Arc Light bombing by B-52 bombers and friendly artillery fire against the enemy area. The company commander had warned of the bombing in advance: "Dig deep" and "Bend your head and cover your ears." He checked the area after the bombing, which had created many craters with a radius of 30 meters and a depth of five meters. The power of the bombing was enormous. There was a constant smell of rotting flesh, but no dead bodies were found.

On August 1, during the withdrawal phase, Captain Hwang gave the 2nd Platoon leader an order to "look for the wreckage of a U.S. aircraft which was shot down," but he could not find it. Instead, the 2nd Platoon found a village with many huts in the mountain. The village contained about 20 cabins and was known to be where the headquarters of the Viet Cong K-13 Battalion was stationed. It was also a village inhabited by the Viet Cong and their dependents. By that time, the enemy was all gone. The men of the 2nd Platoon found about 50 cows the Viet Cong village had been keeping. They handed the cows over to South Vietnam's provincial office. And then they incinerated the huts and withdrew.

At around 18:25 that evening, his men arrested three Viet Cong suspects dressed in civilian clothing and interrogated them. But the suspects claimed that they were civilians. However, the platoon men found a notebook in a man's pocket that contained some records from during the Viet Cong meeting, so they were presumed to be Viet Cong, not civilians. In addition, from the joint interrogation by South Vietnam's County Office and 3rd Battalion S-2, one of them turned out to be the chief of the local Viet Cong people's committee. On August 2, the 2nd Platoon found an enemy corpse in the Thai Long Valley and buried it. After that, while resting for about 10 minutes, a point man sat off the jungle trail and touched the booby trap tripwire. He was seriously wounded by its explosion but died during the evacuation. It was difficult to detect the enemy's booby traps cleverly installed on the jungle paths. Captain Hwang's heart was broken when he thought of his parents at home, who would grieve over the loss of their beloved son.

In retrospect, no matter how large the scale of an operation might be, the actual battles were fought in the jungles by the junior-level officers below the company commander. Therefore, the instant

The soldiers of the 2nd Bn of the White Horse's 30th Rgt. firing 106-mm RR into the enemy position during Operation White Horse 9 in October 1968. (Source: ROK Army Museum).

judgment and command of these junior leaders determined the success or failure of the combat. Captain Hwang thought that the operation might have gone right if the intervention of senior commanders in the middle of operations had been less frequent. Sometimes Captain Hwang operated with the radio off when fierce battles were in full swing. And he realized that a fluent English-speaking commander could effectively conduct a unit's fighting on the battlefield. In case of casualties that occurred between operations, he would have no problem requesting a dust-off if an officer were fluent in English. Last, Captain Hwang emphasized that you should not have a love of gain in Vietnam. The beginning of all the injustices and contradictions in Vietnam had something to do with this lust for things.[20]

## Chapter 7

# Class of 1969 to 1971: Toward the End

### Class of 1969 (25th Class)

The Class of 1969 entered KMA on February 1, 1965, and graduated on March 28, 1969. A total of 254 cadets entered, and 203 graduated. The attrition rate was 20%, the lowest up to that time. Only the Classes of 1963 and 1969 had graduated more than 200 cadets. When they graduated, the Vietnam War was slowly coming toward an end. After a large-scale Tet Offensive by North Vietnam in January 1968, the U.S. concluded that a military victory in South Vietnam was impossible.

And on March 31 of that year, President Lyndon B. Johnson announced on national TV and radio that the United States would stop the bombing in the region north of the 19th parallel and that he would not run in the 1968 presidential election. In addition, the U.S. proposed peace negotiations with North Vietnam. North Vietnam regarded this proposal as the "surrender of the United States" and accepted it, and peace talks began in Paris in May 1968. Moreover, Richard Nixon, with his presidential inauguration on January 20, 1969, began to accelerate the "Policy of Vietnamization" of the Vietnam War.[1]

Upon graduating from KMA in March, the Class of 1969 was assigned to the front-line divisions as platoon leaders in April. About six months later, these officers who wanted to experience combat in

Vietnam hurriedly applied for duty in the war. From the beginning of 1970, these graduates of 1969 began to go to Vietnam with Second Lieutenant Yoo Gwang-sun as the first runner. And until the end of the war in March 1973, 102 officers had crossed over the ocean. From 1970 on, the number of KMA graduates in Vietnam sharply decreased. This was a natural phenomenon because the Korean military gradually reduced the size of its troops while preparing for the withdrawal. None of these officers served as company commanders. All returned to Korea after serving as platoon leaders or assistant staff members of higher units. Beginning with the graduates in 1969, war participants decreased significantly.

## "Our Platoon Leader is a Phoenix."

Among the graduates 1969, Second Lieutenant Kwon Kyung-seok (Later, two-term ROK National Assemblyman) had probably the harshest combat experience in Vietnam. While serving as a platoon leader in the 10th Regiment of the 8th Division in Korea, he volunteered for the Vietnam War, believing he had to have combat experience as a professional soldier. Though he was first supposed to be a member of the White Horse Division, he was transferred to the Tiger Division as a replacement when he was about to depart for Vietnam. There was too much combat loss of platoon leaders in the Tiger Division. All officers and soldiers going to Vietnam were supposed to be trained at the 7th Replacement Training Camp at Oeum-ri, Gangwon Province. But he was exempt from that process and left Busan Port in May 1970 because the situation in the 26th Regiment of the Tiger Division was too dire. After a week's voyage he arrived in Da Nang, Vietnam, and was immediately put into the operation area by helicopter as the 2nd Platoon leader, the 5th Company of the 26th Regiment. His company commander was Captain Seo Yong-won (Class of 1964), followed by Captain Yoon Hang-joong (Class of 1965).

The 5th Company base was on a low hill of rugged mountain Moh Joh at Song Cau in Phu Yen Province. Second Lieutenant Kwon served only in this company as platoon leader for a year until returning to

Korea in April 1971. He was put into operation at least once a week during his tenure there. The first operation he was engaged in right after being assigned to the company was Operation Tiger. It was a division-level operation that lasted for ten days, starting on July 21, 1970. On D+1, the 5th Company maneuvered to the operational area by helicopter and landed on Mt. Vinh Tanh's highest Hill 975. The company searched the area but found no enemy except one deserted cave.

Two days later, Lieutenant Kwon's 2nd Platoon set up an ambush on a ridge as an AP of the company. While approaching to treat a wounded soldier and returning to his foxhole, Corporal Kim, a medic, was misidentified as an enemy in the darkness and killed by the friendly fire. This incident served as an opportunity to build a strong fighting posture for the platoon and tighten discipline. According to his company commander, Captain Yoon Hang-joong, "Kwon felt a profound responsibility for Kim's death as a platoon leader. He saved his battle allowance and sent it to Corporal Kim's bereaved family in Korea, whom he had never met. It's a touching story of what could happen in war."[2]

Lieutenant Kwon mentions the "Battle of Re Wen village," which occurred in early September 1970, as the fiercest battle that he will never forget. Re Wen village, located 3-4 kilometers north of Song Cau, was a so-called "hostile village" controlled by the Viet Cong. Also, in the nearby jungle, Special Forces of the North Vietnamese regular army hid and threatened the friendly units. The ROK-U.S.-South Vietnamese coalition forces launched a joint sweeping operation against the Re Wen village. Lieutenant Kwon's 2nd Platoon (12 men, one-third of his Platoon), the 5th Company's advance party, surrounded the outskirts of the village and set up an ambush to block the North Vietnamese forces from entering the village.

Suddenly, at around 7:15 in the evening, he heard a bird chirping in the forest, and at that moment, the North Vietnamese troops appeared. In an instant, bombs exploded here and there, and the firefight started. Initially, the enemy, who suffered severe damage, seemed to retreat for a while, then enveloped Kwon's platoon with reinforced strength and poured concentrated fire on Kwon's position. It continued throughout the night. The outnumbered platoon was sandwiched between the North Vietnamese forces trying to enter the village and the Viet Cong, who ran away from the village, chased away by the South Vietnamese forces who had swept the village. Kwon estimated the enemy forces on both sides to be over two companies.

To make matters worse, the ammunition of Kwon's platoon was running out. Three of his men were seriously injured during the engagement and evacuated by a U.S. helicopter around midnight. Lieutenant Kwon had been half in doubt about whether the chopper would come to pick up the wounded at such an hour. But it came. He was deeply moved by the dedication of the American helicopter pilot who flew into the area where a fierce battle was occurring. Artillery fire by Korean and American forces continued until 5 a.m., with the aerial illumination lit all night. In particular, Lieutenant Ahn Byung-moo (Class of 1968), the 4.2-inch mortar platoon leader attached to the 5th Company from the Regimental Combat Support Company, shelled mortar fires all night.

Lieutenant Paik Nam-hwan (Class of 1969), the 1st Platoon leader of the same company, was ordered to provide the struggling Kwon with ammunition and break through the enemy's encirclement. Paik and his men managed to reach Kwon's position. The two classmates fought together. The fierce battle ended around 5:00 at dawn, and the surrounding areas were pocked with bombshells. If only two shells had landed in the platoon's position, all his men would have died. As the day dawned, Kwon's men found many firearms left by the enemies, including AK rifles. Lieutenant Kwon and his platoon members who survived the hellish trap were all awarded the Inheon medal, the fourth-highest medal for distinction in combat.

Kwon said in an interview later, "I had to fight even until two weeks before I left for the home country." "I encountered and engaged the enemy countless times, and it was a miracle that I survived. If I had been killed, I might have been evaluated as one who had taken reckless actions. I only did my best; that's it. I thought that the result would be something beyond human ability. Since then, I have kept the motto in my heart, 'Do your best and wait for Heaven's Mandate.'" Through many battles, he survived. From that time on, Lieutenant Kwon became a fatalist. His men would refer to him and say, "Our platoon leader is a phoenix."[3]

2Lt. Kwon Kyung-seok (left, Class of 1969), quenching his thirst by drinking coconut water with 2Lt. Paik Nam-hwan (Class of 1969) after the battle of Re Wen village in Sept. 1970. The torn pants on Kwon's right knee indicate how fierce the battle was. Both were the platoon leaders of the 5th Co, of the 26th Rgt, Tiger Div. (Courtesy: Kwon Kyung-seok)

In a sense, Lieutenant Kwon's story is the story of Lieutenant Paik Nam-hwan, who shared the same combat experience with Kwon as a platoon leader in the same 5th Company. Upon graduating from KMA on March 28, 1969, Second Lieutenant Paik Nam-hwan (Later, the 68th Infantry Division commander as a one-star general) was assigned as a platoon leader in the 32nd Regiment of the 2nd Infantry Division on the Korean front. Around January 1970, Major Lee Cheol-ho (Class of 1960), an assistant to G-1 Personnel of the division, recommended that he go to Vietnam because an officer has to have war experience. Major Lee had been a tactical officer of Paik's 2nd Company of the Corps of Cadets at KMA when he was a junior cadet. Some claimed that the soldiers who went to Vietnam were American mercenaries, but he did not care. At that time, all graduates of KMA thought it natural to participate in the war to gain combat experience as professional soldiers. Upon arriving in Vietnam in June 1970, he was appointed the 1st Platoon leader of the 5th Company with the 26th Regiment.

Although it was usual for the incoming junior officers to go through the chain of command—division, regiment, and battalion— for reporting before being assigned to their platoon leader positions, Paik had to go directly to his company because it was out in the jungle. After taking a helicopter from the battalion to the operating area, Lieutenant Paik immediately arrived at the company's position in the jungle and descended on a rope from the chopper. At that time, Operation Tiger was going on. Lieutenant Paik was welcomed by the 2nd Platoon leader, his KMA classmate Kwon Kyung-seok, who had been there some months earlier.

In Operation Tiger, Paik's baptism of fire was painful. He lost a point man during a search on the first day of his appointment, two more during a firefight days later, and suffered severe casualties even before he could adapt to the jungle. The biggest battle he had fought in Vietnam was the Battle of Re Wen village, where he had to deliver ammunition to Kwon's platoon enveloped by the NVA regular forces. The fighting started in the early evening and continued until dawn the next day. It was a battle that lasted long enough to require illumination by aircraft.

In particular, since the enemy was judged to be of company size, it was almost impossible for Lieutenant Kwon's platoon alone to stand and escape the trap. Paik succeeded in penetrating the trap in which Kwon's platoon struggled. Paik told Kwon, who had more experience than Paik in Vietnam, to take the unified command of the two platoons. Paik got a deep impression of the bravery and determination that his classmate had shown during the fighting. After returning home, Lieutenant Kwon quit his military career to serve as a public official in the government. But Paik always thought that Lieutenant Kwon was one who should have remained in the military. Even now, on Memorial Day, Paik visits his subordinates lying dead at Dongjak-dong National Cemetery and begs forgiveness for his incompetence leading to their deaths.

During the Vietnam War, one of the things that always came to his mind was what the children of South Vietnam were trying to get something to eat whenever they met Korean soldiers. They used to follow the Koreans, shouting with their thumbs up, "Daihan No. One! Daihan No. One!" Indeed, when he was a boy, Paik saw his peers doing the same thing in the Korean War. He had lived in a house next to the railroad tracks near Daegu. When a transport train of American soldiers passed, they immediately rushed to the railroad tracks and shouted, "Hello! Hello! Giv'me chocolate!"

He could not even fill the C-type return box, the smallest one every officer was allowed to take home when he left Vietnam. He filled half of it with only C-rations. When he arrived in Busan, five or six women vendors followed him from the pier, even to his house. When he asked why they were following him, they said they were business traders who bought the items in the box from the returning soldiers and sold them at a higher price. Paik couldn't sell them the electronics they favored because he had no such appliances with him, and instead, he sold half of the C-ration to them at a giveaway price.

In June 1971, when he boarded the returning ship after a one-year tour, he lay on the cabin's bed and reflected on what he had learned through his combat experiences in Vietnam. First, violence is not the solution to a problem between individuals or countries. There should be no war or violence in any case. Second, he realized how happy the country of Korea was. Before being deployed to Vietnam, he had often quarreled with those around him over trivial matters and deceived his superiors not by doing what he didn't want to do. But after returning home, he had a mind to make concessions and complete the given task, even if he stayed up all night. Finally, as a person who overcame the fear of death, he gained confidence that he could do anything.[4]

## The Troublemaker of the Cavalry Regiment

Almost all of the 1969 class members went to Vietnam as platoon leaders, but some served as staffers from the beginning. One of them was Second Lieutenant Oh Hae-jeong, who went to Vietnam in June 1970. After a voyage of five days and four nights aboard the USS Geiger, Lieutenant Oh arrived at Quy Nhon. Upon arrival, he was assigned as an assistant to S-3 Operations, the Cavalry Regiment. He was an English interpreter and an aide to the regiment commander. The regiment had Lieutenant Colonel Mullens from Texas as a liaison officer between the ROK and U.S. forces. Lieutenant Oh and LTC.

Mullens closely cooperated in the matter of maneuvering troops by helicopter.

In August 1970, two months after arrival, Lieutenant Oh got on a helicopter to assist his regiment commander in Operation Tiger, the division-level operation. In August in Vietnam, the temperature on the ground exceeded 40 degrees Celsius, making it difficult for people even to breathe, while inside the helicopter, it was too cold at an altitude of 2,000 meters. In less than 30 minutes, he had the urge to urinate. Having no other option, he took out the canteen he was wearing on his waist, opened the lid, and went to work. But, Alas! The opening of the canteen was so narrow. He managed to do his business inside the vibrating helicopter, but half of the urine flowed out. He couldn't help laughing at this mishap in the first helicopter operation. He felt a bit sorry for those grunts who had come to Vietnam together with him, wearing a helmet, carrying an M16 rifle and heavy combat gear, and leading a platoon through a rugged mountain jungle. In contrast, he was always comfortable onboard the helicopter.

While he was grateful to U.S. helicopter pilots for their dedication in supporting Korean operations, he occasionally had conflicts with them. In April 1971, he visited the adjacent 1st Regiment with his regiment commander and operations staff by helicopter. However, the helicopter pilot who took his group there said he could not stay long enough for them to finish their business and would have to return immediately to his unit. When Lieutenant Oh protested that he had never heard of such a thing in advance, the pilot insisted on returning. Oh was so angry that he drew his pistol and threatened the pilot. Yet the helicopter went away, and his party had to rent a jeep from the 1st Regiment and return. They had to risk encountering Viet Cong attacks and traveling tens of kilometers to their base.

Due to this incident, the American pilots called Lieutenant Oh, the troublemaker of the Cavalry Regiment. The Deputy Division commander called Lieutenant Oh, rebuked him, and ordered him to return home immediately. The Tiger Division, supported by the U.S. 129th Helicopter Company, was forced to take tangible measures in response to the U.S. protests. However, it was a matter of formality because he was scheduled to return to Korea two weeks later, and this incident caused him not to receive a medal for combat distinction. At that time, he justified it to himself that he committed such a reckless act just for the safety of his regiment commander.

Now, he believes that threatening a pilot of an allied nation with a pistol was an irresponsible act caused by a young officer's rashness. He returned to Korea aboard the *USS Upsher* in June 1971, and shortly after that, he applied for a job to serve as an aide de camp to the Commanding General of the Eighth U.S. Army in Korea but was rejected. The incident in Vietnam was heard of even by the Eighth U.S. Army in Seoul. Lieutenant Oh couldn't help but admire the intelligence of the U.S. military. Even here in the U.S. units in Korea, he was known as the guy who had threatened the U.S. officer with a pistol in Vietnam and, thus, was one that had to be avoided. How would you expect them to select someone like you as the U.S. commander's aide de camp? Oh was too naïve.

Lieutenant Oh's experience in the Vietnam War continued to play a role even after the war. Lieutenant Oh left the army and served as a government official for the rest of his public career. In December 2013, 38 years after the war's end, he was selected as a mid-to-long-term advisory group of KOICA based on his experience as a government official. He was dispatched to Lang Son Province in Vietnam to provide policy advice. Oh's job was to impart know-how to SMEs and investment policy. He did not say that he had fought in the Vietnam War more than 40 years ago. He didn't do it on purpose. He didn't know how they would react, and he didn't want to remind the Vietnamese post-war generation of the pain of war. However, he tried to help them with a sincere heart by repaying what he had done wrong.[5]

## "Do not listen too much to the braggart of veteran soldiers"

"The 2nd platoon leader with the 9th Company, the Cavalry Regiment, Second Lieutenant Won Jong-gwan (Class of 1969) detected a group of the enemy, killed seven of them, and captured some weapons including two AK rifles, two pistols, one SMG, and one B40 Rocket Launcher on a long-range ambush operation." Thus writes Asadal No. 68, dated May 1971.[6]

Known as a model cadet who had shown sacrificial spirit in an intramural sports event at the Academy, Lieutenant Won went to Vietnam in June 1970. While in Vietnam, he served only as a platoon leader throughout his one-year tour until his return to Korea in July 1971. His company commander was Captain Cho Young-gil (Later, ROK National Defense Minister), who showed outstanding leadership in managing the company and conducting combat operations.

In Operation Fierce Tiger 16, he saw many casualties occur in the adjacent 10th Company because of the mistaken fire by the friendly artillery. Lieutenant Won could not forget his fight on Nui Hon Quit on D+6. His company continued to operate without a significant result for six days, and on April 29, 1971, around 15:00, his platoon found a natural cave on the left ridge of the mountain. While his men were moving to the valley searching for water, they found traces of the enemy boiling and eating ants. Lieutenant Won told two squads

Defense Minister Jung Rae-hiuk who visited ROKFV, inspects the enemy weapons captured in Operation Eagle 70-1. Right next to him is LTG Lee Sei-ho, CG of the ROKFV. (Source: Institute for Military History, MND)

to block all sides of the cave. Then, he himself entered the cave with one squad connected with ropes each other and threw gas bombs and grenades. Seven enemy bodies were found dead. Lieutenant Won's squad captured some weapons inside the cave, including three small weapons, one revolver, one M79 grenade launcher, and a P-10 radio. In the Vietnam War, it was a usual practice to bury the enemy corpses and only sometimes to leave them unattended to continue operations. Won's platoon buried the dead enemies.

False reports in the Vietnam War were frequent occurrences among the Korean troops toward the end of the war. The division sometimes sent helicopters to transport the enemy dead to prevent false reports. Lieutenant Won highlighted three lessons from the Vietnam War. First, do not listen too much to the braggart of veteran soldiers; second, commanders must show great conviction and exert firm control during operations; and third, since small units mainly fought the war, more study should be focused on small-unit warfare tactics.[7]

The following are just a few representative examples that most graduates of the 1969 class encountered on their Vietnam tour.

Second Lieutenant Yoo Gyo-jun was the 3rd Platoon leader of the 2nd Company of the White Horse Division's 28th Regiment from June 1970 to July 1971. During his tenure, he participated in Operations Dokkabi 18, 19, 20, and Eagle 71-1. The battle he will never forget is the battalion operation carried out around April 1971 at Nui Hon Ba. He searched Hill 350 and Da Bia Valley, and there were a lot of booby traps installed. Advancing to avoid the booby trap, he encountered a rock formation. The point man in front, followed by the leader of the point squad and Lieutenant Yoo, was climbing a rocky area when he met an enemy, and a firefight erupted. The point men were killed on the spot. The enemy was firing in the cave, and his platoon injected 66 LAW fire and gas bombs into the cave. As a result, his platoon killed two Viet Cong inside the cave and captured one M16.[8]

On the other hand, First Lieutenant Jang So-gil (Class of 1969), 2nd Platoon leader of the 1st Company of the White Horse's 29th

Regiment, had many combat achievements. Recognizing his combat performances, the regiment commander allowed him to take a week-long leave to Saigon. However, thinking of the suffering of his platoon members, he felt guilty about his comfort. He came back earlier than he was allowed to, went out on an operation against the enemy in advance, and was killed by an enemy sniper during road reconnaissance. His battalion commander LTC Paik Myung-yoon (Class of 1957), recalls Jang's sacrifice later: "He was a true soldier and showed courage and love for his men as a KMA graduate would go on the battlefield."[9]

## "Enemies always step on your tail"

First Lieutenant Cho Geum-joo served in Vietnam for 16 months, from March 1971 to July 1972. While serving as the 3rd platoon leader of the 10th Company of the 30th Regiment, the White Horse Division, he participated in Operation Dongbo 17, conducted in July 1971. On the third day of the operation, the company commander asked, "I think this area is the most difficult to take. Which platoon leader will go ahead?" Lieutenant Cho volunteered, saying that his 3rd Platoon would do it. As he advanced through the jungle, he found a fleeing enemy, and the point man attacked first. After a brief pause in the shooting, Lieutenant Cho, and two of his men set out to pursue the enemy. During a lengthy pursuit, he heard a bird chirping on a tree and instructed the soldier to go and check it out. Suddenly, the Viet Cong jumped out from under the tree and ran away.

Lieutenant Cho's rifle was locked, and the enemy ran away when he unlocked it. Later, Lieutenant Cho emphasized the logic of "don't be afraid of a safety accident, and you should leave the rifle unlocked." He could have gotten the Viet Cong if he hadn't locked the rifle, but he let the enemy go. Another lesson he learned was, "Enemies always step on your tail." His company commander always used to say to the platoon leaders: "You need to be continually vigilant and search even during breaks. In particular, be careful of the enemy's attack from the rear during movement. After receiving a re-supply, the friendly troops throw away cans, biscuits, and leftovers they did not eat. The enemy often follows to pick up those snacks."[10]

First Lieutenant Lee Gang-eon (later 1st Infantry Division Commander as a two-star general) served as the 2nd Platoon leader of the 8th Company of the White Horse's 28th Regiment from April 1971 to June 1972. The 8th Company was deployed on Hill 38 and had strong combat power. In front of the company, there were four avenues of approach judged to be enemy infiltration routes, and the company was tasked with controlling them. About two battalions of the Viet Cong were believed to be nearby, and Ong Sang Mountain in this area was dangerous. He participated in Operation Dokkabi 21, which started on August 10, 1971, and made significant achievements on D+3. On this day, the 2nd Platoon took on the role of advance party and went through the jungle of Ong Sang Mountain in a single file.

At this time, Lieutenant Lee's platoon discovered the enemy's natural cave and fired at the hiding enemy. Inside the cave, his platoon killed five enemies and captured three firearms. Those killed were buried locally. Lieutenant Lee targeted a bird called the Viet Cong and trained the soldiers to shoot. The bird was slightly smaller than a pigeon, and with the number of birds they caught, he evaluated the firing competence of an individual soldier.[11]

## Class of 1970 (26th Class)

The Class of 1970 entered the Academy on 24 January 1966 and graduated on 28 March 1970. The initial entrants were 258 elite young men selected out of 95 high schools in small towns and cities throughout the country, and only 183 were commissioned as second lieutenants. The attrition rate was 28%. Upon graduation, all of these officers were sent to various combat infantry divisions as platoon leaders or staff officers in the DMZ area. Among these men, 94 graduates went to Vietnam. Only with one year of experience as platoon leaders or staffers in the homeland they jumped into the jungles of Vietnam. Among the 94 participants, two were killed in action, and one other died of the war's traumatic after-effects some years after returning home.

The Class of 1970 was the second from the last among the KMA classes who went to Vietnam. Their tour of Vietnam began with Second Lieutenants Kim Sang-soon and Chang Yang-woon in late 1970, the year of their graduation, but most men went from mid-1971 to 1972. After a one-year tour in Vietnam, all these men returned home by March 1973, mainly as the last element of withdrawing troops. Indeed, the Vietnam experience with the Class of 1970 began with its first year at the Academy. When they were freshmen in 1966, some Vietnam veterans who had been in the first troops to be deployed the previous year returned and assumed the tactical officers of the cadet companies. One was Captain Lee Chun-geun (Class of 1959), the hero of the Battle of Duc Co. The other was Major Park Dong-won (Class of 1958), the 5th Company commander of the 1st Regiment, who had shown bravery in Operation Fierce Tiger 5. They were the objects of cadets' awe and wonder.

While listening to their heroic stories, the Class of 1970 still could not feel the reality of the war. Even when the statue of the late Major Kang Jae-gu (Class of 1960) was erected on the campus in 1966, the war in Vietnam seemed none of their business. It was one to be fought by others in far-off places. Since the summer military training of 1969 in their senior year, the Vietnam War entered their hearts and began to locate itself there. The primary training programs for the seniors were organized to be used for conducting operations in Vietnam.

And the OBC, where new second lieutenants were supposed to serve a three-month apprenticeship before being sent to their first commands, was offered to the Class of 1970 during their senior-year summer military training. This process meant they might be directly deployed to the units in Vietnam right after graduation. This arrangement of offering the OBC to the senior cadets during their summer training was planned by the ROK Army authorities, who feared that the Vietnam War would be extended for quite a long time and the need for junior officers would be increased. This measure was also reflected in the number of new cadets selected in 1971. The number of incoming cadets of that year, the Class of 1975, doubled compared to former classes. The initial entrants of the Class of 1975 amounted to as many as 520, twice the number of usual recruits.

The contents of the summer training were composed of those

tactics and techniques to be practiced in Vietnam, mainly of counter-guerrilla-warfare exercises, which included: ways of attacking pillboxes in the crevices of the rocks, making frontal assaults against the sun-browned hills, detecting the enemy's primitive booby traps scattered in the jungle trails, searching for labyrinthine VC tunnels, locating the VC's underwater hideouts, planting M18 claymore mines, experiencing ghost caves to overcome fears of the VC's nightly surprise attack, battering the VC into submission with the Hammer and Anvil Movement, and so on. Some veteran instructors gave them nerve-racking combat stories with the elusive VC. Thus, Vietnam was the land of horror to those who knew nothing of war. Yet, simultaneously, it was the land of adventure calling out to them with a challenge. To those who had had no chance of going overseas, Vietnam was just a romantic land, far off to the south, with the Southern Cross twinkling on starry nights and thickly wooded with palm trees.

The first person who went to Vietnam among the graduates of 1970 was Second Lieutenant Chang Yang-woon, a platoon leader of the 6th company, the 1st Regiment, the Tiger Division. In four months since he reported to the regiment, Chang participated in Operation Fierce Tiger 15, jointly with South Vietnamese forces. Although he had been on some ambushes and reconnaissance patrols since he came to the regiment, this was the first large-scale operation he would be in combat as a platoon leader, so he wondered if he could make it. While waiting for the battle, he was overwhelmed with doubt and fear. Two separate helicopters loaded with his platoon took off and landed in the jungle of Nui Hon Che. Still, the lead pilot misread the coordinates of the LZ and unloaded his platoon at a place about 2 km from his company command post. He was at a loss about what to do in this unexpected situation. But he made a desperate effort to grope his way toward the company through the thickly forested jungles in the pitch-dark night. It took as long as three hours to get back. His platoon bivouacked for the night there and jumped into action early the following day.

As the firefights gradually intensified, a few casualties occurred in the adjacent platoon. Having received information about the enemy's hideout from the regiment S-2 Major Lee Jae-tae (Class of 1960),

Chang's company commander designated his platoon as an advance party. Upon this order, Chang divided his platoon into two parts, led by himself and his platoon sergeant, and went on patrol along the jungle trails and ravine. On the way, he detected some VC and opened massive fires with all his platoon's weapons, including M79 rifle grenades and machine guns. After a fierce firefight, he approached the enemy position with an adjacent unit and discovered it was the enemy's training center. His platoon collected dead VC bodies and captured AK rifles, machine guns, 122-mm mortar shells, food, and raincoats. It was a significant achievement in his first battle since he had come to the unit. He was proud that he had overcome the fear of death that had haunted him. In this operation, he was awarded the Hwarang medal.

During the operations, his battalion was informed about the presence of a North Vietnamese battalion and ambushed the enemy in the reed forest of Nui Hon Che. Then around two o'clock at night, there were crashing sounds of claymore mines and machinegun fire. In the morning, however, the "enemy" proved to be a mountain goat as big as a bull caught by the adjacent company. This was a usual occurrence around the company's tactical base. For example, this happened in the early period of the Korean troops being stationed in Vietnam. On a pitch-dark night, the soldiers on guard had caught the sight of a massive movement forward of the company's perimeter and delivered a concentrated fire on it. But the movement was made by some twenty water buffaloes of the villagers roaming around that night. In the morning, the villagers swarmed to the base in groups and protested about what the soldiers had done. The Korean army authorities apologized and paid them for all the dead animals.[12]

## A platoon leader, the primary target of the enemy sniper fire

Another Hwarang medal awardee was First Lieutenant Kim Pil-soo (later, Commander of National Security Command as a three-star general). Kim's Vietnam experience was a bit different from his other classmates. It was usual for most junior officers to get transferred to other staff jobs of higher commands after a certain number of months

Some members of the Class of 1970 in the Tiger Division: From the rear left, First Lieutenants Song Yeol-jae, Kim Soon-shin, Kim Pil-soo, Lee Heung-yeol, Chang Yang-woon, Chun Yong-min, Kim Chang-ho, Chung Il-seok; From the below left, Choi Young-chang, Kim Tae-geun, Kim Tae-woong, Uhm Myoung-chul, Shin In-sun, Kang Hang-joong (Courtesy: Kang Hang-joong)

of service as a platoon leader. It was not so for Kim, though. Even when such a chance to move on came to him, Kim was excluded from being considered. Moreover, nobody told him the reason. He underwent unfair personnel management for the first time since he came to the field. Deprived of the chance to move on, he had to stay in the platoon leader position throughout the thirteen months of his Vietnam tour. A platoon leader became the primary target of the enemy sniper fire in Vietnam.

Upon arrival in Vietnam in April 1971, Kim was assigned as a platoon leader of the 5th Company of the First Regiment of the Tiger Division. He replaced the platoon leader killed with his platoon sergeant in Operation Fierce Tiger 16. After two months as a platoon leader, he was committed to Operation Fierce Tiger 17. From June 25 to July 11, 1971, this operation was the last division-size operation in Vietnam conducted by the Tiger Division. It was to destroy the North Vietnamese 12th Regiment and rear elements of the VC in Binh Dinh Province and secure the region's stability. Up to that time, he had not engaged in any combat action, and thus as a platoon leader, he was overly concerned about what his performance would be like in a real firefight.

The operation was carried out around Nui Vo Ca and its vicinity. On the third day of the operation, his platoon was fired from the enemy bunkers inside the natural caves on the top of Nui Vo Ca. An intense firefight ensued for 20 minutes, and two men in his platoon were shot by the enemy's rifle fire. At first, he was so embarrassed that he had no idea what to do, but he came to his senses immediately and evacuated the wounded soldiers by calling the dust-off. After that, his company received massive fire support from artillery and Huey gunships to neutralize the enemy positions for forty minutes. Then the company made an all-out assault, and his platoon finally seized the enemy positions. Among his company's achievements were eleven dead enemies, some captured weapons, including 82-mm mortars, machine guns, small arms, revolvers, North Vietnam's national flag, the 12th Regiment's colors, and many medicines.

The enemy position was estimated to be at least above a battalion-size unit headquarters when analyzed from the things captured. In contrast to the enormous combat gains, his platoon's damage was very slight, with just two wounded. However, if soldiers inexperienced in jungle terrains with limited information about the enemy had made such a reckless attack, they might have paid a tremendous price. This was the lesson that Kim had acquired from that operation. Although Kim's baptism of fire, that soldier's sacrament in the Vietnam jungle, was stern and severe, it became a valuable asset for his professional military career in the future. At the same time, he witnessed unfair leadership practiced by his superiors and the discrepancies between appearance and reality in war and the military. All these things allowed him to know how to distinguish what a leader should do from what they should not.[13]

## "What do I look like in this situation?"

First Lieutenant Oh Hyun-koo (later, Commander of the First Corps and Aviation Operations Command as a three-star general) initially went to Vietnam as a White Horse Division member in May 1971.

When he arrived in Vietnam, Operation Fierce Tiger 16 was ongoing, and some junior officers were killed or wounded. Thus, nine officers were transferred to the Tiger Division to fill the vacancies from amongst those new arrivals. The next day, Oh was put into the operational area to replace a platoon leader killed in action. His position was a platoon leader in the 12th Company of the 1st Regiment. He was immediately sent to the jungle without an official report to his immediate superiors.

For the first time, Lieutenant Oh had the experience of lethal bullets flying toward him. The shots were live ammunition, not dummy ones. In evacuating the dead body of the former platoon leader by helicopter, his platoon was exposed to the enemy and received concentrated fire. He wondered, "What do I look like in this situation?" Strangely enough, his mind was calm, and he came to himself. He could identify the direction and position that the enemy fired with a clear sense and ask his company commander to deliver artillery fire on it. The operation ended with his unit's victory and no casualties in his platoon. Indeed, Oh had a grueling debut upon arrival in Vietnam. After completing the operation, Oh could meet with his battalion and regiment commanders. Three months later, Oh moved to the position of helicopter officer.[14]

First Lieutenant Kim Soon-shin (later, the 26th Infantry Division commander as a two-star general) arrived in Vietnam in April 1971 and served for two years. During the days of his service as a platoon leader in Vietnam, he was continually full of tension. On a reconnaissance patrol one day, he led his platoon into a booby-trapped jungle trail, and several of his men, including his platoon sergeant, stepped on mines and were injured. After evacuating the sergeant, he spent the night holding another of his wounded point man in his arms. After a one-year tour as the platoon leader, he was ordered on short notice to attend the Vietnamese language school in the Tiger Division.

After months of language training, he was dispatched as a liaison officer to South Vietnam's 20th Infantry Division. While there, he participated in many engagements against North Vietnamese troops and had some near-death experiences. In early 1973 the Korean forces were planning to withdraw from Vietnam, and the company tactical bases of the Korean forces were being handed over to South Vietnamese troops. While doing these jobs, Kim went through indescribable horror scenes, which still haunt him with nightmares now, even after 50 years have passed. Yet, the two years of his Vietnam experience became a guidepost that has ensured he never forgets his true mind—to stay on the right path--for his long years as a career army officer.[15]

Likewise, First Lieutenant Park Heung-keun (later, the 3rd Infantry Division commander as a two-star general) still can't forget the love and warmth his men showed him when he was hospitalized after being seriously wounded on a patrol mission. Park came to Vietnam as the 2nd Platoon leader, the 10th Company, the Jae-gu Battalion with the First Regiment. At first, he was mainly engaged in patrols and ambushed forward of his company's tactical base and vicinity. In January 1972, Lieutenant Park's platoon killed three VCs in the division operations to free the Go Boi plains from Viet Cong control. While searching for the villages' enemies, Park encountered some VCs and received their massive fire. In this skirmish, he and one of his men received a penetrating wound to the leg. Immediately he and his man were sent to 106 Evacuation Hospital in Quy Nhon by the dust-off helicopter.

After a month-long hospitalization, Park returned to his platoon to serve the remaining period. Looking back later years, he felt he was lucky that no one in his platoon had been killed, and he had been awarded the Inheon medal, thanks to the dedication of his men. He still cannot forget the love his men showed him while in the hospital. In their free time in the company base, when there was no action, he used to teach them some songs he had learned in his cadet days, including the theme songs of the movies "Stagecoach," "Railroad Song," etc. While he was in the hospital, his men recorded the songs they sang in chorus on a cassette tape, wishing for his fast recovery, and brought it to him in the hospital. When he met with them with the tape, he felt a sudden gush of affection that he had never felt for anyone. The invisible threads that bound his platoon together pulled at him. Park still remembers vividly that he and other hospitalized officers had listened to the tape with their eyes full of tears. Such devotion, simple and selfless, the sentiment of belonging to each other, was the one decent thing the soldiers found in a war otherwise full of betrayals and horrors.[16]

The Class of 1970 had fewer artillerymen than those of other classes. For this reason, the number of artillery veterans among the 1970 class was relatively small. Among those few artillerymen is First Lieutenant Kang Gwang-seok (later, the 36th Infantry Division Commander as a two-star general), who went to Vietnam in June 1971. Though he volunteered for Vietnam in early 1970, he was not allowed to go because his battalion commander didn't want to lose him. There were few KMA graduates in the frontline units at that time. Many commanders wished to keep the junior officers fresh from KMA in their units. However, the order of his dispatch came down from the Army Headquarters. So eventually, he could go to Vietnam and was assigned to A Battery, the 10th Artillery Battalion, the Tiger Division.

Kang was in Vietnam for less than a year. During that short period, he went through all positions a junior artillery officer had to assume, including FO, Fire Direction Officer, and Firing Battery Commander. He participated in all sizes of operations, from small to large ones, and contributed to the victories of the infantry units with strong fire support. During his tour, he fired over 25,000 explosive shells and is still suffering from the aftereffect of the war with chronic symptoms of tinnitus. One thing that Kang feels regretful about is that he was not awarded any commendation medal. Although he was mentioned as a potential candidate for medal recipient every time an operation had ended, he was never selected. Yet he still feels proud that he overcame the fear of death with his courageous young men in the backcountry called "Killer Valley." Those men of his organized an "Artillery Veterans Friendship Club" when they returned home and still share their memories. Unfortunately, some of them are suffering from long-term effects of exposure to the defoliants.[17]

## War does not always progress, as written in the book.

Any soldier's academic background or social origin had nothing to do with combat performance in the face of death on the battlefield. Among those who had such an experience was First Lieutenant Kim Won-gyung (Class of 1970). He arrived in Vietnam as a platoon leader of the 6th Company, the 28th Regiment of the White Horse Division, in October 1971. While engaging in many reconnaissance patrols and ambushes, he participated in a large operation for the first time, Operation Dokkabi 22, in March 1972. He discovered that the "combat competence of the soldiers had no connection with their educational background." In combat, bullets do not discriminate over who they kill. War does not always progress, as written in the book. As shown in much American literature, those with higher education were likely to place their words before actions, arguing about war ethics, morality, rules of engagement, etc.

Although this was not always true of all the armies fighting in Vietnam, Lieutenant Kim Won-gyung saw that the soldiers who took the lead at the forefront of his platoon were, in most cases, the ones socially underprivileged. Many of these men were those who had not been duly treated in the society and communities to which they belonged. But they had overcome the fear of death with indomitable will and developed comradeship in the jungles of Vietnam. Through this experience of getting to know these young men, Kim learned the importance of the "leadership of serving," not reigning over others to pursue his career as a professional soldier. When he returned to his homeland and was reassigned to the field units, he made it a rule to pay more attention to the socially underprivileged.[18]

## "I don't want you to think you're a John Wayne."

First Lieutenant Han Chul-yong's (later, the 8th Infantry Division commander as a two-star general) Vietnam tour was a journey to learn about life and leadership. He went to Vietnam as a platoon leader of the Reconnaissance Company of the 28th Regiment. Before he arrived in Vietnam, he heard that a platoon leader in the war was considered consumable or throwaway. But he would question the implicit meaning of the words, especially the military jargon that a lieutenant's combat life expectancy is measured by minutes, not hours. Yet here in the war zone, he had to face reality. Half in doubt of these words, he stayed overnight at the Division headquarters and directly flew to the 28th Regiment headquarters at the foot of Nui Hon Ba by helicopter.

After reporting to the regiment commander, he came to his platoon immediately. From the beginning, when he had first thought of going to Vietnam, he had decided that the best thing to do in Vietnam was to survive and return home safely. "Our primary objective is to survive and go back home safe," he declared before his 40 men, "I don't want you to think you're a John Wayne." And yet he emphasized that "a single drop of sweat in training is same as a drop of blood in real fighting," which he had heard innumerably in his cadet days. Fortunately, he had no KIAs, only some wounded men in his platoon during his tour. He still feels proud of that.

In the summer of 1971, there was a regiment-size operation in Suoi Cai Valley and its surrounding area, some 20 km northwest of Tuy Hoa. His platoon got three soldiers wounded from the enemy's surprise attack in the operation. As to the occurrence of the three casualties, he blamed himself for his carelessness not to have cool-headedly judged the situation as a leader. According to Han, his platoon was advancing through the jungle. At that time, his point man found a case of the enemy's 155-mm rocket round and was excited that it might be a sign of VC's hidden weapons and ammunition storage site somewhere nearby. Capturing the VC weapons was considered a significant combat achievement. His men dug in the ground, hoping to find something. However, they found just a case full of rice and some 50 to 60 more filled with provisions. Still, being sure of a weapons storage site somewhere there, his men insisted they should go on searching.

Indeed, at this point, he had to keep his composure as a platoon leader, but he didn't do so, for he was also excited. The image of medals was flickering in front of his eyes. Eventually, he ordered his men to continue searching. In the meantime, the enemies had built an entirely defensive position, and they were waiting for the approach of his men, including placing claymore mines and other booby traps. Into the mouth of this rat trap walked his platoon. At that moment, the enemies' claymore mines exploded, and three of his men fell. Immediately he requested gunships to be sent by radio and barely rescued all the wounded men. Going over the incident later, he thought that a leader should not get quickly excited but keep calm. That was a valuable lesson he learned in his first engagement as a platoon leader in Vietnam.

A farewell party for Col. Park Gap-ryong (Class of 1955), the third from left, Chief of Staff of the White Horse Div, who returns to Korea after a one-year tour. This was hosted by the White Horse Chapter of KMA Alumni Association. Some members of the Class of 1970 posed with the colonel: from left, 1Lts Noh Yoon-sup, Lim Joong-bok, Col. Park, 1Lts Park Sung-joon, Lee Seung-gwan, Yoon Seok-tae. (Courtesy: Lee Seung-gwan)

At the time, the 28th Regiment of the White Horse Division had suffered from small and large operations failures for the past few months. Once, a battalion command post was overrun by the VC's night surprise attack, and heavy casualties occurred. The result was fifteen friendly KIAs, and the battalion commander was dismissed. Not long after that, an armored vehicle of the regiment was attacked by the VC's antitank grenade launcher (RPG-2), and a platoon leader and his crew were all killed, with the vehicle destroyed by the fire. As a result, the regiment commander was removed and sent back home. Although inevitable war situations caused those failures, the two commanders were excellent and competent men. Lieutenant Han still feels sorry for their dismissal.

On the contrary, removing a battalion commander was a good example of how vital a commander's leadership role was in combat. This battalion was waging an operation on the rocky hills west of Vung Ro Bay near Nui Hon Ba. Then, the weather suddenly deteriorated, so the battalion could not continue the operation. The battalion commander ordered his troops to withdraw on foot because scores of helicopters couldn't fly in. The problem was, though, that he alone used a helicopter. He had the aircraft land at his command post for a second to pick him up. Of course, the battalion commander was also dismissed. He had disregarded the fundamental principle that a commander should be with his troops in any place and at any time.[19]

## An MP Officer at Cam Ranh Bay

KMA graduates must select one of five combat arms—Infantry, Artillery, Armor, Engineer, and Signal—when they graduate. However, after two years of service in their initial branches, they can change their branches to other combat service support ones if they want. Among the Class of 1970 graduates, First Lieutenant Noh Kyung-hwan was one of the few officers who changed his initial branch of service to Military Police. Before being transferred to the MP branch, he was a platoon leader of the 99th Regiment of the 32nd Infantry Division as an infantry officer. When he became an MP officer, Lt. Noh wanted combat experience like his classmates who volunteered to go to Vietnam. Eventually, as a member of the MP Company of the White Horse division, he went to Vietnam and was appointed as the head of its MP Detachment at the port of Cam Ranh with some forty members under his command.

As the first and foremost strategic base of South Vietnam, Cam Ranh harbor was full of military installations and busy with the arrival and departure of lots of military vessels and supplies. As a law enforcement officer, he was in charge of conducting investigations of any wrong-doings of Korean troops coming and going during his tour of duty. When the Korean troops withdrew from Vietnam in 1973, he was the security officer for Lt. General Lee Sei-ho, Commander of the ROKFV. Four years after he returned from Vietnam, he left the army and served as a government official for the rest of his public career. From the start, he chose the Supreme Prosecutors' Office in the government based on his service as a law enforcement officer in the military. Noh served in various capacities in the Office until his retirement, including as a U.S. FBI Special Agent for Korean Government.[20]

## Class of 1971 (27th Class)

The Class of 1971 was the last class of KMA graduates who went to Vietnam. The Vietnam War was coming to a close as the United States pursued the Vietnamization Policy from 1969 onwards. The policy stated that the South Vietnamese would be able to defend themselves by dramatically enhancing the self-defense capabilities of its forces. Then the U.S. would withdraw its troops. At the beginning of 1969, the U.S. military maintained the highest 550,000 strength. According to Richard Nixon's withdrawal plan, which began at the end of that year, the U.S. was to withdraw its 50,000 troops by April 1970. After General Creighton Abrams took command of USMACV, ground operations were changed from active Westmoreland's Search and Destroy to the passive Clear and Hold operation. The former was to search, fix, and destroy the enemy, increasing the body count. The latter was to clear the populated areas and secure them to support the implementation of pacification plans.[21]

Moreover, in early 1971, the Détente mood was created through ping-pong diplomacy between the U.S. and China, and it reached its climax with Nixon's visit to China in February 1972. Under these circumstances, the Spring Offensive of the North Vietnamese Army started in April 1972 to lead the Vietnam War to a victorious finale. The Tiger Division fought the final battle in Vietnam at An Khe Pass on Route 19. The ground battle ended on March 23, 1973, when the ROK forces withdrew completely.

The Class of 1971 graduated from KMA just two years before the Korean withdrawal from Vietnam, yet many graduates wanted to gain a war experience in Vietnam. Still, they had to complete their tours as platoon leaders in the front divisions in Korea for a certain period. Thus, they could only go to Vietnam in early 1972. The point was that the Korean government was no longer receiving any applications for going to Vietnam. Therefore, the Class of 1971 had only a limited opportunity to go to the Vietnam War.

Nonetheless, some officers who had applied as early as possible from the front had a chance to go to Vietnam. Officers who had been fortunate to go to Vietnam in early 1972 became the members of the withdrawing units after a one-year tour. Although the least number among all the classes of KMA had been to Vietnam, the 1971 class graduates were proud that they were the members of the last elements of the Korean army to end the war, which had been fought for eight and a half years.

The Class of 1971 entered KMA on January 24, 1967, and 188 out of 256 were commissioned as second lieutenants on March 30, 1971. The attrition rate was 28%. Of these graduates, only 22 officers could go to Vietnam a year after graduation. The long journey of the KMA graduates--the Blue White Line--to the Vietnam War ended with the Class of 1971, eight and a half years after the first Korean troops began to be deployed in 1964. The 22 officers, all dispatched in 1972, consisted of 12 Infantry, 9 Artillery, and one Engineer officer assigned to the Tiger Division, White Horse Division, and First Logistics Command.

First Lieutenant Lee Gyu-sup went to Vietnam in August 1972 to name some of these officers. A platoon leader in the Tiger Division, Lee killed two VCs during an ambush operation. Though he was not awarded a medal for his combat performance, he was only allowed to go on a leave of comfort to Saigon after being cited for his achievement. First Lieutenant Lee Sang-don served in an American unit in Quy Nhon as an Engineer officer in charge of supplying equipment. First Lieutenants Kim Gwang-won and Lee Sang-tae fought as platoon leaders in the First Regiment of the Tiger Division. First Lieutenant Oh Yong-geun, who arrived earlier than his classmates, fought as a platoon leader of the White Horse Division, but the Viet Cong's sniper fire injured him during a patrol mission. Perhaps Kim Hyung-suk and Sung Sa-hyun would be the last KMA graduates with combat experience toward the end of the Vietnam War.

## At An Khe Pass along Route 19

Among the Class of 1971 graduates, First Lieutenant Moon Dhu-shiek was the one who had the harshest combat experience. He arrived in Quy Nhon on 16 April 1972, just one year before the official ending of the combat action. There should have been a welcoming ceremony with a military brass band when disembarking at the Quy Nhon port. But, at that time, the Battle of An Khe Pass, the biggest battle of the Vietnam War the Korean troops had ever fought, was going on, and the Korean army was in a bad situation. Moon was appointed the 1st Platoon leader of the 11th Company of the Cavalry Regiment of the Tiger Division, replacing the 1st Platoon leader killed a few days before. The 11th Company was then engaged in the battle. Before entering the fight, he was ordered to clip his fingernails, toenails and cut off some hair as DNA samples, just in case.

While waiting to board the helicopter, he saw an enemy shell explode in a nearby artillery battery base, killing 4-5 soldiers whose bloody corpses were immediately transported to the heliport. He went to the 11th Company CP by helicopter and had to jump out of the air because the plane could not land. The bullet-proof jacket with iron plates on the front and back, food for three days, ammunition, grenades, and two canisters amounted to over 30 kg, so it wasn't easy to jump out of the air carrying the heavy load. Moon arrived at the company CP but could not see his company commander. He was out in the field, leading the troops.

After some time, Moon met with the company commander Captain Cho Seung-ok (Class of 1965), who came to Vietnam from KMA, where he was teaching Philosophy. Captain Cho had been fighting for several days, and his face was black with blue lips. Moon was the 1st Platoon leader, and his duty that night was to guard the company CP. When he checked his platoon, he found only eight people. The rest had either been killed or evacuated because of battle wounds. Protecting the company CP entailed guarding around a barbed-wire fence of about 20 meters. Anti-personnel mines were planted all over the area. The battlefield seen from the company CP was bright all night, with the sound of gunfire intermittently and the jungle lights lit by the artillery fire, both far and near.

The next day he was sent to the battlefield. Route 19 from Quy Nhon to the Cambodian border through An Khe passed Hill 638, and his platoon was to set up an ambush on the ridge of Hill 638. It was night, and he couldn't stay calm because of the sound of gunfire and grenades exploding nearby, making him shudder in horror. The following day, his platoon's mission was to collect the bodies of the fallen soldiers. Disposing of comrades' bodies who died in the valley was horrific. The swarms of flies on the bodies were disgusting. He could not help smoking for a while because of the rotten smell, even though he had never brought a cigarette to his mouth. Ambush at night and search and patrol continued during the day, and shells and bullets were flying in, but there were no casualties.

One day, a soldier was stung by a scorpion on his forearm during an ambush. He screamed in pain all night, cursing his platoon leader badly. When transported the following day, his forearms were as swollen as his calves. After about 15 days, the battle was over, and all other units withdrew, but only his company remained in the combat zone for longer days. During the time, he saw an engineer soldier killed by a shell fired by the enemy. The soldier of the Engineer unit was restoring Route 19, filling the pocked holes, and paving the road.

About a month later, Moon's platoon was on guard on an unknown hill around Mang Yang Pass beyond An Khe Pass toward the Cambodian border. Although his company went out for a search operation, his platoon remained away from the company headquarters to build an independent guard post on Route 19. It took about 15 days to make the camp on a low hill to protect the base from the attack of enemy mortars. After working hard for 15 days, the soldiers began to collapse from exhaustion and fever. Despite the need to evacuate the patients, he rebuked their laxity of discipline and forced them to practice Taekwondo in the narrow space available. It was unlucky for his platoon members to have such a vicious leader. But he got a fever. He thought that his illness was retribution for what he had done to his men. He tried to overcome it with his will, but it was impossible. He was evacuated to the regimental medical unit, but it was so painful that he was about to burst into tears. At the end of treatment, he was transferred to the regimental operations staff as its assistant. The five-month period as a platoon leader was over.

Around August 1972, he was transferred to the position of helicopter officer. His job was to plan and implement a schedule to deliver personnel and supplies by helicopter to each region's company and platoon bases daily. An American officer, Lieutenant Colonel Levandowski, was dispatched to his regiment as a liaison officer. He treated Moon like his son and made him feel very com-fortable. Moon was not good at English, so he always kept Min Byung-chul's book, Living English, to communicate with the Colonel better, but it didn't work well. A liaison officer's secretary was a Vietnamese high school graduate. She spoke much better English than he did as a college graduate. That made him a little discouraged.

One day, a platoon leader of the 11th Company, conducting a search and reconnaissance mission, requested artillery fire on some 30 suspicious people moving outside the CCL. It was rare for 30 enemy troops to move in broad daylight, so Moon decided to check with the Binh Khe County Office. When civilians went out of the CCL to cut trees, they were required to obtain permission from the county office, which should, in turn, inform the Korean regiment. But there was no such pro-cedure. Moon asked the Governor of Binh Khe to check their records, but the Governor said it was okay to shoot.

Although hesitant, Moon directed the Artillery battery to fire on them as requested. As those people were bombarded, they descended to the plains out of the mountain. Onboard the helicopter with

1Lt. Moon Dhu-shiek (right, Class of 1971), 1st Pt Leader, 11th Co. of the Cavalry Rgt, with his orderly in front of the "Victory Monument Stone" erected at the battle site of An Khe Pass.

LTC Levandowski, Moon called two gunships to fire on the civilians. The scene where the gunship followed and fired upon the civilians fleeing from the paddy fields was not one to see. After firing artillery shells and gunships for about 30 minutes, a civilian rushed to the front gate of the 3rd Battalion and asked to stop shooting because they were civilians from nearby villages. Eventually, the situation ended with one person killed and one injured.

On February 8, 1973, the United States and North Vietnam agreed to a ceasefire. It was a truce and the beginning of a new war. From that day on, the Korean army began to withdraw. On February 10, Lieutenant Moon took off with a DC-10 from Phu Cat Airfield on his regiment's last withdrawal plane and arrived at Dong Cheon Airfield in Daegu, Korea, at dawn the next day. He returned home without a regimental commander's commendation, let alone a combat medal. Still, his one-year tour to Vietnam was a valuable war experience and became a great asset to his career as a professional soldier.[22]

# PART III

# CONSIDERATION: ISSUES

Chapter 1

# The Vietnam War as a War of Platoon Leaders

"The Vietnam War was a war of the platoon leader,
by the platoon leader, for the platoon leader."
Kwon Kyung-seok (Class of 1969)

The Korean troops were stationed in Vietnam for eight years and six months, from September 1964 to March 1973. While in Vietnam, they conducted as many as 576,302 small unit operations below the company level. These operations were all carried out by two Army Infantry Divisions (Tiger and White Horse) and one Marine Brigade, which comprised about 93 companies with approximately 372 platoons. It has been calculated that these platoons were involved in 213 engagements every day throughout the period. If so, every platoon is said to have conducted two meeting engagements or skirmishes every three days.[1]

From October 1965 to June 1968, for example, the Tiger Division alone conducted operations, amounting to as many as 69,338 ones, among which the small units carried out 69,238 from squad to platoon, accounting for 99.8% of them.[2] This data reveals that at least two squads in all echelons of combat troops were committed to operating daily in various ways. During the period, the number of officers dispatched was 22,339, of which 7,689 were captains, and 9,857 were lieutenants, accounting for 79% of the total. This number shows that many junior-level officers were needed to carry out the small operations in Vietnam.[3]

Many people have testified from experience that the Vietnam War was a war fought mainly by small units, which was indeed the case. "The man, who carries out the daunting task of attack order at the forefront of the battle, is a platoon leader or rifleman," writes Vietnam veteran Shin Jung (Class of 1964) in his book on the Vietnam War, *The Wall of Jungles*.[4] As a platoon leader of the Tiger Division, the first Korean combat unit deployed to Vietnam in October 1965, Second Lieutenant Shin fought many small battles and returned home in October 1966 after a one-year tour. A platoon is typically the smallest military unit led by a lieutenant. According to him, the Vietnam War was a war of platoon leaders because most combats were conducted at a platoon or squad level. Lieutenant Colonel Lee Sang-hoon (Class of 1955), the 1st Battalion commander of the 26th Regiment of the Tiger Division, argued that "about 80% of all operations conducted by the Korean forces were small unit actions."[5]

Even Kim Hyungseok (Class of 1965) titled his Vietnam memoir as *A Platoon Leader of the Vietnam War*.[6] Lieutenant Kho Jeong-sang (Class of 1970), a platoon leader of the 30th Regiment of the White Horse Division from April 1971 to June 1972, even asserts that "the Vietnam War was a small-scale battle below the platoon level. The combat in Vietnam was nothing more than a battle fought by three or four men in front of a platoon leader. In a search operation in the jungle, a point man stands at the head, then the point squad leader, then one or two soldiers of the squad, then the platoon leader, followed by the remaining platoon men."[7]

In defining the nature of the Vietnam War, many platoon leaders who participated in the actual battles emphasize that the Vietnam War was a war fought by small units. Lieutenant Won Jong-gwan (Class of 1969) asserts that since most of the Vietnam War consisted of platoon operations, a proper judgment of the situation by the platoon leaders was the key to victory. As a platoon leader who was mainly involved in the operations, he says, "Even now, I still believe that the Vietnam War was a platoon's war."[8] In addition, Captain Lee Jun (Class of 1963, later ROK National Defense minister), who served as the 6th Company commander of the 1st Regiment of the Tiger Division, recalls that "I respected the opinions of the squad and platoon leaders because all combat activities were centered on their activities."[9]

Captain Byun Hong (Class of 1965), who served as the 10th Company commander of the 1st Regiment of the Tiger Division, was asked about his feelings after completing Operation Fierce Tiger 16, conducted on April 24, 1971. He responded that he had gained confidence that the Vietnam War was a war for platoon leaders and that their conviction was necessary. The 11th Company commander of the 1st Regiment, Captain Lee Eui-ho (Class of 1966), agreed with Captain Byun's argument, saying that the Vietnam War was a war for the squad and platoon leaders. This was a great lesson the Korean military learned from the Vietnam War.

The same was true of the American war in Vietnam. Philip Caputo's widely acclaimed memoir A Rumor of War describes an experience of a Marine platoon leader. Tim O'Brien, an American fiction writer of the Vietnam War, focuses on a platoon of American soldiers fighting on the ground as the object of depiction in his best-seller novels *Going After Cacciato* (1975) and *The Things They Carried* (1990).[10] A 1986 American war film, *Platoon,* follows a U.S. Army volunteer soldier assigned to an infantry platoon of the 25th Division in Vietnam.

All these observations of the officers about the war may be summed up in the remarks of Lieutenant Kwon Kyung-seok (Class of 1969), who had the harshest battle experience as a platoon leader among his classmates. He said, "The Vietnam War was a war of the platoon leader, by the platoon leader, for the platoon leader,"[11] parodying U.S. President Abraham Lincoln's famous remarks in his Gettysburg Address in the American Civil War.

The tour of duty for Korean officers and soldiers in the Vietnam War was one year, like that in the U.S. military. The 7th Replacement Training Camp was established at Oeum-ri, Gangwon Province, in March 1966 to train all replacement troops for the returnees. After the Tiger Division was first dispatched in October 1965, replacement troops began to go individually, not by the unit, from the following year. Therefore, except for the White Horse Division deployed in October 1966 in its entirety, all replacements went to Vietnam on a rotational basis. The troops departed almost every month. In the case of KMA graduates, most of the combat arms replacements were lieutenants assigned to platoon leaders. After completing their tour as platoon leaders at the frontline units in the homeland, they joined the ranks of Vietnam again as platoon leaders.

Of the 2,069 KMA graduates who had been to the Vietnam War, 53%, 1,092, served as a platoon leader or their corresponding level positions. When the first deployment troops were being organized, the class eligible for platoon leadership was the Class of 1964. After that class, all the KMA classes were assigned primarily as platoon leaders. Of course, those who had missed going to Vietnam as junior officers were promoted to a higher rank and had a chance to go to Vietnam, this time as a company commander or staff member for higher units. But most graduates volunteered to go to Vietnam as second or first lieutenants. This contributed to their belief that the Vietnam War was a platoon leader's war.

What, then, were the factors that made the Vietnam War a war of platoon leaders?

First, the Vietnam War was a war of counterinsurgency against guerrillas. The guerrilla war is an unusual form of military combat that often utilizes raids, ambushes, sabotage, and other irregular tactics. The battle involves small groups of men, armed civilians with limited training, attacking traditional and larger military groups. The guerrilla war is fought virtually on no fronts against unseen enemies. There is no clear line distinguishing a hostile area from a friendly one. Most combats were carried out by small groups, using the landscape to avoid open battles with the larger enemy and launch raids and surprise attacks before disappearing into the hideouts. The Vietnam War became mainly a war of minor skirmishes, not a battle of proclaiming victory by erecting a flag on a hilltop or occupying any territories.

Second, most of the combat engaged in by the Allies were to pacify the villages or hamlets of peasants. However, the pacified villages or hamlets were likely to again fall under Viet Cong control, and you had to conduct another operation in the same place. You would have to repeat the same combat you had fought in the same village. For example, the villages in the Go Boi plains in Tiger Division's TAOR were battlegrounds for the entire war period. Countless actions were repeated in the same villages. A fierce firefight conducted by a platoon at a village in November 1965 was repeated at the same place seven years later in 1972. Enemies would come back to the village with which they had a connection. In other words, the guerrilla war in Vietnam was a war of hide-and-seek. Thus, as mentioned above, the soldiers described the Viet Cong tactics as "the enemy is all around us biting our tails" or "the enemy steps on our tails."[12] In a similar vein, Michael Herr, a Vietnam war correspondent, talks about the tactics of both the U.S. and the VC in his book *Dispatches* (1978), "The VC had an ostensibly similar tactic called Find and Kill. Either way, it was us looking for him looking for us looking for him, war on a Cracker Jack box, repeated to diminishing returns."[13]

This meant that you had to be vigilant even during searches and breaks. You had to pay special attention to rear vigilance while moving; the enemy would follow you after you left. After receiving resupply, you had to throw away remnants like cans and biscuits that were not eaten. The enemy would often come to find some things thrown away by the Allied soldiers. He would take those things with him. The enemy often made a booby trap in the shape of an empty can by putting TNT in the C-ration can that the allies ate from and then threw away.

This booby trap was dangerous because it was thrown away like a can by the friendly forces after eating, so it was difficult to identify whether it was a threat. And so, almost nobody paid attention to them because the soldiers thought it was simply an empty can. However, if you took a closer look, you would find a tripwire. And if the can weren't stained with food scraps, you would have to treat it as a booby trap. Even the slightest carelessness could lead to misunderstandings, so the soldiers had to be especially careful. The Viet Cong could be understood as germs. The germs will not dare to attack the muscular bodies but attach themselves to the spot of a weak body. Once they stick to weakened areas, they aggravate the entire body quickly if not adequately treated. The same is true of the Viet Cong. The quarters where security was loosened and the posts where guards slept were the perfect places for the Viet Cong to infiltrate.[14]

Third, the military strength and tactics of the guerrillas made the Vietnam War a war for the platoon leader. There were three kinds of Viet Cong. The main force of the VC was composed of relatively well-trained professional soldiers. The top cadre included regular officers from North Vietnam and was organized into companies, battalions, and regiments. Another component was the local force composed of the VC from the rural villages. But the significant commanders and staff were the officers of the NVA. The provincial government operated this force independently and often reinforced the main strength. Before the arrival of the Korean troops in early 1965, the combined total of the main force and local force was nine regiments consisting of 30 battalions, 189 companies, and 110 separate platoons, estimated to be the strength of about 45,000. The third component was the militia, often called local guerrilla. Disguised as a civilian during normal times, this personnel appeared for action when necessary. This force helped the central and local units by doing chores like guiding and carrying supplies. They have spread all over South Vietnam.[15] These forces did not wage massive operations against the Koreans but used the tactic of "Hit and Run" with a small party. Therefore, the Korean troops opposed them with a small force below the platoon level against their small-scale tactics, especially to find their stronghold or hideout.

Looking back on the various battles he had fought in Vietnam over the past year, First Lieutenant Bae In-ki (Class of 1963), the Weapons Platoon leader of the 11th Company, the 1st Regiment, said, "Most battles have been guerrilla fights in expanding and clearing operational areas within the TAOR. First of all, the enemy was reluctant to fight openly with us. In addition, with extreme limitations in support from the rear and overall supply, the Viet Cong tried to achieve their goals using Hit and Run tactics. Therefore, *we had no choice but to respond with small units at all times,* and we had to improve the tactics of small units and strengthen our combat power."[16]

Fourth, due to the topography of Vietnam, large-scale operations were impossible. Coordinated search and destroy operations between the units were difficult because the visibility in the densely forested jungles was quite limited to the extent that you could not see things, sometimes even five meters away. Though you divided large areas into small ones and conducted coordinated operations, finding and eliminating the enemy was impossible in such places. Therefore, you had to organize the troops into small units like squads or platoons. Further, when you moved a unit larger than a platoon through the thick jungle trail, movement in a line formation was almost impossible. A column formation in a smaller unit was the only effective way to get through the jungle. Correct map reading was essential in Vietnam's densely forested jungles. However, there was a marked difference between the map and the actual topography. In many cases, you marked your location wrong and requested artillery fire based on the wrong data, which led to imposing damages on the friendly troops, often in large-unit operations. When you flew into the jungle by a helicopter, sometimes the American pilot misread the coordinates of the LZ and unloaded your unit far from the designated place.

Finally, the concepts of operation of both the U.S. and Korean forces became factors that turned the war into one fought by the platoon leader. Above all, as Commanding General Chae Myung-shin once said, "At the present stage of the Vietnam War, there are no objects to seize at the risk of taking enormous casualties."[17] That meant a large-scale operation needed to be revised. There was no massive body of enemy troops to defeat. Although the Korean military waged corps- and division-level operations like Ojakgyo Bridge, Hong Gil-dong, Fierce Tigers, and White Horses, all of those operations were to expand its TAOR and pacify the broader area of South Vietnam. The small units carried out over 90% of operations and actions during the presence of the Korean military in Vietnam.

Referring to the "Search and Destroy" operation, an operational concept of the U.S. military, the Korean army developed it into a three-stage idea of separation, interception, and destruction to fit the nature of guerrilla warfare. The tactic of the U.S. military was to search the area where the guerrillas were believed to be hiding, find them, and then destroy them with powerful military force. The most critical task in carrying out the three-stage operational concept adopted by the Korean military was to separate the Vietcong mix from the residents and prevent them from contacting the residents. To this end, the Korean military built a tactical base on the outskirts of the residential areas. At the same time, the concept of a tactical base was to contribute to increasing the pacified region. Hence the idea of the "company's tactical base."

## Company Tactical Base

In the Vietnam War, the company tactical base was essential for waging counterinsurgency warfare through small unit operations. This base concept made possible the repeated battles in the same places, a distinctive feature of guerrilla war. This base idea comes from Major General Chae Myung-shin. He had adopted the concept as the starting point of all Korean operations since he arrived in Vietnam in October 1965. The company's tactical base was a camp for the operations and daily life of the rifle company. It was heavily entrenched so the company could conduct independent operations and sustain the enemy's regimental-sized attacks for at least 48 hours. In the early period, the company base was not fixed. It was moving base. Once the company pacified an area, it moved to other places where it again engaged in the same projects.

These continuous position movements went on for quite a while, but did not cause any problems because it took only three days to construct a company's tactical base. The main point of building the base was to dig foxholes, trenches, and dugouts and set up quadruple barbed wire entanglements around the perimeter. The claymore mines were first introduced and placed around the base with trip flares in the war's early phase. From 1969, however, the concept of a company's tactical base was changed from a temporary base to a semi-permanent one, constructed as a fortified position using lots of materials and heavy equipment.

A fundamental concept in constructing the company's tactical base was to build it not in the form of a frontal defense but in the shape of a perimeter defense so that fire distribution could be appropriately made. Above all, it was to establish a complete defensive system for night combat. And two-person bunkers rather than individual foxholes were needed so that two soldiers could mutually depend upon each other. A trench had to be composed of two or three two-person bunkers constructed in a vertical line, not a horizontal line, to be connected to the squad leader's foxhole to the rear. Even if any squad or platoon was overwhelmed by the enemy, other platoons could keep resisting.

Artificial constructs were indispensable in stopping the enemy's

MG. Chae Myung-shin, Tiger Div. Cmdr, showing a Tiger Division's company tactical base to American and South Vietnamese generals in the Go Boi plains in the early period of its deployment in 1966. (Source: ROK Army Museum).

approach by placing obstacles outside the base. Barbed wire entanglements were installed, and antipersonnel mines and trip flares, too. And a killing zone was designated for self-defense of the base, including fire net construction and the setup of claymore mines on the enemy's avenue of approach. For the measures of signaling for night activities, various kinds of sound and flashlights were needed in case the enemy's attack cut off communication. Finally, the base was supported by artillery fire. Against the enemy's massive attack, it needed to be within the range of artillery fire regardless of general or direct support. If the base did not get artillery support, the base might be devastated by the enemy, and the men demoralized. Therefore, the company's tactical base was not to be constructed beyond the reach of the artillery fire.[18] In sum, the company's tactical base was the starting point for all platoon actions.

Chapter 2

# Combat Motivation: What Makes Them Fight?

## Fear, Courage, and Comradeship

What makes the soldiers charge forward on a battlefield where bullets pour in from all directions? What makes them run towards the enemy position without the fear of death? Is it an act of courage? Is it an act of patriotism? Or is it a product of the unconscious? What else could it be? Sigmund Freud concludes that the thought of "Nothing happens to me" makes the soldiers do such an action.[1] In reality, it is said that the soldiers unconsciously rush forward when they see their comrades, who were alive until a moment ago, falling, or when they have confidence in their platoon leader who has taken care of them in earnest.

In his American Civil War novel, *The Red Badge of Courage* (1895), Stephen Crane defines courage not as something that has anything to do with "all larger purposes of war" but as the expression related to "paroxysm" or "frenzy" or "sublime recklessness" coming from the choking atmosphere of war and the following sense of rage. Amid chaotic war, his protagonist feels himself "a savage religion-mad."[2] In the novel, the writer subverts the traditional concept of courage. Likewise, in "Speaking of Courage," a story in the Vietnam novel *The Things They Carried* (1990), the writer Tim O'Brien suggests that the award, a military recognition of courage, is not conferred in the accurate measures of heroism. In other words, "many brave men do not win medals for their bravery, and others win medals for doing nothing."[3]

However, this is not always true when talking about KMA graduates. These men, who were commanders and leaders in Vietnam, recognized courage or heroism as something related to honor or the larger purpose of the war. Yet most of them understand that courage comes, to a considerable degree, from a sense of duty or self-sacrifice. In his book *The Wall of Jungles* (1967), Second Lieutenant Shin Jung (Class of 1964), who was a platoon leader of the 11th Company of the Tiger's 1st Regiment, defines his reason for fighting:

> It has been a month since the Tiger Division landed in Vietnam. In the meantime, about 28 of our soldiers have been killed. Of course, some people died during the fighting, but more died due to safety accidents. What does this mean? One of my KMA classmates also suffered an unfair death when a mine exploded due to his subordinate's mistake. Some say, "We came to Vietnam purely for our selfish interest, regardless of whether it was to have combat experience or personal development." As a platoon leader, however, when I risk my life and wander through the jungles with my men amid the rain of bullets, I think, "Still can I say that I came here for myself?" You can call us crazy. I want to place a much higher value on the concept of "for others" than "for me." Here, "others" may be referred to as "country" or "nation." There is nothing to be ashamed of, even calling it "world humanity." We are fighting for others.[4]

The same was true of Captain Lee Jae-tae (Class of 1960), the 11th Company commander, the 1st Regiment of the Tiger Division. He says, "the soldier's view of life and death is closely related to the duty of self-sacrifice." He continues, "If only we are determined to sacrifice ourselves to honor the military and the nation, we will willingly dedicate ourselves to the nation's just cause. It is the military spirit. In particular, soldiers choose a place worthwhile to die. Caesar said he would choose to die in war rather than die from illness, and the warriors in the East thought it a shame to die in pursuit of safety during a war. Soldiers on the battlefield have something like 'mysophobia about death.' 'When we die, we wonder whether we have done anything wrong to feel ashamed about. Is there anything embarrassing about your body?' I think about it all the time."[5]

Captain Lee tells the story of his experience when he saw one of his subordinates charging toward the enemy's position. Sergeant Lee Sang-shik, a member of the 3rd Platoon, was a troublemaker. One day, when his platoon was on duty, he went AWOL and appeared at a honky-tonk in Quy Nhon. While drinking beers, he quarreled with an American sergeant and knocked him out. He was summoned back to the company at Cu Mong Pass. He also made some trouble here, but Captain Lee always condoned him.

One and a half months later, in the Chi Hoa battle, the 1st Platoon to which Sgt Lee belonged was the last element to withdraw. While passing beside Captain Lee, Sgt. Lee saw his company commander seriously wounded. All of a sudden, his face got contorted and trembled with anger. He rushed into the enemy position in the blink of an eye, saying, "Sir! I'll pay them back for what they did." Seeing him refuse to come back, the squad ran toward him, followed by all men of the 1st Platoon. A fierce firefight ensued. Captain Lee recalls years later, "I still can't forget his contorted facial expression full of anger. I felt Sergeant Lee's true loyalty from the depth of his heart."[6]

As mentioned earlier, any nation in war must have a just cause for which to fight. The reason is that the nation must persuade and mobilize its people and materials to conduct war. At the same time, all kinds of rhetoric are needed to accomplish the nation's war aims. However, the soldiers in combat are not interested in such an idea of a country or nation. What matters the most to them is to survive the war and return home alive. Their platoon leader and fellow soldiers are the most reliable for the soldiers isolated in the jungle. More than anything else, many officers agree that comradeship or loyalty to one's superior is the most crucial factor that makes men fight. Such a factor is something that unites the members of a unit. Here the individual soldier realizes the meaning of one important myth—the myth of army buddies. Such values as comradeship, loyalty, sacrifice, responsibility, and courage come from this legendary bond.

Why do soldiers fight? They fight for their survival. In the bushes, patriotic slogans do not help them withstand the fear of death. No matter how thoroughly a soldier may be armed with ideological and physical training, he is instinctively dominated by the fear of death. No one can alleviate a soldier's fear of death on the battlefield. Only the comrades who get shot and bleed together beside him, thus sharing joy and sorrow, become a source of comfort and strength. The platoon leader takes pride in his men, and the men give their confidence in him. This solidarity is possible because the men firmly believe their leader will save them from the dangers of death.

## The Bond between a Platoon leader and his Men

On the battlefield, the smallest tactical unit commander is the platoon leader, and the soldiers' dependence on the platoon leader is absolute. It forms the basis of their motivation for combat. The strong bond between the platoon leader and his men is irreplaceable on the battlefield. A leader is a guardian who takes care of his soldiers. Thus, the parents who sent their sons to the battlefield frequently wrote letters to the platoon leader or company commander, asking for their son's safety. And the soldiers completely trust their platoon leader, who will keep them safe. Where else in human society can we find such a relationship? But in war, no one can guarantee the safety of anyone from the bullets flying randomly. On August 19, 1971, during Operation Dokkabi 21, conducted by the White Horse Division's 28th Regiment, Sergeant Kim Choong-hwan of the 3rd Company was killed. Before leaving for the operation, Kim's company commander received a letter from the soldier's father, saying, "He is my only son, so please take good care of him and bring him home safely." And the commander promised he would. But he did not keep his promise. "Why did I bring him into battle?" It remains as an indelible debt in the heart of the commander.[7]

Soldiers fight for their platoon leader and comrades. The presence of the platoon leader motivates the soldiers to fight more than any cause. A female nursing officer at 106 Evacuation Hospital in Quy Nhon in October 1966 recalls, "When the surgeons operated on the soldiers who had been in the hospital after being seriously injured in battle, no soldiers looked for their father or mother, even amid great pain. They looked for platoon leaders, squad leaders, and comrades. They shouted that they should get well fast, return to the unit, and fight. No one wanted to evade the battle. They said, 'we must go back quickly and get revenge on the Vietcong.' I felt that the strong bond formed on the battlefield between men is greater than any love in the human world."[8] This is not just apparent in the Korean military. In an American war novel about Vietnam, *Fields of Fire* (1979) by James Webb, the platoon leader Lieutenant Hodges states that "man's noblest moment is the one spent on the fields of fire."[9] Comradeship is the unique value found between men only on the battlefield.

Second Lieutenant Shin Jung (Class of 1964) contracted malaria and was taken to the evacuation hospital. Malaria, which he didn't take care to prevent, left him in hospital for a month. As a platoon leader, he felt sorry for his men and was disappointed with his own helpless appearance. Then he received a letter from the Platoon Sergeant commanding the platoon on his behalf. "You were exhausted from fever for several days. You pushed forward through the jungle at the head of the platoon, but eventually, you fell into the woods. When you were evacuated by helicopter, we all were stunned. The platoon without you seems to be empty. When will we be able to see you? All the platoon members are praying that you will recover soon and we will be able to go to battle together again."[10] Where else can you see the love between people like this, except in war? It is what makes the subordinates charge toward the enemy's position.

Lieutenant Koh Tae-soon (Class of 1970), a platoon leader of the Tiger Division, witnessed how much his men depended upon him while engaged in an operation at Go Boi plains. During the operation, his platoon was separated from his company. Further, a lack of water in an utterly isolated jungle drove all his men to extremity. Amid all the supplies cut off, they desperately struggled even for a sip of water to quench their thirst. Right then, a man brought Koh a canteen with some water. Koh could not withhold tears welling up in his eyes while holding the canteen in his hand. He still does not forget the man saying, "You're the person who should protect our life, so you need to drink, not us." But Koh insisted they all would share even a drop of water, and the water problem was solved immediately with the supplies airdropped by helicopter. But this incident became an opportunity for him to be aware of how important it was to win the men's hearts and minds on the battlefield.

The same was true of Lieutenant Kim Soon-shin (Class of 1970) and Lieutenant Park Heung-keun (Class of 1970), who were keenly aware that comradeship between a platoon leader and his men matters most in combat. Once in a battle, Lieutenant Kim had sat up all night embracing his wounded man to his bosom. Kim remembers he had felt the man's warm heart as his body temperature was transmitted to his. The wounded man was talking to him through his heart. He must have been grateful to his platoon leader, who had protected him all night. Likewise, Park Heung-keun felt that the man who motivates his men to fight is the platoon leader. According to Park, soldiers can acquire methods and skills for fighting through military training. But it depends on a leader's ability to motivate them to fight at the risk of their life on the battlefield. The platoon leader is the one who makes it possible to exert such knowledge and abilities. The mighty strength of the platoon leader comes from not only his leadership competence but also his heart-to-heart love toward his men—comradeship.

When a comrade who had been with you was injured and evacuated to a field hospital or his homeland, you might feel his vacancy, and he would be deeply missed. It was not because he possessed qualities that made him special but simply because he had been *one of them*. Indeed, the most crucial point that motivates men to fight is comradeship. The problem was, as Philip Caputo says in his book *A Rumor of War*, "at times, the comradeship that was the war's only redeeming quality caused some of its worst crimes—the act of retribution for friends who had been killed. Some men could not withstand the stress of guerrilla fighting: the hair-trigger alertness constantly demanded of them, the feeling that the enemy was everywhere, and the inability to distinguish civilians from combatants created emotional pressures which build to such a point that a trivial provocation could make these men explode with the blind destructiveness of a mortar shell."[11] Indeed, most civilian massacres in the Vietnam War occurred under these circumstances, as shown in the My Lai massacre in March 1968.

## Temerity and Cowardice

Who was the most courageous soldier in the field of battle? Some officers say: In Korea, the soldiers were encouraged to be "brave," but there was no need to emphasize bravery in Vietnam. When they saw their comrade fall, they became brave. They ran forward on their own initiative. So, a commander even went so far as to say that there was no need to emphasize bravery when training. According to Captain Kim Dae-heon (Class of 1958), a company commander of the Cavalry Regiment of the Tiger Division, the soldiers, who were usually brave and raging, were feeble in battle. Still, rather introverted and calm soldiers fought bravely and well in actual combat.[12] As mentioned above, a soldier in *The Red Badge of Courage* named Wilson is a braggart who is "loud" in front of the recruits before being committed to the battle. However, when the battle begins, he is terrified and does not know what to do. He is a coward. Instead, the quiet recruit in ordinary time bravely dashes forward and becomes a hero.

Captain Kim Jeong-jo (Class of 1965), the 7th Company commander

A company commander of the 29th Rgt, White Horse division checks a maneuvering route with his men during Operation Bat 2-6. (Source: Institute for Military History, MND)

of the 29th Regiment of the White Horse Division, had a similar experience to Captain Kim Dae-heon. He participated in Operation Bat 26 and searched the jungle on June 22, 1971. His regiment commander often said, "The brave lives, and the coward dies." A person who had been usually brave was not always courageous on the battlefield. At that time, a few gangster soldiers were in his company before joining the army, and Kim tried to send them to the front line as point men. But when he saw one of them fire into the air in the presence of the enemy, he felt that he was just a braggart. His words didn't accord with his deeds.

On the other hand, there was a squad leader named Sgt. Ji Yong-gil, a short and lean person who was quiet and shy. Before coming to Vietnam, he had already married and had a child. Outwardly, he was a gentle and kind model soldier, so Captain Kim doubted about whether such a person could go out to the battlefield and lead the way. But in actual battle, he was different. He didn't mind making him the point man of his squad. He stood up bravely more than anyone else and rushed forward courageously. "As a result," Kim said, "rather than the boasters in normal times, it is necessary to discover soldiers with calm and inner courage and entrust them with the important tasks."[13]

In combat, temerity is prohibited. Sometimes temerity or being eager for fame can lead to tragic results. Second Lieutenant Kim Hyung-seok witnessed an incident where his KMA classmate, Lieutenant Suh Soo-geun (Class of 1965), was lured and killed by the enemy while chasing him recklessly. Of course, this resulted from Lieutenant Suh's immaturity of experience. But it was an incident that showed what results temerity could lead to. On December 24, 1967, Lieutenant Suh, a platoon leader of the 6th Company of the Tiger Division's Cavalry Regiment, encountered two VCs at sunset in the deep valley of Nui Phu Cat during Operation Fierce Tiger 9. The enemy disappeared into the jungle, and the 6th Company commander ordered Lieutenant Suh to search the enemy's hiding place. Suh and his men fiercely chased the enemies without delay. However, the enemies lured them into a concealed position and poured a massive fire on them.

At this time, the size of the enemy increased twice that of the platoon. Lieutenant Suh asked the company commander to reinforce troops while attempting to escape to the right side. But he suffered a penetrating wound in the thigh from enemy fire and received emergency treatment from a medic. Still leading the battle, he again took an abdominal piercing injury. While he handed over the platoon's leadership to his sergeant, he only fought against the enemy with grenades. Eventually, his life ended in the jungle of central Vietnam. Some of his men escaped from the spot with severe injuries and reported to the company. However, this case was often discussed in the Korean army headquarters in Saigon as an example of a battle caused by arrogance and inexperience combined with the search for glory. Lieutenant Suh was an officer with a sense of justice and bravery since he was a cadet and one of Vietnam's most courageous platoon leaders. Lieutenant Kim does not necessarily view his death as a catastrophe caused by arrogance. Suh was a soldier.[14]

Quoting some taboos that officers should evade from Sun Tzu's *The Art of War,* Lieutenant Kim Hyung-seok briefly differentiates courage from temerity. Firstly, if an officer fights desperately, he may eventually die. You shouldn't jump into unnecessary or urgent things and force yourself to death. In other words, it means that officers should thoroughly examine the situation, judge the importance and timeliness of work, lead the way, and act heavily with self-respect. It is deplorable for young officers to be obsessed with the cause of loyalty and courage or to die while rushing to the enemy camp on impulse without discernment due to ambition or fame. True dedication and courage are to risk your life only when it is judged that you cannot save your subordinates and troops without sacrificing yourself.[15]

When fighting, you have to understand the psychological state of your soldiers. The fear of death is a human instinct. In other words, it is natural to have a fear of death on the battlefield. First Lieutenant Bae In-ki (Class of 1963), the Weapons Platoon leader of the 11th Company of the First Regiment, directly experienced how soldiers expressed their fear on the battlefield. There was a case where his platoon set up an ambush during the day in an area where enemies frequently appeared. Sergeant Son, whose return home date (DEROS) was just around the corner, saw eight armed VCs moving about 20 meters in front. The Sergeant did not report it to him though he saw them. Lieutenant Bae found out the reason later, but it was simple. "If he had told me, there would have been a battle. If so, he might have been injured or killed. He didn't want to make trouble. He tried to avoid dangerous things. He didn't want anything wrong to happen before his return home in a few weeks."[16]

In addition, Lieutenant Bae found out that even if he ordered a soldier to reconnoiter a spot up to 100 meters in front, the soldier reported that nothing had been wrong when he returned to the platoon position. However, the reconnoitering soldier had been lying face down, only 30 meters from where his platoon was. For this reason, a leader needs to understand each soldier's psychology and avoid making a big mistake at a critical time. Of course, it is difficult to define this as cowardice. In his Vietnam memoir *Dispatches* (1977), Michael Herr named the psychology of these soldiers "The short-timers syndrome."[17] In fact, many American novels deal with short-timers' psychology. In the American novel of the Korean War, Michael Lynch's *An American Soldier* (1969), the protagonist, black Corporal Francis, wants to pass the time, with a month left on his return home. "Count our days. We're short-timers. We can take it easy."[18]

There were several ways to avoid battle. It seemed legal, but in reality, it was a cowardly act. Most of the battlefields in Vietnam were low-lying grass and low shrub zones, so mosquitoes were rampant at night. The struggle for survival with poisonous insects became a wild game rather than a fight against the enemy. The night ambush operations were the time for the soldiers to apply mosquito repellent to their exposed faces and skin. Mosquitoes continued to swarm in at dawn even after using the repellent. Mosquitoes sucked human blood and, in return, transferred malaria fever to their hosts. However, some platoon soldiers did not apply mosquito repellent during the night ambush. Instead, they applied greasy butter on their faces and waited for mosquitos to come. After a few days, they developed a physical fever and reported it to the platoon leader through a medic when it rose to 39 degrees centigrade. Then, the unit took them to the hospital, treated them, and returned them to the unit. In the meantime, a series of exhausting combat activities continued in the unit. When the fever rose again, they were evacuated again. As such, there would be 2-3 soldiers in each platoon who traveled back and forth between hospitals and units as habitual fever patients and ended up boarding the return-home ship.

Even a few platoon leaders have been rumored to have had a similar patient life. They couldn't overcome fear and physical pain in the operational area and used such mean tricks to relieve self-inflicted conditions. After returning home, they somewhat exaggerated their combat experiences, stole other soldiers' stories as if they had been theirs, and pretended to be heroes. Kim Hyung-seok recalls: "I discovered the variety of the human beings and felt sad at their ugliness,"[19] as the American writer Mark Twain had found such "miserable human beings" in a supposedly incorruptible town of "Hadleyburg" some centuries ago.

## Chapter 3
# Dehumanization of War

In the Special Forces A Camp at Me Phuc Tay there was a sign that read, "If you kill for money, you're a mercenary.

If you kill for pleasure, you're a sadist. If you kill for both, you're a Green Beret."[1]

Michael Herr, *Dispatches*

"Even though I walk through the valley of the shadow of death, I will fear no evil, for I am the evil."

A parody of the verse Psalm 23

War is characterized by randomness, cruelty, and death. The factors leading to such violence in the Vietnam War differed from those of other wars. The deaths mainly came from snipers, mines, boobytraps, and ambushes. For example, Philip Caputo writes of the nature of death in Vietnam: "In the sodden world they inhabited, the mere act of walking, an act almost as unconscious as breathing, could bring death."[2] For this reason, as Peter L. Stromberg says, "death has a quality different from other wars" in Vietnam.[3] The fear of death coming from unseen sources and the contradictory mandates under which the soldiers were subjected to fight brought to their mind frustrations, anger, and hatred. The mental state was well reflected in their frequent question: "If we're not here to shoot these people, what the hell are we doing here?"[4] At this stage, their moral sense was blurred, and they, in turn, degenerated into atrocity producers.

In Vietnam, many killings and brutalities were committed directly, face-to-face. In such cases, the individual soldier can see what he has done. For this reason, as Robert Jay Lifton said, “the psychological aberration and the moral disintegration are very advanced.”[5] In the process, the soldier thinks he is no longer morally and psychologically sane; he recognizes himself as evil. Caputo writes, “Perhaps the war had awakened something evil in us, some dark, malicious power that allowed us to kill without feeling.”[6]

Indeed, the Vietnam War is remarkable for recognizing the individual as a source of evil. When a green soldier engages in his first battle, he is often overwhelmed by fear of death. Not only that, but he also has difficulty firing at point-blank range. Perhaps, for this reason, there was an unbelievable story from Vietnam that some novitiates had been ordered to disfigure an enemy’s dead body to arouse a killer’s instinct in their minds. Soldiers in combat are often said to transform from humans into nonhumans. In war novels or memoirs, the soldiers usually identify themselves as incarnations of evil.

In fact and fiction, they self-mockingly compare themselves to evil, as shown in the parodied verse, “Even though I walk through the valley of the shadow of death, I will fear no evil, for *I am the evil*”. Coming from Psalm 23 of the Bible, the parodic graffiti is engraved onto belongings like Zippo lighters or helmets. Man has a destructive instinct that is always ready to jump out. It is the “power full of dark

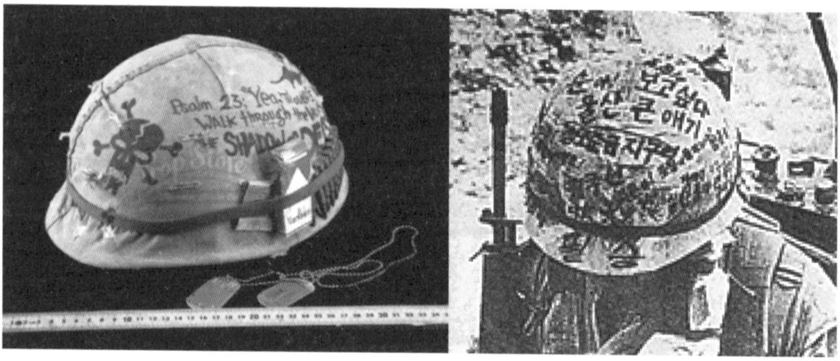


A graffiti on the American soldier’s helmet (left): “Yea, though I walk through the valley of the shadow of death, I shall fear no evil. Cuz I’m the meanest motherfucker in the valley.” A parody of Psalm 23. A graffiti on the Korean soldier’s helmet: Translating, “I miss you, sweetheart,” “No aiming at this spot,” “Absolute Victory,”and “Board the Returning Ship.” The words on both helmets show good contrast between the Korean and American soldiers about what mattered most to each of them. ” (Source: Institute for Military History, MND)

and evil mind" latent deep within a man's unconscious that arouses him to commit the killing. It takes only a split second for a soldier to be transformed from an innocent man to a cruel one. There is only a thin veneer between these two men. In WWII American novel *The Thin Red Line* (1962), the author James Jones says, "there's only a thin red line between the sane and the mad."[7]

This envisioning oneself as evil becomes a factor that makes a man frequently worry about his reentry into the world. A soldier in *The Short-Timers* (1979) by an American Vietnam veteran Gustav Hasford, pessimistically looks ahead to his return to America: "Those of us who survive to be short-timers will fly the Freedom Bird back to hometown America. But home won't be there anymore, and we won't be there either. Upon each of our brains, the war has lodged itself, a black crab feeding."[8]

Why is it possible for men to kill each other in war? Once in a war, men are not humans. How can such an average person with reason kill another person? In the Vietnam war novel *Fields of Fire* (1979), written by Vietnam veteran James Webb, almost all the soldier characters have no real names but are called nicknames. Their real names remain unknown throughout the novel. But their nicknames are un-human ones like Snake, Cannonball, Homicide, Water Bull, Phony, Bagger, Crazy, Cornbread, etc. In the novel, the nicknames are given because the "grunts" in the Vietnam War felt abandoned or thrown away. After all, they cannot commit the brutalities in a normal state of mind because they cannot meet the horrors of such a cruel war with human names. With such words, the soldiers are not human; they symbolically lose their standard human identities. In WWII American novel *Catch-22* (1969), the author Joseph Heller raises a more fundamental question, "the man who has not gone insane cannot kill other men in war, but how come the insane man can go to war?" There seems to be no answer to this question. War has continued, and so many human beings have been killed.

On the other hand, the enemy should also be a non-human being. To kill the opponent in war, you must downgrade him into an animal or non-human being, for it is difficult for a man to take another man's life away as long as he remains a human being. In WWII, the Imperial Japanese army (Unit 731) in Manchuria used humans for experimentation on biological and chemical weapons and live vivisections, dismemberments, and other experiments. It called the subjects of investigation a particular project code-named "Maruta," which meant "log" or "tree trunk," not human. In all events, such a brutal action was an unethical and anti-human crime.

In Vietnam, the Viet Cong and North Vietnamese soldiers were called "gook." But later, this word was applied to all Vietnamese people as a derogatory slang term. As mentioned above, the name "Mere-Gook-Rule" (MGR), an unofficial policy of the American army in Vietnam, was used as a means for soldiers to be prosecuted very leniently for harming or killing "gooks," even if the victims turned out to have no connection with the enemy. In the guerrilla war in Vietnam, the rationale for this rule was that the soldiers had a tough time determining which Vietnamese were civilians and which were enemies. Such a rule like MGR was a kind of legal action to instill an idea of taking human life very lightly in a soldier. Americans and Koreans only fought against the "gooks," not humans in Vietnam. Both sides, hostile and friendly, should be non-human beasts or animals on the battlefield.

Having landed in Vietnam in early 1971, First Lieutenant Koh Tae-soon (Class of 1970) participated in more battles than anybody else among his classmates. For his successful performances, he received several medals, including Inheon medal and South Vietnamese decorations. This meant that his platoon had produced a lot of enemy casualties. When he had no action, he used to jot down what had happened. It was not all about combat actions but included writings of transformed human appearances in a war situation and heart-breaking stories in which his platoon had to kill others. He hated war. Indeed, it is the young soldiers that hate war most, for they are the ones who should go out to the field and fight. A man of gentle mind in his cadet days, Koh named himself a "brutal fucker" after having undergone a violent baptism of fire, the primary sacrament of war. He was no longer an infant(ry). He might have had to show his men what he should be as a leader. That is, he had to mask his gentleness with surface roughness. A moment of paternalism might bring disaster to his men. You cannot kill the enemy or win the battle if you don't become a brute. He had to train his men to be cruel. Indeed, they were also innocent young men, typically feeble enough not to kill even a small creature like an ant.

Koh's platoon participated in a Tiger Division's operation conducted in the Go Boi plains for some days. His platoon made an all-out assault and killed all the enemies they encountered. He ordered his men to ease the pain of the still alive enemies by thrusting bayonets into their bodies to end their suffering. His men did only what they were told to do. Amid the bullets pouring down, the men's reason was paralyzed, and they were like animals, automatically responding to their instincts for survival. They seemed to have suddenly grown wild with an enthusiasm of unselfishness. It was as if they had been driven by something unknown. What had been written in innumerable literary documents about men during the war was realized here in actuality. Koh's confession: "Many lives were lost by my hand. It's now my true feeling that I was only faithful to my duty at the time. Making an insensible excuse for what I've done, I consoled myself by saying that my brutality was inevitable in that situation and thus should be understood from that perspective. We all were in a state of frenzy."[9] In the war that the country ordered him to fight, he did his best to fulfill his mission. Yet he concluded that there should not be war.

According to some Vietnam veterans, Korean troops committed atrocities against the enemy's corpses. Although such an atrocious act toward the dead could not be imaginable, the Korean soldiers often practiced it in the early phase of the war under the pretext that it would give the soldiers guts to kill. The recruits did not even dare to face the enemy in combat.

One day in February 1967, on his way back to the company, Captain Lee Jae-tae (Class of 1960) met his 2nd Platoon leader Lieutenant Kwon Woo-heon (Class of 1964). With his guidance, the company commander Captain Lee took a quick look at the area where the platoon had conducted its operation. Several soldiers were eating C-rations in front of the enemy's trench beside a fence surrounded by bamboo forests. In front of them was the bloody dead body of an enemy killed a while ago, and the squad leader was stabbing the body with a bayonet. The company commander was surprised by this horrific act. The squad leader said, "I'm nurturing the audacity of the recruit."

Some recruits of the platoon who had recently arrived stood there with dumbfounded faces.

The platoon leader, who witnessed this cruel act, stood aghast at this scene. Captain Lee immediately left the scene without a word. A soldier's primary virtue is to be polite to the dead on the battlefield. Even if it is an enemy body, it should be buried with as much respect as possible. This is because killing is not to satisfy a personal grudge but for the cause of the nation that one fights for. He condoned the atrocities. He always felt guilty for having used the bodies of VC as bait to lure them into his ambush site on Route 1 in the Phu My area. He pretended that he had not seen the horrible act. But he saw it clearly and evaded the scene. He still has painful remorse for that deed.[10]

One day in February 1967, Captain Park Jong-sik (Class of 1959), the 11th Company commander of the 28th Regiment of the White Horse Division, told the soldiers to cut off the left ear of the dead enemy they killed to prove what they did. They did what the company commander told them to do and hung the cut-off ears on the branches of a tree in the company base. Some American war correspondents came to his base and saw the ears threaded into the wire. He worried that it would become proof that the Korean troops had been committing cruel acts. He sincerely asked them not to report the issue of cutting ears. The American reporters had the humanity to keep their promises. In the article they sent to their home country, the reporters removed the phrase "Cut off their left ear" and published an article only praising the bravery of the Korean military.[11]

Lieutenant Kim Hyung-seok (Class of 1965), the 1st Platoon leader of the 6th Company of the 30th Regiment, witnessed such abuse of enemy dead bodies in his company. On April 28, 1967, the 3rd Platoon of the 6th Company went on a night ambush operation. At around 10:20 p.m., the 3rd Platoon reported that five suspicious people were pulling carts and heading toward Hill 287. According to the 3rd Platoon leader, the area was prohibited for residents. Nevertheless, they were heading toward the mountain in groups at night, perhaps to deliver supplies to the enemy. So, the 3rd Platoon members fired and killed all of them. However, no guns were found in the carts, except for food and drinks. The battalion headquarters instructed them to cut off the ears of the bodies to secure solid evidence.

That evening, the 3rd Platoon did what they were told to do, mixed them with salt and brought them to the unit. The 6th Company had not had such a remarkable combat result up to that time as it had eagerly wanted, but this was a tangible success, so the company commander was pleased and praised the 3rd Platoon leader for what he had done. But the issue was whether the five people were enemy combatants. The officers under the company commander judged that it was evident that the five people who moved to the mountain at that hour were enemy military personnel and that there was no possibility that they were innocent civilians. Upon reaching this conclusion, the company commander and the 3rd Platoon leader seemed relieved. Then, the next day, an official from South Vietnam's local government came to the battalion and protested that the Korean soldiers had killed five innocent woodcutters. At this, the battalion's Civil Affairs officer claimed they were VC agents disguised as woodcutters. So, the Korean soldiers killed them because they had ignored the regulations that stipulated not to move around at night.

These inhumane acts sometimes occurred in Vietnam. Rumor had it that some units had returned to their base even with the head cut off of the enemy's body to instill the killer instinct into the soldiers. A company commander of the White Horse division told his men to cut off the dead enemy's toes to confirm their alleged combat achievement. However, he could not bear to see the toe of the enemy's corpse being cut off, so he turned his head away. Why did this phenomenon happen? Was it because only the results of battles mattered? This dismemberment of enemy bodies has a precedent in the history of Korea. In the 16th century, the Japanese invaded Korea (then Joseon Dynasty) and cut off the noses of dead Koreans. They took the noses back to their country and built nose tombs. It was a shameful period in the history of Korea. Under any circumstances, the mutilation of the dead body should not happen. It's a war crime.

The battalion decided to return the ears and then visited the victims' village to express condolences to the bereaved families and talk with local government officials to cover up the incident. On top of that, the battalion paid additional compensation for damages for each victim.

This incident, which took place in the 3rd Platoon, was later included in *The History of the Korean Forces' Deployment to the Vietnam War*, Vol. 2 as a case of civil dispute between the South Vietnamese government and the ROKFV.[12] Captain Hwang Won-tak (Class of 1962), the 10th Company commander of the 1st Regiment, the Tiger Division, acknowledged the Korean troops did a lot of inhuman acts to increase the soldiers' guts, such as mutilating the corpses of dead enemies. He outrightly banned these practices committed by some soldiers in his unit. He ordered his men to bury the enemies killed in an appropriate place to honor them.[13]

## A man's life was cheaper than a single gun

Dehumanization is a phenomenon that can affect any soldiers' behavior depending on the particular war situation, but institutions and policies can also contribute. Although you killed many enemy soldiers in the war, your action might be regarded as having killed innocent civilians if you didn't capture weapons. In the Vietnam War, the standard view of the body count was that "even though you killed the enemy, you still killed the civilians." Therefore, the level of combat achievement depended upon the number of captured weapons which proved the result.[14]

To the Viet Cong, weapons were considered very precious, so if their comrades were killed during the battle, they left their bodies behind, but usually recovered their rifles. While carrying out the operation, Korean military commanders learned that they had to capture the enemy's gun rather than kill it. This policy was the opposite of the American military's "body count" tactics.

The Korean army in Vietnam had a unique method of evaluating a unit's operation based on the number of captured weapons. It was the only active way to reduce the fighting power of the Viet Cong, who were not well-supplied with weapons. In battle, the Viet Cong always retrieved the weapons of their killed or wounded comrades, and even when captured, they hid their weapons well, so the Korean troops could not take their guns away. The Korean army ruled that one weapon was equal to three enemies killed, and there was even a saying, "one medal for one gun." In addition, the integrity of the subordinate units' combat reports was judged by the number of weapons captured. It was considered unreliable if any combat report indicated no guns for ten kills. It was challenging to capture personal firearms like this, let alone crew-served weapons. Rumor had it that some soldiers would buy rifles on the black market to win medals.

The rule "one gun equals three lives" was the one that Korean soldiers had followed as an unwritten rule in the Vietnam War. Sometimes, the troops who went out to battle inflated their combat achievements, so bringing the enemy's weapons was proof of fighting. Why did this rule exist? The troops often killed innocent civilians while trying to kill the Vietcong. In other words, the rule indicated implicitly that a man's life was cheaper than a single gun. What a sad irony! On the battlefield of Vietnam, humans were nothing more than throwaway things.

Captain Lee Jae-tae (Class of 1960) recounts his battle experience ironically. Once during combat in 1966, his company killed six enemies and captured several firearms. To celebrate the combat achievement, they held a beer party at the company headquarters. They were to recognize and celebrate a definite act of murder. Since Lee had engaged in some battles, he had become insensible to humanity even after committing the killing, as if it were nothing. In other words, the men of Lee's company belonged to authorized murder licensees. The reason that his soldiers wandered into the jungle every day was to hunt or kill human beings. Ironically, a few days before, a mountain goat stood on a rock before the camp as if saying, "Shoot me," but the company guard did not shoot. If it had been an enemy, the sentry would have shot. The price of a man on the battlefield was lower than that of an animal.[15] War is contradictory. *If you kill a person in society in peacetime, you will be executed, and if you kill a person in a war zone, you will receive a medal.* Initially, humans are considered unreasonable, but aren't they still created in the image of God?

Lieutenant Noh Nam-seop (Class of 1967) saw how men in war become wretched. The fighting soldier, enemy or friendly, was nothing but an expendable thing when he was dead. Let me quote his somewhat long story here:

> A scene that always made me sad during my presence in Vietnam still

sometimes comes to mind—looking at the helicopters flying in the sky, the memories of catching dragonflies with a butterfly net in the countryside as a child spring to mind. However, that was only a brief recollection. He often witnessed the helicopters falling in the jungles with their propellers broken by the Viet Cong's anti-aircraft gunfire or crashing into the fields. The helicopter pilot and all soldiers on board were killed or seriously injured, and the survivors became prisoners of the Viet Cong.

Meanwhile, there were some casualties after the Allied forces completed the sweeping operation to find and eradicate the Viet Cong. The wounded were hurriedly evacuated to the hospital by the dust-off, but the dead were transported by ropes hanging from the plane. Why do the strings carry the deceased under the aircraft? The U.S. Army ruled that 'corpses cannot be loaded into helicopters with living personnel on board. Sometimes, I saw one or two corpses transported by the helicopter tied to a rope hanging down. Each time, my heart ached at the tragedy of the battlefield.' It was incomprehensible that the dead, who deserved due respect, could not even get into the aircraft and were transported in such a way. It reminded me of my childhood memories of catching grasshoppers, stringing them onto foxtail stalks, and wearing them around my waist. So, my mind was confused with the thought that 'If you die on the battlefield, you are treated like a grasshopper.'

Flamethrowers usually caused the Vietcong's death during cave-clearing

A Tiger division soldier attacking a VC cave on Nui Phu Cat using a flamethrower during Operation Fierce Tiger 6 in September 1966. (Source: ROK Army Museum)

operations, so there were many cases where it was difficult to identify their faces because of their dark black color. In the case of the Vietnamese, their mothers usually came out to find their sons among the corpses on display. However, they didn't cry, perhaps because death seemed a usual affair in the lives of Vietnamese who had been at war for over 30 years. There might be no tears left for them. Their tears may have dried up.

I visited the Battalion headquarters in Cam Ranh for business and returned to our battery. On my way back, I saw 12 Viet Cong corpses lying near the wooden bridge of Dien Khan. Seeing the mothers looking for their beloved sons seemed so pitiful and painful. The carcasses showed how fierce the previous night's battle had been.[16]

## War is hell: Winning is the be-all and end-all.

The American Civil War General William Tecumseh Sherman coined "War is hell." When Sherman ordered evacuation and burning of Atlanta in 1864, Confederate commander General Hood protested his plans. Sherman replied, "War is cruelty, and you cannot refine it, and those who brought war into our country deserve all the curses and maledictions a people can pour out."[17] In Sherman's view, "war is entirely, and singularly the crime of those who begin it, and soldiers resisting aggression (or rebellion) can never be blamed for anything they do that brings victory closer."[18] The meaning implicit in the words "war is hell" is that "soldiers fighting a just war can do anything at all that is useful in the fighting….The war rights of the just are total, and any blame their actions entail falls upon the leaders of the other side."[19] This argument can be extended to the logic that soldiers who are told to fight for their country (or for the cause) can do anything to accomplish its war aims. In other words, they are only fighting the war, not by choice but by necessity. The phrase has been commonly used by those war realists who had committed inevitable or willful atrocities in war.

Due to the nature of the guerrilla war, where the Viet Cong and innocent civilians could not be distinguished, many civilians were sacrificed, and General Sherman's words were summoned each time.

And such civilian sacrifices used to be justified or condoned in the name of war. As mentioned earlier, unwritten rules such as MGR or TOG applied among the American soldiers justified the sacrifice of civilians or actions that violated the rules of engagement. The same was true of the Korean soldiers. On 30 November 1966, Lieutenant Kim Hyung-seok (Class of 1965), the 1st Platoon leader of the 6th Company of the 30th Regiment, found a woman during a patrol mission working in the field while pursuing an enemy running away.

He saw his men aiming at the woman. He shouted at the men not to shoot. He kept calling, "Stop shooting," but it was too late. His men fired at her. The woman struggled with the last gasp of her life and then drooped to the ground. He was seized by severe guilt. She may have just been doing her farm work. Yet he thought the woman must have been staying there to detect the activity of the Korean soldiers under the guise of a farmer living in this Viet Cong-controlled area. If he hadn't killed her, she might have given information about them to her Viet Cong authorities, which would lead to their death. He tried to talk himself out of guilt, taking their killing for granted. But Lieutenant Kim struggled to justify what they had done to her.[20]

The cost of a single bullet sometimes led to tremendous results. According to Captain Lee Jae-tae (Class of 1960), the 11th Company commander of the First Regiment, if an enemy fired a single round at a friendly or shot at a passing helicopter, the artillery barrage flew in less than five minutes, and the village became a sea of fire in a moment. In the end, the village became the target of the friendly forces' ground attack and was burnt down. The whole village was often obliterated, and numerous lives were lost. When the general public read this story, they thought the Allied forces were reckless and ruthless. But soon, they were persuaded by the military authorities, who justified that the bombing was inevitable due to the atrocities committed by the enemy. When the enemy went into villages and hid amongst the inhabitants, the friendly forces tried to separate the enemy from the residents in various ways, using flyers and broadcasts, but all was in vain. Sometimes the enemy resisted to the last by taking the residents as hostages. In the end, the village received indiscriminate retaliation.[21]

In connection with the issue of "war is hell," the massacre of civilians was a hot topic during and after the Vietnam War. While some activist groups condemned the killing actions of the soldiers as massacres, the soldiers defended the actions as inevitable ones carried out during the operation.[22] Captain Park Jong-sik (Class of 1959) probably had a more profound understanding and experience of the massacre of civilians than anyone else. In his Vietnam memoir *The Forgotten War and Dollar: A Memoir of an Infantry Company Commander*, Captain Park, the 11th Company commander of the 28th Regiment, recounts what he underwent in the war. He wrote the book under the pseudonym of Captain Park Soo-gyu and treated such scenes as having happened in dreams.

Captain Park Soo-gyu, the commander of the 11th Company, who was resting at the Tan My base after completing Operation Ojakgyo Bridge Phase 2 in April 1967, fell asleep one afternoon. In the book, he said, "I had been sleeping for a while, but the Viet Cong People's Committee Commissar, captured in the valley of Mt. Ong Sang a few days ago, appeared in a dream and attacked me. He said, 'The Korean soldiers tied me tightly to a large, dead tree, held my daughter before me, and 13 men under the 2nd Platoon leader Lieutenant Kang gang-raped her.' And then the Chairman beat me hard with sticks. I fell out of bed trying to avoid the stick, and it was a dream."[23] Perhaps part of this incident was genuine. In fact, at first glance, this seems to have nothing to do with the operation to seize an enemy target. Although she was the daughter of a high-ranking enemy official, there is no mention that she did any damage to the Korean military. However, it is estimated that Captain Park's unit has been damaged several times by the Viet Cong before. And the unfortunate incident that occurred on the inhuman stage of the war was "treated as dreams not to hurt the public sentiment."[24] Perhaps it was an act of retaliation against the Viet Cong cadre. And yet, could it be justified in the name of war?

## "Kill everything alive, destroy everything standing, and set it on fire."[25]

The Company commander Captain Park Soo-gyu did not say these words directly. But he implicitly acquiesced to their silent request by saying, "OK." He knew how much his subordinates were angered by what the enemy had done to their comrades in the past several months. And it's common in memoirs for war veterans to represent a horrendous thing as a dream or a third-party detour technique. The following is an example of what occurred to Captain Park's Company in Operation Plum Flower on March 8, 1967.

Captain Park's 11th Company was ordered to penetrate 20 kilometers into the Lac My village before dawn and secure the LZ for the helicopters and artillery positions. All company members crossed the 007 Bridge on the Darang River in two rows and approached the village. For some time before this operation, the company had been attacked several times by the enemy and suffered some casualties. At the start of the operation, the company lost Sergeant Lee Sang-deuk, a deputy squad leader of the 3rd Platoon. He was a competent soldier loved by all his fellow soldiers. Besides Lee, three others were also injured. Sergeant Lee's death was tragic. While moving forward, Lee and his squad received the VC's grenade attack. Sergeant Lee embraced the grenade with his body when the soldiers had no time to avoid it. If Sergeant Lee had not done so, all nine squad members would have been killed.

Captain Park couldn't believe that Sergeant Lee had died in such a way. Overwhelmed by the thought of revenge, Park temporarily put aside the sadness of his loss. He ordered the platoon leaders to assault the enemy position, but they didn't move at all. The reason was that, as the villagers were mixed with the Viet Cong, they could not distinguish one from the other. They said they wouldn't attack because they knew there was a firm instruction from Commanding General Chae Myung-shin, "Protect one resident even if you miss a hundred enemies." Despite all this, Park decided to choose the latter option—to go ahead--with the determination to willingly risk punishment for what he did. The moment he made such a decision, he felt his life collapse as if it were coming to an end.

He shouted at the soldiers' request. "Ok!"

The soldiers made an all-out assault at his command, killed everything alive, destroyed everything standing, and set it on fire. The soldiers erroneously interpreted the commander's "OK" to be that those people in Lac My village could be VC or VC sympathizers. Instead of guerrilla fighters, they found about 100 unarmed villagers, mostly women, children, and older men. They acted with extraordinary brutality and destroyed the village. After some time, Captain Park was shocked by the cruelties that his men had committed. "How could this be possible?" In one basket of a *Ganh Gao*[26] was a thin baby with its left arm cut off. The baby's mother put the Ganh Gao on the ground and sat down, staring at the soldiers with panic-stricken and vacant eyes, and her face was also bleeding from a shot. Captain Park first called in the medic, instructed him to provide first aid to the baby and mother, and told all the men to stop shooting. "If anyone shoots or fires recklessly, I will shoot him with my own hands," he barked.

This was a war crime. He couldn't believe that his men did this. It was not an act committed by human beings. He was at a loss for what to do. He mumbled in his mind the hymn "Amazing Grace."

Amazing grace! How sweet the sound!
Thou saved a wretch like me!
I once was lost, and now am found;
Was blind, but now I see.

They retreated from the village to their field base and stayed up all night with open eyes. Captain Park fell asleep for a while and had an ominous dream. In his dream, the 3rd Platoon leader entered Lac My Village and captured a young baby's mother alive. And the Platoon leader, Platoon sergeant, and Squad leaders committed rape. But only Sergeant Chung, the 3rd Squad leader, refused. Sergeant Chung reported this fact directly to Captain Park. He said they first killed her mother and mother-in-law with three grenades. Then they raped her by rank and killed her and the baby. Captain Park was surprised and pinched his forearm to see whether this was a dream. He called the 3rd

Platoon leader by radio to check for anything.

The 3rd Platoon leader said that he discovered a cave in Lac My village and that the 2nd Squad rifleman Private Park Young-soo boldly pursued three Viet Cong who were running away and killed them all. Captain told him the story of the ominous dream, and the platoon leader said, "Don't worry, sir. How could I do such a terrible thing as a human being?" Operation Plum Blossom began on March 8 and lasted for five days. The company returned to its Tan My base. Captain Park wondered if such acts of rape and the massacring of civilians that they committed were of help to achieve the purpose of war. It was the retaliation for the death of a comrade, but could it be justified as a legitimate act? What kind of explanation could the apologetics provide for that act? Could General Sherman's dictum "War is hell" condone all immoral and inhuman acts committed in the name of war?

What is the limit at which humans become vicious? Perhaps Caputo's words would be comforting. "It was the dawn of creation in the Indochina bush, an ethical as well as a geographical wilderness. Out there, lacking restraints, sanctioned to kill, confronted by a hostile country and a relentless enemy, we sank into a barbaric state. Most American soldiers in Vietnam could not be divided into good men and bad. Each possessed roughly equal measures of both qualities. I saw men who behaved compassionately the Vietnamese one day and then burned down a village the next."[27] Indeed, the war in Vietnam was one fought in the ethical wilderness.

Chapter 4

# Appearances and Realities in the Vietnam War

War is a disorder by nature. Combat does not always progress as planned or expected. A moment's conversion of judgment is sometimes needed. For this reason, flexibility is considered one of the essential traits of leadership. War is the wilderness where all moral values are distorted and subverted. Sometimes, injustice becomes justice, untruth truth, and cowardice courage. It is where the existing value system does not have a binding force anymore. Hence the phrase, "The first casualty of war is truth."[1] A coward in combat is later transformed into a hero who has shown courage. In fact and fiction, such stories frequently appear of veterans who do not condone the cowardly acts they did in the war and who, feeling a sense of guilt, commit suicide. For example, in American fiction, the protagonist Sgt. Marty Randi in Francis Polini's Korean war novel *Night* (1961)[2] and Corporal Norman Bowker in Tim O'Brien's Vietnam novel *The Things They Carried* (1990)[3] belong to that category. Both veterans of the respective wars are highly praised as the soldiers who showed courage for their comrades in battles, but they are the ones who betrayed or abandoned their comrades in combat. They suffer from survivor's guilt in their hometowns. And in the end, they do not overcome their guilt complex and end their lives by committing suicide.

The just cause of war conducted in the name of freedom and peace sometimes causes pain and damage to the people for whom they were told to fight. Americans and Koreans in Vietnam were supposed to fight under the strict rules of engagement, but it was challenging for them to observe the rules because of the frustrating war situation. Captain Kim Jin-gyu (Class of 1958), an artillery battery commander of the Tiger Division's Cavalry Regiment, writes in the book *The Crusade of Hwarang Warriors* (1967): "In retrospect, all of my 444 days in Vietnam were not always proud and honorable as they should have been."[4] The implication is that the Vietnam War was not necessarily a war worth fighting and that all he did was not always justifiable.

Many KMA graduates also witnessed what should not have been done in the military. Justice and truth are the two most cardinal moral virtues of Academy graduates. They recite *The Cadet Creed* day and night for four years; the third line reads, "We will choose the harder path of justice over the easier path of injustice." Not only that, but *The Cadet Code of Ethics* also dictates what we should do. The first line of the *Code* is "Cadets always tell the truth." Moreover, the fundamental ethical principle of *The Honor Code* is "Cadets do not tell a lie." They should always tell the truth. Sometimes they are tempted to condone the injustice in the name of defending the nation. Those mandates above dictate that even if the majority is against something, you must have the courage to stand up for a cause that you believe is correct.

Let's take a look at some realities in Vietnam during the war era. At that time, the South Vietnamese government did not win the confidence of its people because of the frequent coups and the ensuing political unrest; rather, Ho Chi Minh of North Vietnam was revered as the leader of the whole Vietnamese people. According to the memoir of General Chae Myung-shin, most staff offices of the ROKFV in Saigon hired one or two female students of the University of Saigon as typists or clerks. One day, the General asked a student in his office, "Who do you most admire in Vietnam?" and got the answer, "Ho Chi Minh," without hesitation.

When he was appointed as the commander of the Tiger Division, General Chae had already concluded that he could not win the war in Vietnam and that, if so, he would jump into the war as a loser. A counterinsurgency war will likely be fought against the civilian population because the guerrillas reside there. For this reason, moral philosopher, Michael Walzer, argued that "the war [in Vietnam] cannot be won, and it should not be won."[5] Michael Maclear expressed this

At a battle in the jungle near Hue in April 1968, an Alpha Company sergeant of the U.S. Army 101 Airborne Division guides the dust-off helicopter hovering in the air to evacuate some wounded soldiers while other fellow soldiers move them to a pickup place. (Source: U.S. National Archives)

American dilemma as the "Albatros around the nation's neck."[6] It is known that then-President Park Chung-hee gave a secret order for the General to minimize the sacrifice of the Korean soldiers and focus more on pacifying projects than engaging in reckless actions. Perhaps, for this reason, the Korean troops tended to request inordinate air and artillery support from the American military, even for the small operations about which the American side used to complain. Further, in the early stage of its deployment to Vietnam, the Korean military mainly focused on constructing its tactical bases and stabilizing regional security. Hence, they had to hear harsh criticism from their American counterparts.

To the American soldiers in Vietnam, the war was a nightmare. The total strength of American forces in Vietnam amounted to as many as 540,000 in 1968. And the total death toll was over 50,000 by that time. Some of the main tactics and rules that the American military adopted in Vietnam included "Search and Destroy," "Body Count," "Rotation Policy," "Winning Hearts and Minds," "Rules of Engagement," "Free

Fire Zone," and so on.[7] Indeed, all these tactics, policies, and rules were the best ones that the American military could come up with in the Vietnam situation. However, these sometimes became the main complaint factors to American soldiers and South Vietnamese people. As we all know, the Vietnam War was a guerilla war with no clear-cut battlefronts against unseen enemies; thus, the first objective of American military tactics was to separate the guerillas from the civilian population.

Hence came the concept of "search and destroy." Although it was the most reasonable tactic in such guerrilla war, the individual soldiers saw it as "dangling the bait," the implication of which was "expose yourself and draw fire." In other words, "once you get shot by the enemy, then you fire back." As Captain Park Dong-won (Class of 1958), the 5th Company commander of the First Regiment, said earlier, it was tough to determine when and where they would fire and fight. Because of this, in ordinary cases, the battle started when the friendly took damage in the first place.[8] There was no territory to seize in the Vietnam War. Almost the whole country was the battleground. Even if you took and stabilized any village or area, you would have to conduct another operation to retake it because it would fall again into the hands of the VC. You had to repeat numerous battles in the same places.

Likewise, the program of winning the hearts and minds of the civilian population, though it started with the best intention, paradoxically produced a population dispossessed from their home villages. And the Free Fire Zone declared around villages under guerrilla control frequently yielded innocent civilian casualties. One of the unique features of the Vietnam War was the body count. The concept of winning in previous wars was primarily based on territorial occupation, but in Vietnam, it was based on decimating the number of enemies. Thus, a unit's effectiveness tended to be measured by the body count it produced. So, a unit commander made every effort to increase the body count, that is, the enemy corpses, the number of which would be the measure of victory.

Philip Caputo, a Vietnam Marine veteran, criticizes the U.S. Army's strategy of body count in this way: "General Westmoreland's strategy of attrition had an important effect on our behavior. Our mission was not to win terrain or seize positions but simply to kill: to kill Communists and to kill as many of them as possible. Stack 'em like cordwood. Victory was a high body-count, defeat a low kill-ratio, war a matter of arithmetic. The pressure on unit commanders to produce enemy corpses was intense, and they in turn communicated it to their troops."[9]

These concepts erroneously contributed to the increase of innocent civilian deaths, frequently overlooked under the "Mere Gook Rule," whose implication was "If he's dead and Vietnamese, he's VC." According to some unidentified sources, it was said that a public cemetery was dug up by artillery fire, and the parts of dead bodies were included in the body counts. Further, a unit's achievement was evaluated by the number of captured weapons, so some soldiers even bought the enemy's rifles from the black markets. If you seized a gun, it was equivalent to one enemy killed (three killed in the Korean army), but a crew-served weapon was equivalent to several enemies. This practice caused a heated rivalry between units or individuals and sometimes hampered the solidarity in a unit. From the military perspective, the body count strategy was the one that produced the result in which you won the most battles but lost the war.

While the Korean military recognized the American tactic of body counts and tried to follow suit, it questioned whether the body count strategy would help accomplish the war's aims. General Chae Myung-shin was afraid that the high kill rate would become an excuse for the massacre of civilians, so he emphasized that each unit would stick to this strategy but not boast about high kills. There was no village without Viet Cong. If someone said, "All villagers, including children, died, but we had only one casualty," then this was an act to ruin the Korean army. It gave an excuse for the propaganda of the Viet Cong. The U.S. military might also question, "Isn't the Korean military killing many civilians?" The people of South Vietnam might think, "The foreign guys are here to kill all of our people, aren't they?" Therefore, "under the circumstances of Vietnam, the low kill rate is something to be rather proud of."[10] There was a nuanced difference between the U.S. military and the Korean military regarding the issue of body count.

Another policy the American military adopted was the rotation policy. Until WWII, there was no such a system in the U.S. military.

In WWII, an American soldier had to stay in the army throughout his tour of duty, 36 to 40 months, without any hope of returning home. However, the rotation policy was first introduced to the military during the Korean War. In Korea, the U.S. Army adopted the point system, in which enlisted men needed to earn forty-three points to be eligible for rotating back to the States, while officers needed thirty-six points. The soldiers on the front lines got four points a month, the rear lines three points, and further back two points. So, if you served only a unit on the front lines, you could go back to the States in nine months.

During the Vietnam War, the U.S. Army adopted a period system, rotating soldiers on one-year tours. Officers also spent a year in the country but with six months in a troop command. Upon arrival in Vietnam, American soldiers knew precisely when they would return to the States, in an acronym, DEROS. The term FNG needs to be mentioned. Usually, but not always, the term refers to recruits fresh from the U.S. who joined the units in Vietnam. Every unit had some FNGs. These recruits had a higher attrition rate than experienced soldiers, and the small units of veteran jungle warfare troops saw them as a liability. The rotation policy ensured a continuing supply of low-quality, inexperienced soldiers at the most significant stress points in combat units.

The Korean troops, both enlisted men and officers, adopted this policy. When they stepped on the ground of Vietnam, the soldiers knew when they would return home. From that moment, the war to them became a fight with time in which survival was their highest objective. The rotation policy was a defense mechanism to overcome the fear of death. By giving the soldiers hope that they could go back to their homes after a certain period, the policy worked positively as a factor that motivated them to do their best during the period of their tours. Paradoxically, this policy could have been more productive regarding combat effectiveness. When their time to return home approached, some soldiers beat their brains to leave the war alive and tended to be unfaithful to their duties. Thus, this rotation policy virtually foreclosed the possibility of establishing fighting units with a sense of identity, morale, and strong cohesiveness.

According to retired Major General Han Chul-yong (Class of

1970), then a platoon leader in Vietnam, such a negative aspect of rotation policy was not so grave a matter in the case of the Korean soldiers. But it was customary that the soldiers who were a few weeks away from returning home were exempted from participating in combat or some other dangerous activities. Lieutenant Yang Chul-seok (Class of 1970) heard a cautionary remark, "100 day-jinks," from some veterans before leaving for Vietnam. In other words, "Be careful 100 days after arriving in Vietnam and 100 days before leaving Vietnam." Indeed, the concept of rotating in the war led to an institutional climate where a soldier might shrink from positive action at the last stage of his tour. This policy was dysfunctional and was not the best system to develop soldiers' combat motivation.

In Vietnam, American and Korean forces were fighting under strict rules of engagement. In his book *A Rumor of War*, Philip Caputo, a Vietnam veteran, says "it was morally right to shoot an unarmed Vietnamese who was running, but wrong to shoot one who was standing or walking."[11] Several questions arise here: Why should the act of running identify someone as a Communist? What if we shot a Vietnamese who turned out to have a legitimate reason for running? He continues, "It was wrong to shoot an enemy prisoner at close range, but right for a sniper at long range to kill an enemy soldier who was no more able than a prisoner to defend himself; it was wrong for infantrymen to destroy a village with white-phosphorous grenades, but right for a fighter pilot to drop napalm on it. Ethics seemed to be a matter of distance and technology."[12] The contradictory dictates of this war—to abide by the rules of engagement and yet to produce high body counts—aroused frustration and a sense of absurdity in the minds of American soldiers. Despite the war being executed in the name of just cause, the war in Vietnam was subjected to many debates and criticisms regarding its means of conduct.

Another problem in Vietnam was the question of false combat reports and unfair distribution of medal awards. In his memoir *Vietnam War and Me*, General Chae Myung-Shin mentions the feat the Korean forces achieved when they first arrived in Vietnam. When the 3rd Battalion of the Tiger's First Regiment bivouacked in a village north of Quy Nhon on October 28, 1965, some VC infiltrated the front of the

3rd Platoon of 3rd Company, and a soldier on guard caught sight of them and fired, killing two VC. That was the first combat achievement made by the Korean troops since its arrival in Vietnam. General Chae wanted to take this opportunity to boost the morale of all the Korean troops by enhancing this man's achievement and making him a model for all soldiers. So, in front of all the troops assembled at the Regiment drill ground, General Chae decorated the soldier with the Hwarang medal. Since the awarding ceremony, Chae writes in his memoir, "Soldiers' will to kill the enemy rose high throughout the Division, and every time the enemy attacked, they repelled or killed the enemy."

And the General continues, "The combat medal is the highest honor to a soldier. Even if he's an enlisted man, the medal awardee becomes a man of national merit. Even after being discharged from the army, he is supported by the government as a patriotic veteran."[13] What is the value of a medal? A soldier receives honor by giving his life, this symbol of honor is the medal. It is an indication of gratitude and recognition of the country for the sacrifice that he made. The medal is a humble piece of iron, but, in a sense, soldiers throw away their only precious life to get a lump of it. Yet this was a means to heighten their combat motivation. However, the excessive desire to get a medal caused the soldiers to fall into the enemy's trap while in action and significantly damaged their whole unit, let alone themselves. In evaluating the individual soldier's performance, some ruptures emerged between superiors and subordinates or fellow soldiers. Discord among the unit members sometimes harmed its cohesion and unity.

Shin Jung (Class of 1964), a platoon leader, was forced to withdraw, with only five men injured in a surprise attack by the enemy. His battalion commander severely rebuked Shin at the company base camp. "The disgrace of our battalion is because of you." He had no choice but to quietly bow his head and listen to the battalion commander's harsh criticism. Sometime before, another platoon leader of his battalion was awarded a medal despite the three men dead and eight injured he had taken in combat. The reason was that he had some significant results despite his loss. Shin wondered which one was more important, his men's life or enemy's neck.[14]

False reporting was a frequent occurrence. Captain Kim Jae-gil, the

7th Company commander of the 2nd Battalion of the 1st Regiment, testified that his company had no damage or combat achievement during Operation Fierce Tiger 17 in 1971. Strangely enough, the detailed combat report was that his company killed four enemies and captured two AK rifles. It was a false report. Such a report may have been good for him and his unit then, but it would not be suitable for the later generation of officers. He was afraid of what those officers would think of them in the future if they came to know the truth later.

Things like snatching combat achievements from others occurred in Vietnam, and even inflating an individual or unit's combat performance was usual practice. The pressure of higher command to emphasize better results made its subordinate units feel on edge. It was almost impossible to obtain an expected outcome from the search and destroy missions because hidden enemies were not easily detected and disappeared even if caught in sight. For this reason, the Korean soldiers, as well as the Americans, could not achieve what they wanted. This led them to exaggerate, capturing the enemy's weapons. As mentioned earlier, if you seized a rifle, it was equivalent to three enemy KIAs and a crew-served weapon like a machine gun or mortar to the number of its operators killed. It was to hit the jackpot if you found an enemy's weapons storage facility or arsenal. The phrase "Estimated Kills" went around in the circles of Korean forces in Vietnam.

For example, let's suppose that when a company reached an enemy's position after a fierce firefight with the support of massive air strikes and artillery fires, it found almost no enemy dead bodies. On the contrary, the company suffered more casualties than expected. In this case, the commander is tempted to make a false report, for he made no remarkable result despite the enormous expending of firepower and friendly casualties. Hence the notion of "Estimated Kills." The point was that such a report of "Estimated Kills" was acquiesced by higher authorities, for the inflated number of "Estimated Kills" had the effect of offsetting the wasted ammunition and loss of friendly lives. A similar way of calculating the enemy dead was practiced in the U.S. military. Once, an American marine platoon seized a hill with the support of massive firepower but found only four dead VC. There were a lot of body pieces and blood trails around the dead. So, the Americans reported eight estimated VC KIAs to the higher authorities. Asked how that figure had arrived, they said they counted up the arms and legs and divided them by four.

The testimonies by many Korean commanders frequently question the credibility of the official combat reports in the Vietnam War. Such a phenomenon was widespread in the Korean military at the time. Why this phenomenon? Colonel Lee Sung-soo, the 28th Regiment Commander, even quoted a report in the South Vietnamese newspapers saying, "Korean soldiers came to Vietnam to obtain medals."[15] Excessive competition for such medals resulted in false battle rewards. Lieutenant Colonel Kim, the 3rd Battalion commander of the 30th Regiment, minced no words, "some combat reports written at the wartime had been framed in advance."[16] According to Lieutenant Kim Hyung-seok (Class of 1965), a platoon leader of the 30th Regiment, the neighboring unit of his company was the 10th Company of the 29th Regiment. And one day, the 10th Company discovered the arsenal of Khanh Hoa Province's Viet Cong units and captured three 60-mm mortars and 61 firearms.

As mentioned above, capturing a weapon was considered equal to three enemy kills. The reasons were that the enemy might be civilians, some enemies were unarmed, and there was no way to confirm the enemy dead in the jungle battle except by relying on honest reports. The 10th Company was reported to have made tremendous achievements by discovering many weapons in the enemy's arsenal. This calculation method created nonsense that forced soldiers to focus more on finding hidden weapons without contact with the enemy than bravely risking their lives to kill enemies. This also led to side effects of inducing inordinate rivalry for combat achievements between units and falsely exaggerating the combat performance. Ultimately, this ridiculous phenomenon resulted in devaluation or distrust of official combat reports.[17]

Forty-five years later, Kim Hyung-seok (Class of 1965) was surprised to find that his combat story in 1967 was distorted in *The History of the Korean Forces' Deployment to the Vietnam War* under the title of "Lunar New Year at Nui Hon Ba." He remembers the situation so vividly. And further, the content was so different. The fighting of his platoon in the cave was exaggerated, and the book recorded the captured firearms as five. SFC Ko, the 2nd squad leader, embellished his seizure of rifles to five from the initial two by adding three more and reported to the company clerk. Kim confessed through his book. "Strictly speaking, perhaps I'm also an accomplice to have consented to it without identifying the dubious part. His company commander Lee (Class of 1957), a symbol of integrity, may take final responsibility for the false report composed by his company clerk." Kim stressed, "Even if it is disgraceful to an individual who committed such an unjust deed, the wrong records should be revealed and corrected for the future of the Korean military."[18]

Such a practice could not be passed over by the moral KMA graduates who have abided by such values as integrity and truth. Less than two years after graduating from the Academy, most junior officers witnessed such absurdities and contradictions as customary in the war and the military. Seeing those wrongdoings committed in Vietnam, many KMA graduates learned valuable lessons on the discrepancies between appearances and realities. They took them as examples they should not follow while pursuing their long military careers.

Commanding General Chae Myung-shin's policy in Vietnam was

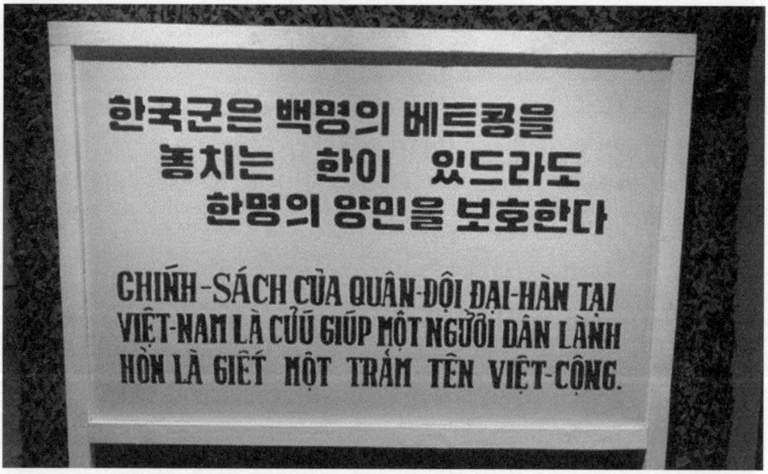


This signpost says, "The Korean soldiers will protect a civilian even if they miss a hundred Viet Cong." General Chae Myung-shin took this slogan as the policy of all Korean forces. This signpost, written in Korean with the Vietnamese translation, was placed at the main gates of all Korean units' bases in Vietnam. (Courtesy: Chae Myung-shin)

to "protect one civilian even if you miss a 100 Viet Cong." According to this instruction, the Korean military conducted active civil operations with all soldiers as psychological warfare agents. He said, "Every soldier is a diplomat. Treat residents with courtesy." Accordingly, the Korean military gave rice to villagers, built bridges and buildings, taught Taekwondo, and provided medical treatment. In Vietnam, "Daihan" (people of the Republic of Korea) was synonymous with kindness. Even if the U.S. military offered the same materials, the Vietnamese refused, but they were grateful if the Korean military offered it. Because of such a policy, the Korean troops amounting to about one-tenth of the U.S. troops could achieve results beyond expectations.

General Chae interviewed Glenn C. Troelstrup, a *U.S. News and World Report* combat correspondent, in July 1967. He responded to a question about the Korean tactics: "Our tactics are not much different from those of Americans. We mainly concentrate on the enemy with superior strength and firepower, enveloping him by surprise movement and wreaking havoc on him. In densely populated areas, we conduct nighttime movement when possible. I emphasize the nighttime operation because the enemy mainly attacks at night, taking advantage of the darkness. We also envelop a given area by doing that. We open an escape route to the enemy. Yet we set up some ambushes in the vicinity of the route. We do not use artillery fire early morning. Instead, we let the villagers know by microphone or leaflets that we aim not to harm them. We inform them of how to take the escape route. At the same time, we prepare food, water, medicine, a tent, and other necessary items for them.

The villagers leave their houses to evade our fire and arrive at the safe places we prepared for them. But the VCs mostly stay in their positions. On the second day, we continue to tell the villagers to leave their houses through the microphone or the leaflets. And at night, we make ambush attacks on the VC who attempt to run away. In actuality, nobody can break through our enveloping net. Anyone who tries to escape in the direction we did not designate will be killed. However desperately the VC may try, they won't find a way out from our net, for our soldiers are well-trained at close combat. Only the time when the enemy approaches within ten yards they fire. The VC within the enveloping net has only two ways, to stay in hiding or to surrender. Otherwise, we force our way through the village and kill the resisting enemy. We will save an innocent civilian among 100 VC instead of killing them all, even though we miss 99. This is our tactic, and the Vietnamese know this well."[19]

The problem was that such a policy led to an increase in Korean casualties. It meant the Korean army risked its soldiers' lives to save civilians. Despite the good intention of the instruction, it caused some complaints from the leaders and commanders. Do they have to put the civilians' lives ahead of their men's? Sometimes people run into pitfalls that lead to confusion.

Chapter 5

# War and Humanism: To Humanize an Inhuman War

"War cannot be humanized. It can only be abolished."
Albert Einstein

War is inhuman. However, men try to humanize the inhuman war. They created the rules of engagement or the Geneva Conventions and urged the fighting soldiers to observe these stipulations. These humanitarian laws were created to limit the suffering caused by war by protecting and assisting its victims as far as possible. War is essentially fought for survival in the ethical wilderness, so it is challenging to abide by rules or laws. Philip Caputo, a Vietnam war veteran who wrote of his experience as a platoon leader of the U.S. Marine Corps, said that the Vietnam War was "the war in which each soldier fought for his own life and the lives of the men beside him, not caring who he killed in that personal cause or how many or in what manner and feeling only contempt for those who sought to impose on his savage struggle the mincing distinctions of civilized warfare—that code of battlefield ethics that attempted to humanize an essentially inhuman war."[1]

This is a critique of the discrepancy between the two premises of conducting war—war's ends and war's means. It is an expression of frustrated soldiers in combat who have to face squarely the realities in which unconditional victory contradicts the just means of fighting—winning and fighting well. War ethics generally emphasizes that a soldier has a just way of fighting, even if he is to kill the enemy in combat. However, it is difficult for combatants to follow such ethics of chivalry. In the presence of this logic, will the rules of engagement ever be observed? Will such a theoretical argument apply to battlefields where only situational ethics makes sense? Thus, most war critics see war as the primary source of disillusionment, therefore focusing on criticizing the futility of war.

And yet, in fact and fiction, many other critics emphasize the contradictory nature of war. Archibald MacLeish, an American poet of WWI, criticizes the partiality of the war writers only focusing on the brutalities of war and insists on a dramatization of the war experience as a totality: there were in the war "speeches, brass-bands, bistros, boredom, terror, anguish, heroism, endurance, humor and death. It [the war] matched great cruelty with great courage. It had its fine sights and its unspeakable sights. It was a human war."[2] These ambivalent natures of war are also well summarized in an American novel of the Vietnam War, *The Things They Carried,* in which the author Tim O'Brien seems to concur with the idea of MacLeish: "War is hell, but that's not the half of it, because war is also mystery and terror and adventure and courage and discovery and holiness and pity and despair and longing and love. War is nasty; War is fun. War is thrilling; War is drudgery. War makes you a man; War makes you dead."[3]

War is the territory of the youth where they share the fear of death. What the soldiers only rely upon on the battlefields are the unit and their comrades fighting beside them. It is this unique nature of war that you can find humanity in such an extraordinary situation that is rarely seen during normal times. War is a horror but a mysterious place where such moral virtues as courage and love are found. War is painful but full of thrills sometimes. War kills a man but makes him who he is. War becomes a place where friendship, comradeship, and solidarity among men that they hardly expected to find in ordinary times are revealed. War is a place where you are in perpetual tension and ordeal, but a place where you experience deep attachment to life. When a battle is over, a soldier remembers his comrades happily, who listened to his agony, encouraged him and saved his life.

After all, this comradeship is the only thing that remains in one's memory. It is a bond that a word, boredom, divorce, or anything other than death cannot break. Ironically enough, the tragic war is where human emotional drama is made beyond our imagination. To Captain Lee Jae-tae (Class of 1960), the 11th Company commander of the First Regiment, war was a place full of contradictions where sadness, pain, joy, and romance coexist. During his service in 1966, Captain Lee experienced boundless camaraderie and pride as he saw his men roaming the jungle in search of their enemies in the heat of the scorching sun. He used to be deeply moved by those soldiers who made assaults on the altar of their possible deaths. Their blunt expressions, like cast iron, were reliable and even attractive. On the other hand, in a village operation, he felt sad to see a young woman standing on the crumbling railing of a thatched house devastated by the fierce firefight and staring into space without expression. And the innocent eyes of an infant in her arms would give him indescribable agony.[4]

In actuality, war is a continuation of tedious waiting. Soldiers in war situations are constantly under stress and tension. For this reason, there are many ways to release their tensions, like R&R or social gatherings. The soldiers won't withstand strain or boredom if no such diversion exists, even in a war zone. According to some 1970 class graduates, many welcoming or farewell parties were held for those who came to and from Vietnam. These parties were usually thrown by their elders. At these parties, various liquors were served, including the Scotch whiskies like Johnnie Walker, Chivas Regal, and French Cognac, which they had never heard of before in Korea. The elders and juniors, who are wide apart in years, drank these expensive whiskies like they used to drink Makgeolli, one of the oldest and cheapest traditional rice wines most Koreans enjoy. If it had not been for the four-year experience that all the graduates at the Academy had shared, such a firm sense of camaraderie and solidarity among them would not have been possible.

Many war books deal with aspects of humanity that emerge on the battlefield. Such an idea is based on the shared consciousness that friendly and hostile soldiers are the contemporary youth thrown upon the killing field and are told to fight against their will. It is said that such a thought sometimes works in the minds of the fighting soldiers. First Lieutenant Shin Sang-gil (Class of 1970), a platoon leader of the 1st

Regiment, was temporarily transferred to the adjacent 26th Regiment for its operation and was engaged with the VC in the jungles of Nui Phu Cat. One of his men stirred up a hornet's nest, and the insects stung many of his men. While they were at a loss for what to do, they detected a VC and pursued him, but he ran away.

Shin was blamed for that by his company commander, but secretly he prayed for the man's safe escape. Just like Shin, he was fighting for his country and its cause. Feeling fear of death moment by moment under extreme conditions, he also desperately fought against the Allied forces. The battlefield is where an indescribable human feeling of compassion that ordinary people in peacetime could never understand sometimes arises. Later, Shin captured a Vietcong alive in the vicinity of his company tactical base and regained the honor he had lost a while before. Yet the image of the escaped VC soldier who may have been killed in another battle remained in his mind for quite a long time.

We can find similar stories in American novels and memoirs of the Korean War. James Hickey's Korean war novel *Chrysanthemum in the Snow* (1990) deals with the Battle of Heartbreak Ridge on the central front of South Korea, where a platoon leader Lieutenant Robertson is a protagonist. Here the lieutenant finds an older man in a hut treating a wounded North Korean soldier. He gives the older man food, firewood, and even the coat he is wearing so that he can better care for the enemy soldier. In another scene, Robertson is caught in a trap to be killed by a Chinese soldier. Surprisingly enough, the enemy soldier waves his hand, gesturing to run away quickly to a safe place. It is a touching story. Sometimes the battlefield becomes the common space of death shared by friends and foes. Thus, both parties feel pity for each other as poor companions in the same situation.

The war story of First Lieutenant Kim Tae-woong (Class of 1970) is indeed an emotional human story deserving to be a theme of war literature. The 2nd Platoon leader of the 11th Company of the Tiger Division's First Regiment, Kim was fighting against the North Vietnamese Army regulars in the mountainous area of Kim Son Valley in Binh Dinh Province in July 1971. Though the division operations ended, Kim and his 24-member platoon were ordered to stay there for one more week of ambush. They waited almost a week in the jungle, but the enemies didn't appear. One night, however, when their patience was about to be exhausted, they finally found some enemies approaching stealthily to their position. Immediately the claymores exploded, followed by fierce firefights. Kim's platoon killed five NVA soldiers in the engagement, and capturing five AK-47 rifles and ten hand grenades.

The following day, one of his men brought a note to him, which was found inside a dead enemy's pocket. The letter appeared to be from his sweetheart back at home, and it was written in pretty handwriting in Vietnamese on a small-sized paper about seven centimeters in width and length. Translated by his civil affairs soldier, it read as follows:

> "My darling, Lhong! How much scared and painful you are! When this war ends, we'll be able to live together, never separated, won't we? Until then, you must stay alive, please! I love you!"

Kim was shocked as if he got a heavy blow by something to the head. Though leveling a gun at each other, he was also a human who had left a beloved woman behind. He was a young man like him, fighting for his country's cause on this battlefront, far from his hometown. The woman's desperate hope to see him back was shattered because of a Korean officer. "Alas! What the hell I have done to her!" He couldn't suppress his feelings of regret. He thought of the NVA soldier dead in the jungle on that moonlit night. "Can't we avoid the war in which humans kill or are killed this way? What's war? What's Vietnam to us? Why we're fighting here?" He was immersed in these questions.

The previous month, he had lost one of his men to the enemy's booby trap. He'd had a firm determination to revenge his loss. However, he was lost for words in front of the gruesome dead bodies. At that moment, the thought of his hate and revenge vanished. What kind of enmity makes us kill each other? He felt his heart aching too much for the woman who would have to live on, missing her lover who would never return to her just because of him. Even after returning home from Vietnam, he was haunted by the thought of the dead enemy soldier and the letter of his lover, and it remained a trauma in his memory for quite a long time.

The awareness that the vicious opponent to be killed on the battlefield is also a fellow human frequently appears in combatants' experiences. These men implicitly agree with the humanists that war is not a machine war but a human war. First Lieutenant Moon Dhu-shiek (Class of 1971) was committed to the battle of An Khe Pass in April 1972, the fiercest one toward the end of the Vietnam War. He was a platoon leader of the 11th Company of the Cavalry Regiment. One of his men brought Moon a letter kept by the dead NVA platoon leader killed in the battle. The letter, translated by the Civil Affairs man, was the one he wrote to his mother in the North. Lieutenant Moon was deeply moved by the letter and recorded it in his diary. Although he was an enemy, he was also a human who missed his loving mother. The letter read as follows:

Dear Mom,

Do you remember that bright, sunny morning? For the first time since I was a little boy, I walked out the door of our bamboo house. When I left home, I hid my sadness. Our troops came down south via Laos. And I continued to walk for days with my comrades struggling with nosebleeds. The march was arduous, and the soles of my feet were all swollen and cracked. I persevered and kept walking. While coming down south, I kept fighting the enemy.

The only memory that greets me when darkness falls at twilight is the quiet silence of the front yard in Tram Shon village. When I think of my hometown, my eyes get wet. I miss the blue mist from the stream, cute pumpkin vines full of green leaves, and butterflies flying in the wind. I miss the beauty of my hometown, where thatch-roofed houses are peacefully gathered. When I return to the village, I want to meet many people and neighbors who will warmly welcome me.

But now, everything is a dream. Coming to the South, the faces of the people are unfamiliar, and the customs are slightly different. Still, this is our country. After the sun sets, the water buffalo cry and slowly return to their pen. When the shepherds are herding their cattle and sing a sad song on their bamboo flutes, it reminds me of the young shepherds of my hometown. Ah! So what is hope for me? In my fear and disappointment, I quietly contemplate what we should do. Everything around the southern land is new and strange. People go happily in the market, and the paddy fields are as vast and green as the sea. The sound of bells and the monks preaching the Buddhist scriptures at a nearby temple. At school, children sing with great joy. The yellow flowers of the cabbage field are in full bloom, and butterflies are leisurely playing.

From end to end in the village, a hopeful dynamism springs up and becomes lively. Why should we burn and destroy the South in the name of liberation? Why? Why? Why? Do I have to pull the trigger, making such innocent children orphans and happy mothers widows? When I buried a mine that would kill those like my mother, relatives, and people of the same blood, my hands used to tremble like a bamboo swaying in the wind.

Dear Mom, shall I be able to see you again? Every night in bed, I suffer from countless nightmares. Tears are flowing like blood.[5]

The NVA officer's letter to his mother in the North represents the minds of all soldiers who have gone to war. They go to fight on the state's commands, but almost all young men, either enemies or allies, are skeptical of the causes of war for which they are told to fight. When you read such a story, war seems to be far away. Once again, we ask, "Why war?"

Captain Lee Jae-tae (Class of 1960) obtained a heartbroken letter from a VC husband to his wife. Toward 10 p.m. on 23 February 1966, an ambush patrol team of his company returned with three Viet Cong suspects, two women and one man. When one of his company men frisked them, he found a letter on a young woman's body. It was sent by her VC husband fighting in the jungle. Translated by the South Vietnamese 2nd Lieutenant Tuoy, who was with Lee's company as a liaison officer, the letter read as follows:

My Darling!

The wind is strong, and it rains, so I feel sad. Good luck with your health and family for the Lunar New Year. I'm fighting against our enemies, the armies of South Vietnam, the U.S., and Korea, to defend our country. There's nothing to eat or wear right now. Please send rice, farmers' clothes with long and short sleeves, and blankets to cover when sleeping. All I have is a chick, so please send them as soon as possible.

Darling! I hope you stay healthy. We will meet again when the revolution is completed, calling 'hurrah.' Please be patient until that day. You can send them to the address written on the envelope.[6]

The woman with this letter was the 26-year-old wife of the sender, a guerrilla; the aged woman was said to be his mother, and the 30-year-old youth, their relative. After reading this article, Captain Lee realized that parting was everywhere, and his heart was broken. What in the world is the ideology that makes them fight each other? Why should there be a war where loved ones break up? Despite suffering from hunger, rain, and severe cold, Lee feels a deep admiration and unspeakable compassion for the VC husband's willingness to fight the world's strongest troops. Lee gave the captives food and a bed, trembling in surprise. All lovers in the world have their sadness. The great sorrowfulness will be the separation, as shown in the letter. It broke his heart to think of her husband, who firmly expressed his affection for his country, revolution, and love for his family, even under miserable conditions.

The same is true of a story by Lieutenant Kim Hyung-seok (Class of 1965). It gives us a pause of silence in thinking about human love. One day in 1967, he found a pocket notebook in the belongings of an enemy officer who had run away during Kim's cave search operation. Kim carefully examined whether there was any piece of essential military information. It was the daily diary of the VC surgeon on the battlefield. A beautiful woman and a cute baby's face were drawn on its pages. The font shape drawn with a colored ballpoint pen was lovely. It was like a collection of an aesthetic man who loved literature and art. Attracted by the work, which seemed to hide a lot of lovely stories, Kim kept it himself and put it deep in his backpack, thinking that he would return it to the enemy officer someday.

Of course, keeping an enemy's possession was a clear violation for the soldiers in combat. Yet Lieutenant Kim kept the notebook for decades, only losing it sometime after discharge from the military. He felt such a great sense of loss. At that time, the Viet Cong officer was on the run to avoid the search of Kim's platoon. Was he still alive because he succeeded in escaping? It hurt Lieutenant Kim's heart. He and Kim were all young men fighting for their respective countries. We all had a family and a reason to live. Kim hoped he would live a happy life with his loved ones. Although it was a battlefield where they had to kill each other, the consideration and love for the people in trouble moved human minds regardless of whether they were friends or foes. That is why we can see a ray of hope even amid tragedy.

On July 15, 1967, Lieutenant Kim Hyung-seok discovered a natural cave area two kilometers east of Su Oi village. One of his men captured an enemy alive while searching the area. The enemy, drawn out of the cave, was as small as a middle school student. He didn't carry a gun and didn't even wear shoes. He looked so pitiful. Lieutenant Kim thought about killing him, but he didn't kill the boy. Kim gave him a cigarette and made him drink water by giving him his canteen. He began to feel relaxed as his hardened expression loosened. The platoon leader reported the capture to the company commander, and soon after, the helicopter arrived and took him away. The platoon members waved at him as he left. The boy responded by waving his left hand to the platoon men. Lieutenant Kim also sincerely prayed to the prisoner that he would feel happy and live well. Lieutenant Kim recalled later: "I thought about why we waved at the enemy soldier we fought against. Perhaps it was an expression of sympathy for the young and sick prisoner or recognition of a fellow human being thrown away in the jungle."[7]

A similar story is also found in Philip Caputo's *A Rumor of War*. Caputo, a platoon leader of the U.S. Marine Corps in Vietnam, seizes a VC stronghold and kills some enemies. He finds a small packet of letters and photographs in their bunker. They were all sent from their girlfriends or wives. He surmises that the notes written in the corners of those things will probably be an expression of love and fidelity. He doesn't like to think of those women, dreaming of returns that can never be, waiting for letters that will never come. "What we had found gave to the enemy the humanity I wished to deny him. It was comforting to realize that the Viet Cong were flesh and blood instead of the mysterious wraiths I had thought them to be." Caputo feels "an abiding sense of remorse." "These were men we had helped to kill, men whose deaths would afflict other people with irrevocable loss."[8] One of his men expresses what may have been a collective emotion, "They're young men. They're just like us, lieutenant. It's always the young men who die."[9]

It was customary for the Korean military to bury the dead enemy. During Operation Eagle 71-1, the 2nd Platoon of Captain Kim Seong-hoi (Class of 1963), the 9th Company commander of the White Horse 29th Regiment, found two VCs carrying food in an ambush in Suoi Cai Valley near Tuy Hoa on February 26, 1971, and exploded claymores, killing them all. After temporarily burying them, Captain Kim placed some security guards and held a simple ancestral rite for the dead with the liquor they carried. Captain Kim offered the following condolences during the ceremony. "You are dead, but your descendants should not live in these mountains but find freedom and live well." Although they were enemies, he treated the dead equally courteously as fellow human beings. This was a common practice of the Korean military since it had landed in Vietnam.[10]

The following is another painful story in which the Korean soldiers tried to save a wounded female VC but had to pay homage to her dead body.

## "A burning incense for a VC woman"

First Lieutenant Shin Sun-gyun (Class of 1964), the 1st Platoon leader of the 9th Company of the Cavalry Regiment, led his platoon on a reconnaissance patrol in a village. His men found four Viet Cong and two women running away. They violently fired at them. Shortly afterward, Staff Sergeant Hyun, a squad leader, heard a woman moaning from the enemy's hiding. There, he found that a woman had been shot and collapsed. She raised her hand and kept saying something. Sgt. Hyun could not understand well, but he thought it would mean, "Save me. Please don't kill me."

The platoon leader called in a medic and instructed him to provide

A Korean soldier helping out a VC woman from a village hideout. (Source: ROK Army Museum)

emergency treatment. A bullet had pierced the woman through her hip from left to right, and she was bleeding from both sides. The medic tried to stop the bleeding, but blood continuously flowed from her bottom. The problem was that Vietnamese women did not usually wear underwear. The medic hesitated to open her skirt. But the platoon leader screamed, "Open her skirt. It's a matter of life at stake. What the hell with her genital revealed?" The medic tore her clothes open. It was the blood coming out of the woman's genital area. But the bleeding didn't stop, and she died.

Lieutenant Shin reported her death to his company commander and said he would bury her. He told his men to dig a hole and bury the woman with all their respect. And he put an incense burner in front of her tomb and lit incense to comfort her soul. The smoke of the scent was spreading through tropical evening sunlight.[11] Although they were enemies who aimed their guns at each other on the battlefield, the incident would be an excellent example that they were human beings who showed respect and love for each other even in a brutal war.

## "A Tiger's humanism planted in the Vietnam warfront"

This is a heartwarming story about two opponents: a platoon leader of the Tiger Division and a Viet Cong rifleman, both of whom were wounded on the same battlefield on the same day. They were hospitalized in the same room for weeks, eventually winning each other's hearts through a human bond. The Korean platoon leader was 27-year-old First Lieutenant Kim Ji-young (Class of 1963), the

Weapons Platoon leader, the 3rd Company, the 1st Regiment, and the Viet Cong was 24-year-old Ngo Van. This story was listed as "A Tiger's humanism planted on the Vietnam war front" in *The Combat Histories of Korean Army in the Vietnam War.*[12]

Ngo Van grew up as an orphan in a village in An Nhon County, Binh Dinh Province. After he joined the Viet Cong regime, he participated in many battles against the Korean troops. During Operation Typhoon No. 7, his unit was attacked by the 3rd Company (Commander Capt. Jang Sei-dong, Class of 1960) of the 1st Regiment in the village of Tinh Hoa, 19 kilometers north of Quy Nhon. He fled along the river with his body injured by a grenade thrown by a Tiger soldier. He got seriously wounded in his stomach. Left alone on the riverbank, he was found during a search by the Tiger soldiers, taken prisoner, and transported to the 6th Evacuation Hospital. After an hour-long operation by the surgeon, he regained consciousness, but his demeanor was still cold.

In the same room, Lieutenant Kim Ji-young, who was injured in the right eye by a machine gunshot by a Viet Cong on the same battlefield on the same day, was lying down. Kim had led his platoon across the bridge in front of Tina Hoa village to support the 2nd Platoon in a fierce battle. But at that moment, an enemy bullet pierced his helmet and passed through his right eye. As a result, his face was full of blood, and he couldn't open his eyes. He lost consciousness, leaving his men with the words, "Do not enter the front. Take a detour to the side." Afterward, Lieutenant Kim was transferred to the 6th Evacuation Hospital and underwent surgery. However, he fell into deep despair when he heard he could not open his right eye forever. His dream of becoming a competent commander and his hope of being an excellent Vietnamese speaker by learning the language in a Vietnamese language school had been shattered.

However, Lieutenant Kim accepted the reality. Then, his heart became lighter. He heard the moans of a Viet Cong prisoner lying next to him. In Lieutenant Kim's heart, there was a feeling of pity rather than hatred. Kim approached the Viet Cong and warmly held his hand. The two eyes of the Viet Cong and one of Lieutenant Kim met. Soft words passed between the two of them. Tears welled up in the Viet Cong's eyes, and the frozen Viet Cong soldier's heart melted because of Lieutenant Kim's warm heart. In the end, Ngo Van confessed everything to the Korean interrogator. And he said that he would enlist in the South Vietnamese army and fight the Viet Cong for freedom if accepted after he recovered. That happens in war.

## The Battle of An Khe Pass and Its Implications

One of the bloodiest battles of the war in Vietnam was conducted from April 11 through 26, 1972, at the An Khe Pass. It was to defeat the NVA forces that blocked Route 19 and reopen it. The Tiger's Cavalry Regiment and some elements of the 26th Regiment were committed to this operation. The An Khe Pass was a group of rocky hills, 400 to 650 meters above sea level, in the Central Highlands, halfway between Quy Nhon and Pleiku, about 55km from each of those critical cities of the country. This 7.5 km-long pass was on Route 19, the only supply route leading to the 2nd Corps of South Vietnamese Army stationed in Pleiku, the largest city of the Highlands, bordered by Cambodia. If this critical terrain fell into the enemy's hands, the South Vietnamese forces in Pleiku would be in a disastrous situation. So, the Korean troops were asked to keep their eyes on this area 24 hours a day and report on all enemy movements and activities sighted.

Then in March 1972, during their Spring Offensive, the 12th Regiment of the NVA's 3rd Division came down south and seized Hill 638 and its surroundings, including some unknown hills and major strong points from which they could command the entire area. The enemy troops were firmly entrenched on these hills and waiting for decisive action. The enemy's total strength was 2,280, armed with a considerable number of 122-mm mortars and 191 crew-served weapons, including machine guns. On the other hand, the Tiger Division was preparing to withdraw from Vietnam, so it could not afford to pay much attention to the enemy situation around the An Khe Pass and its vicinity. Yet while the Cavalry Regiment was heightening security measures for its tactical bases, an enemy sapper squad infiltrated into the frontage of its 1st Company on April 11, 1972, which led to "the Battle of An Khe Pass, the longest and fiercest operation since the arrival of the Korean forces."

The battle, which started as a platoon-scale operation, kept expanding gradually into the scale of a company, a battalion, and a reinforced regiment and continued for sixteen days. Consequently, the battle ended as the victory of Korean forces by taking Hill 638 and some neighboring unknown hills, but it was a "pyrrhic victory." The Korean victory in this battle opened Route 19, through which transporting supplies to the South Vietnamese 2nd Army Corps in the Kon Tum-Pleiku area near the Cambodian border became possible. Still, both sides yielded the most casualties in single combat throughout the Vietnam War, and compared to the enormous efforts committed, the result was not as satisfactory as expected. According to an unofficial casualty report, 615 North Vietnamese were reported killed, against 45 Koreans dead. Ultimately, this battle was recorded as having projected a negative image on the Korean forces as an always-winning machine.

Perhaps the person who saw the whole process of the operation at the closest distance was First Lieutenant Kang Hang-joong (Class of 1970), a platoon leader of the Tiger Division's Reconnaissance Company. During the operation, he provided various supplies to the combatting 9th Company by helicopter and was able to observe what was happening. What shocked him was that many dead bodies were left unattended and rotted in the tropical heat of the jungles. When they were retrieved later and waiting to be evacuated to the rear base, Kang forgot how to speak at the terrible sight of the decomposed bodies. "When God created human beings, He made them in the image of Himself. Then, how could He leave His creation in such a horrible state?" was Kang's thought, and he says it has become a reason for him to assume a cold attitude towards Catholicism to which he has been committed. Standing before these awful sights, he asked himself, "Do such clichés as 'died for their country or 'sacrificed themselves for all' make any sense? Can their deaths be justified and comforted ever by those inflated and high-sounding rhetoric?"

Kang's strong indictment of the war's brutality reminded me of WWI American novel William March's *Company K* (1933). Having witnessed innumerable scenes of butchering on the Western Front in France, an American soldier of Company K on night duty finds a shadowy figure approaching him and, alarmed at its appearance, challenges him. This is the conversation between the soldier and the figure:

"Who's there?"
"I'm Jesus."
"If you're such an omnipotent God, how did you let this horrible war go on? You have to feel shameful."
"That's what I wanna ask. If you have the answer, please let me know."
And then the soldier and the Jesus figure sat together, weeping endlessly.[13]

## The Tragedy of Cu Mong Pass

In November 1965, the Tiger Division, which landed in Quy Nhon and was stationed in Binh Dinh Province, set up a market at Cu Mong Pass. The purpose was to restore mutual trust by allowing residents on both sides living in Phu Yen Province and Binh Dinh Province to meet to exchange goods freely. It also had the purpose of psychological

At Cu Mong Pass in central South Vietnam, called the Panmunjom of Vietnam, the civilians of two opposing sides met and exchanged goods at a designated time every day. (Source: Institute for Military History, MND)

warfare to collect information and disseminate the policies of the South Vietnamese government.[14]

After one month of staying at Bac Tang base, the 11th Company (Commander Captain Lee Jae-tae, Class of 1960) of the 3rd Battalion of the First Regiment moved to Cu Mong Pass and took over the area from the 5th Company (Commander Captain Park Dong-won, Class of 1958). The rugged mountain pass 300 meters high above sea level was on the border between Binh Dinh Province and Phu Yen Province, on Route 1 at the southern end of the Tiger Division's TAOR. This pass was a miniature version of the current situation in which two sides were severely fighting. The VC controlled the area from this pass to the south. In this area, the VC had levied tax and military duty on the residents. Looking down the pass far away, you could see the VC checking the residents carefully. It was a scene not entirely different from the DMZ area in Korea, so Lee's men sometimes fell into a trance as if they were in Korea.

On the top of the pass was a marketplace where the residents from the area under VC control and those from the South Vietnamese government's area of control came to exchange goods. The market opened once a day under the supervision of Captain Lee's soldiers. The market was open for two hours, from 10 a.m. to noon. The residents from the VC area mainly brought pigs, chickens, and seafood, and those from the other side mostly brought rice. The rice acquired from here flowed to the VC. But the South's government overlooked the act under the pretext that they also had to live by it. Yet, the South's government strictly controlled the quantity of rice sold below three kilograms. What the VC side also needed was medicines. As the medicines were controlled items, the South side residents took them inside their clothes. They traded the medicines at the market five times the price, higher than in the government area.

One thing to note was that men were prohibited from being in the market. Only women were eligible. Captain Lee, in charge of this market, had some difficulty checking the residents coming to the market. He said, "All of them are women, so it's impossible to frisk them. Because they wear a black one-piece dress without underwear, touching them to investigate is to touch their naked body." What would happen if those soldiers who had to live an ascetic life were ordered to frisk young ladies to search for medicine? The women were mainly ladies in their twenties, some in their teens.

There was a story about why this market was only for women. Before the Tiger Division arrived here, this place was occupied by the U.S. Marines. One day, a soldier was fired on by a VC disguised as a vendor. The Americans gave them a strict warning, and, from then on, men did not appear anymore except some aged men and small children. Yet there was no guarantee that all women were not VC. In reality, the VC mobilized a 10-year-old girl as a vendor and approached an American soldier to throw a hand grenade. It was what happened just before the Tigers arrived there. The women from the VC side were about 250, while those from the government side 150. Among those from the opposing side were the VC's wives, and those from the friendly side were the wives of the South Vietnamese army officers. When the market ended, each returned to their respective areas.

Then one day, a woman from the hostile area approached the friendly side. It was not freedom that she aimed to get. She had tears in her eyes, "My mom's living in Binh Dinh Province. I miss her too much. I implore you to let me see her." Deeply moved by her plea, the Korean soldiers treated her well and escorted her to the South Vietnamese police officers, asking them to allow her to meet with her mom. However, she was rejected and sent back to her place. The profoundly disappointing figure returning to her side made all the Korean soldiers on hand sad and reminded them of the tragedy of their own divided country.

Another heartbreaking story was also about a separated family. One day Lieutenant Colonel Park Kyung-seok, Captain Lee's immediate superior, was on an inspection tour to Cu Mong Pass riding his command jeep. Right at that time, a French-made sedan came to a stop before the checkpoint. Getting out of the car was a South Vietnamese lady wearing a blue filmy Ao dai. She was the first ever beauty that he had seen in Vietnam. She came to him and said in fluent English, "How are you, Colonel?" She continued her long story beseechingly in tears. Her story went like this: She had lived in a small village called Ta Cai, 5 km from this pass. When she was a young girl, she was sent to her relative in Saigon, and there she went to high school and continued to study at a university. After graduation, she got married to a businessman there in Saigon. In Ta Cai village, her parents, younger brothers, and sisters were now living. Even in Saigon, she wished she could visit her hometown to meet with them, even if it were just once in her life. With a dream of seeing them, she came to Quy Nhon by Air Vietnam last night. She said she would come back after staying there overnight. She would give LTC Park as much money as needed for him to arrange it.

The battalion commander did not doubt her words. But he had no right to send her to her hometown, but the South Vietnamese government had it. And thus, he told her, "I'm a Korean officer, and I only have the responsibility to protect all the Vietnamese here. Your problem is beyond my power, but the South Vietnamese police officers here will handle it. I'll do my best to persuade them to do you a favor." And he sent her to the police officers. But they didn't accept the lady's wish. She visited the checkpoint daily, asking us to let her see her family. For a while, he thought about if there were any ways it would be possible. At last, an idea came to his mind that he would bring her family to Cu Mong Pass.

After Lieutenant Colonel Park gave sufficient rice to a woman from the VC side, he told her to bring the lady's family to this place. And then he waited, half in doubt. The following morning, the lady from Saigon was wandering around searching for her family like mad. After a while, a weeping sound of a woman came to him. The lady, at last, met her younger sister. Without a word, they were crying hard with their arms embraced tightly. After some time, however, a woman from the VC side pulled the younger one out of their embrace. It was time for them to part. The policeman from the government side came out with the lady, helping her along the way. The Tiger soldiers couldn't do anything but stare calmly at the scene. This was not a tragedy only for the people at Cu Mong Pass. This was the least humanitarian effort the Korean military could make for the suffering Vietnamese.

# Chapter 6
# The Korean Soldiers Reflected upon the Americans and the Vietnamese, and Vice Versa

In his Vietnam memoir *A Rumor of War*, Philip Caputo once said that "the barbarous treatment the Viet Cong and ARVN often inflicted on their own people" is beyond description, but "neither confronts the crimes committed by the Korean division, probably the most bloody-minded in Vietnam."[1] To the Vietnamese or Americans, the impression of the Korean soldiers was, first of all, their cruelty. This rumor spread widely among the Viet Cong, and they tried to avoid engaging with the Korean troops. According to a captured Viet Cong's confession, "Unless you have 100% certainty, you should evade fighting with the Korean troops under any circumstances."[2]

On March 18, 1967, during Operation Ojakgyo Bridge, the 11th Company (Commander Captain Park Jong-sik, Class of 1959) of the 28th Regiment of the White Horse Division captured the messenger of Captain Hong Tam Nhan, the 2nd Company commander of the 6th Battalion of the 95th Regiment of the NVA, and frisked his body. The Korean investigator found the enemy's official document to evaluate the Korean army. It was the instruction of the NVA's Fifth Division commander. One example of the five items read as follows: "Based on the combat experience so far, the combat power of the Korean army is enormous, and the soldiers are cruel. Therefore, avoiding face-to-face confrontation, finding weaknesses, and weakening their combat power with sniper and surprise attacks are necessary."[3]

One day in July 1966, Captain Jo Gyu-seong, the 3rd Company commander of the 101 Security Battalion, Construction Assistance Corps, captured several VCs in a search operation and obtained some information on what they thought about the Korean troops. A prisoner said, "We have lost our will to fight the Korean soldiers because the Koreans set up ambushes so stealthily, fire, and block our escape route." And he responded to why they didn't defect: "The Korean soldiers kill so mercilessly that we can't come over because we're so afraid. If one of us is killed by a landmine or a grenade, they fire another shot at the dead body to ensure it. Then how do you expect us to come over to your side?"[4]

The Korean soldiers' bravery was recognized through their appearance in battle and the spread of Taekwondo. According to the memoir of General Chae Myung-shin, the First Regiment of the Tiger Division once captured a booklet of the VC, *The Capturing of Korean Soldiers Alive,* in which some VC orders were stipulated. The following was an excerpt from the brochure:

> You cannot capture the Korean soldiers alive. It would be best to avoid contact with the Koreans unless a Viet Cong victory is 100% certain. Never defy the Korean soldiers without discrimination, even when they are not armed. You can kill them but not capture them. They are all well-trained in Taekwondo. Most are said to be black-belt degree holders of martial arts. They smash a stone into pieces and fracture men's bones, too. You have to give up capturing them alive in the first place.[5]

Vietnamese people seemed to think all Korean soldiers were Taekwondo black-belt degree holders. Captain Lee Bong-wan (Class of 1960), an intelligence officer of the Engineer Battalion of the White Horse division, once visited a county office to cooperate with its officials for some business. When he and his men came out of the office, some boys approached them with a few pieces of brick and told them to break them by hand. They seemed to think that Koreans, whoever they were, knew how to break bricks. Unfortunately, there was not a man in his entourage who had done brick-breaking. But one soldier thought it a perfect chance to elevate the honor of the Korean soldier. He struck the brick with all his might and barely broke it. Even though he saved his skin, he fractured the bones of his hand and had to stay in the hospital for some time.[6]

Indeed, the martial art played a crucial role in making and spreading the brutal image of Koreans. Korean Taekwondo instructors continually increased at the request of the South Vietnamese authorities and with the determination of Commanding General Chae Myung-shin. By the time of the Tiger Division's arrival in 1965, the Taekwondo population throughout South Vietnam amounted to over 100,000. As of 1969, about 150 to 170 Korean Taekwondo instructors were in action at 70 different places in nine provinces nationwide. The power of Taekwondo was enormous.

When the Korean combat troops arrived in Vietnam, Taekwondo was well-known to the extent that all Korean soldiers were considered black-belt degree holders. Further, the Korean soldiers armed with Taekwondo were considered fearful beings, and the VC were too afraid of them. Once, when the soldiers of the Tiger Division had taken a VC captive alive, the prisoner in the interrogation confessed that Taekwondo was one of the main reasons for fearing Koreans. However, the primary objective of sending the Korean troops to Vietnam was to contribute to winning the war, but more importantly, to execute the "stabilization efforts" to bring peace to Vietnam. Thus, the Korean troops made it a policy to give horror to the enemy but implant recognition of Koreans as benign friends among the Vietnamese people.[7]

A friendly newspaper, *The Saigon Post*, also noted this attitude toward the Korean army. Dated March 20, 1968, the paper carried an article about the Korean troops: "Koreans' killing ratio has been 15:1, and they drove out the enemy from the swamp area of Binh Dinh, where both French and Vietnamese were not victorious during the past 20 years. The essential characteristic of Koreans is their unending fighting spirit. Unsurprisingly, the enemy avoids contact with the Tiger, White Horse, and Blue Dragon warriors. Captured enemy documents indicate that the VC are instructed not to make contact with Koreans."[8]

Indeed, this impression of the cruel Koreans was what American soldiers had experienced in the Korean War. At the time, Americans had seen Koreans as people with an oriental view of man's life as insignificant. Thus, the Americans in Korea felt a sense of fear and horror toward the entire Korean people, especially toward North Koreans. This attitude toward Koreans might have been transferred even to Vietnam some twenty years later.

The bravery of the Korean army was not only noted by the South Vietnamese newspaper, but *Chicago Tribune* reported. Chesly J. Manly, the war correspondent of the paper, covered the Korean night surprise attacks at the Battle of Binh Tan and the Battle of Chi Hoa, conducted by Captain Yong Young-il (Class of 1960), the 9th Company commander, and Captain Lee Jae-tae (Class of 1960), the 11th Company commander on January 17 and February 27, 1966, respectively. Published on March 13 of that year, the article reads as follows:

> The United States is proud that Korea's Tiger Division is the best fighting force in Vietnam. All battalion commander-level officers have completed OAC at Fort Benning or other American military schools. The fighting doctrine of the Korean army follows that of the American army. However, the Korean military has developed a unique fighting doctrine. Viet Cong fears the South Korean army more than the Australian or U.S. army. The captured documents of the enemy well show this fear. The death rate is one South Korean soldier to 15 Viet Cong guerillas.
>
> One factor that led the Korean military to victory was that the Korean army took a break during the day and conducted nighttime battles. Most South Vietnamese troops search Viet Cong in broad daylight before and after Siesta. During the day, however, the Korean army patrol team goes out to get information from village farmers and ambushes outside to catch Viet Cong at night. There is an essential difference between Korean military tactics and U.S. military tactics. When the South Korean military gets fired from the enemy, it immediately counterattacks from two or more directions before the enemy escapes.
>
> Once the U.S. army receives a shot from the enemy, it steps back and requests artillery fire support or air bombing. And then, when it begins an attack again, the enemy already disappears. In two recent attack operations, the South Korean army

launched a night raid on the rearranged enemy position. The South Korean troops made the enemy tremble with fear by shouting "Tiger," beating drums, and blaring the trumpet. In the dark, the Korean soldiers removed the enemy's weapons and smashed them with Taekwondo. One Viet Cong, who died in a battle without weapons, had its skull crushed and pelvis bent. Korean soldiers break a person's neck and back with Taekwondo, which uses hands, knees, and feet. In a night surprise operation, the 9th Company killed 37 Viet Cong, and the 11th killed 41 enemies.[9]

This article was then adopted as a testimony of the U.S. Congress to brief the performance of the ROK army's Tiger Division and was included in the minutes of the U.S. House of Representatives.

## "Learn from the Korean Army."

The Korean army executed a historical Operation Ojakgyo Bridge,[10] the first largest corps-level operation in Vietnam, from March 15 to May 31, 1967. This operation was to envelop, search, and wipe out the enemy in the region from Song Cau to Tuy Hoa, protect its residents, and deny the enemy food and war supplies. Two main combat divisions were committed to this operation. Separated into the south and north of Route 1, the Tiger Division and the White Horse Division operated in opposite directions, from north to south and south to north, respectively. The meaning of this operation was to expand the TAOR of the Korean forces by opening Route 1 from Quy Nhon, the stationed area of the Tiger Division, to Tuy Hoa at the northern end of the White Horse Division, enabling about 400 kilometers of land transport. It was to stabilize the economy by revitalizing industry and trade in South Vietnam by opening Route 1 only in the south of the 17th parallel. In this operation, the Korean army obtained enormous combat results: 874 KIAs, 417 defectors, and 648 firearms captured while suffering only 35 KIAs and 167 WIAs.[11]

When the operation ended successfully, some leading newspapers in the free world reported loudly, "Learn from the Korean Army," "The Success of Operation Rendezvous," and "The Meeting of Tiger and the

A historic rendezvous of MG Ryu Byung-hyun (KMA 7th Class), Cmdr of Tiger division going south, and MG Lee So-dong (KMA 2nd Class), Cmdr of White Horse division going north, at the Hoa Da village on Route 1 during Operation Ojakgyo Bridge on 18 April 1967. The corps-level operation was to connect and open Route 1, running through south and north. (Source: Institute for Military History, MND)

White Horse." *The Saigon Post* carried the article "Learn from the Korean Army."

Korea is neither an advanced country nor a unified country. Like South Vietnam, it is a developing and divided country with a ceasefire between the two Koreas. And historically, it was not a state to dominate but a state to be dominated. More unfortunate about Korea than us is that there was no vicious invader across our southern sea, but there was an invader called Japan across the eastern sea of the Korean Peninsula.

And in South Korea, a more potent and provocative North Korean communist group aims to invade South Korea beyond the Military Demarcation Line (MDL). Nevertheless, South Korea dispatched combat units to Vietnam, including two Infantry Divisions and one Marine Brigade. This fact shows the will and courage of the Korean people against communism. In Vietnam, they achieved a miracle that no country dared to do.

Operation Ojakgyo Bridge is named after the legend that Gyeonwoo, a male star, and Jiknyeo, a female star, meet at a bridge on the Milky Way on July 7 every year. Meeting Gyeonwoo and Jiknyeo means welcome, joy, and happiness. It is natural for them to feel joy and pride as the Tiger Division and the White Horse Division meet on Route 1; they have created a more critical history by giving our Vietnamese pleasure and happiness.

The opening of Route 1 will give the Vietnamese hope and development in many ways, including the economy, culture, society, politics, and the military. It was a shame that our Vietnamese could not open it. Why are we impossible and

Koreans possible? This suggests the following conclusion: 'There is something we need to learn from the Korean military.'[12]

The praise for the operation's success was not limited to the Vietnamese. The U.S. military also dispatched an observation team to this operation to observe the actual battles of the Korean troops. From March 17 to 26, 1967, 14 selected officers, including Captain James Russell of the U.S. 101st Airborne Division, watched the operation by eight companies under the 28th Regiment of the White Horse Division. The purpose of the visit was to discover and correct the operational defects of the U.S. military by acquiring tactics from the Korean army, which was making a superior achievement to the U.S. military, even with poor conventional equipment and firepower. After watching the operation, they prepared an operational observation report and sent it to the U.S. Army Command in Vietnam and the U.S. Department of Defense. The following is a part of the report Captain Russell wrote,

> I sincerely felt that the Korean military is an army that can help us. They were very good people. We thought the Korean military should be equipped with M16 and M60 machine guns in the shortest period. The M1, AR, and M30 machine guns were utterly inappropriate under the terrain and climate of Vietnam. In addition, Korean troops will have to wear jungle combat uniforms. Their conventional clothing had many difficulties, such as being heavyweight and not drying up quickly. The U.S. highest authorities in Saigon may solve these difficulties as long as they want.[13]

General William Westmoreland, the commander of the USMACV, who frequently visited the Ojakgyo operation site and observed the battles, said several times in public, "The most impressive thing in my military career was Operation Ojakgyo Bridge. The Korean soldiers' complete understanding of strategy and tactics and their actions accordingly pioneered the new frontier of the Vietnam War."[14]

Regarding the bravery and fighting spirit of the Korean military, Park Jung-ki (Class of 1958) finds their origins in Chinese classics dealing with ancient Koreans. In *The Later Han* and *Eastern Ok Jeo*, Park says, "The Korean ancestors are described as having large physiques and being very brave. They are simple, honest, good at fighting with lances and talented in infantry combat."[15] And in *Classic of Mountains and Seas,* written by Chinese scholars of the Xian Qin Dynasty, "The country of gentlemen (Joseon) lies to the North, and their attire is modest. Men wear a sword and hunt animals as a main staple for food. On both sides, two large tigers act as servants when they travel."[16] Park concludes that the Koreans traditionally were skilled in infantry combat. The Korean army inherited this tradition of excellence from those periods long ago. "Our ancestors had the large build and were full of valor, and [they] were talented swordsmen and fought and won valiantly in battle." He continues, "so does it not suffice to say that men from other countries may have also met them with a bit of fear?"[17]

Meanwhile, a war correspondent for *The Times,* Nicholas Tomalin, who had stayed with the Jaegu Battalion and 2nd Battalion for ten days to observe Operation Fierce Tiger 5 conducted in the Go Boi plains by the two battalions, returned to England and wrote the following article in its Sunday edition, dated May 29, 1966. "If Korean forces had taken over Vietnam or Americans had learned the 'Go Boi lesson,' the Vietnam War would have already been over. In short, the Korean military understands this war and is fighting with the same coolness and determination as the Viet Cong in every combat. They fight in small units when occupying large areas like the Go Boi plains. They first deliver a dreadful blow but do not forgive anyone who turns out to be Viet Cong. And then they win the hearts and minds of the villagers."[18]

War is not always destructive. For a year in the early phase of its arrival in Vietnam, the Tiger Division brought peace to the fertile land of the entire Binh Dinh Province along Route 1 and Route 19, including the city of Quy Nhon, by carrying out innumerable operations of ambush and pacification. The Koreans did what the French and Vietnamese forces had tried to do in just a year over the last twenty. Furthermore, the Korean troops made it a policy to protect a civilian even if they missed a hundred Viet Cong. This policy was a factor that may have contributed to the Vietnamese's affirmative attitude toward the Koreans, who had been considered cruel. The Vietnamese seemed deeply impressed by the gentle manners of the

Korean soldiers.

For example, when the displaced villagers returned, they found their homes had been well-repaired and cleaned. Not only that, they also saw their village elders well treated by the Koreans. So, they concluded that the Korean soldiers deserved to be men of manners and courtesy. Perhaps, for this reason, the Korean forces were assigned to more missions concerning civil affairs than any other allies by the American authorities. Significantly, the Korean soldiers were considered to excel over other allied soldiers in combat performance, organization, military specialties, pacifying efforts, and so on.

Given Vietnam's politics, culture, religion, and national character, it is well-known that the execution of the war, depending upon the exclusive use of enormous firepower, did not naturally bring peace. That is why the Korean army widely conducted various activities, including psychological consolidation, besides military operations. Despite differences in many aspects of life between Korea and Vietnam, Vietnam belongs to the sphere of Oriental culture in the religious way of life and many other cultural traditions that Korea shares. This geographical and cultural homogeneity brought many advantages to the Korean forces in their effort to execute the war.

On the other hand, there were also negative evaluations of the Korean military. Some American commanders who had worked with their Korean counterparts sometimes were very critical of the latter. Several of their complaints included "Korean troops make inordinate requests for helicopter and artillery fire support of Americans" and "Korean troops do with two divisions what Americans can do with only one brigade." It is said that Melvin Laird, then-U.S. Secretary of Defense, openly criticized the Korean troops as ineffective and even threatened to stop their support toward the war's end. The Americans' criticism made sense, but the excessive request for fire support was partly to minimize Korean casualties. In the guerrilla war in Vietnam, General Chae understood well in advance that almost all combats would be conducted in small-scale operations. There would be "no objects to seize at the risk of taking enormous casualties."[19] The unanimous wish of all commanders, from the platoon leader to the top, was to bring their men back to their home country without any sacrifice.

Koreans are described negatively in some American novels about WWII and Vietnam War. To take a few examples, a bodyguard of Colonel Sato, commander of the Japanese POW camp in Pierre Boule's book *The Bridge on the River Kwai* (1952), is a Korean, a nasty-looking man with a bulky and robust build. And the guards of a Japanese POW camp in Singapore in James Clavell's Pacific War novel *King Rat* (1962) are mostly Koreans. These people are described as villains who earn money through illegal trade with the inmates. In the Vietnam War era, some U.S. mass media would cast a wary eye on the activities of Korean soldiers in Vietnam. And they would make the cold assessment that Korea used the war in Vietnam as a "God-send" opportunity to develop its economy as Japan had done during the Korean War.

In an American novel of Vietnam, Charles Durden's *No Bugles, No Drums* (1976), the protagonist soldier is the most critical opponent of what happened in Vietnam. He ridicules the corruption of the war, saying that American "grunts" and South Vietnamese "slopes" are all thieves "stealin' 'n' sellin' everything they can get their goddamn greedy hands on… but it's nickels 'n' dimes. Half the fuckin' Korean Army is moonlightin' as guards for grunts that're stealin' to sell to us."[20] In the novel, he even says that a U.S. cargo ship loaded with military supplies and equipment arriving at the port of Cam Ranh was entirely headed for Busan in Korea.

When the Korean government accepted the U.S. request to deploy its troops to Vietnam, strong voices arose against the dispatch in the Korean National Assembly because the Korean army was described as being mobilized in a U.S. contract war or as U.S. mercenaries. For this reason, General Chae insisted that the Korean military should have operational control even to quell these objections, against President Park's recommendation that the Korean army's operational control be under the U.S. command.[21] In any case, the Korean army was to be fully supported by the American government in terms of war expenditures. This atmosphere was a recognized fact even among the American soldiers fighting on the Vietnam front. The following episode well represents such arrogance of American soldiers toward the Koreans.

One day in 1967, Captain Park Jong-sik (Class of 1959), the 11th Company commander of the 28th Regiment, witnessed his fellow officer being humiliated by an American enlisted man during Operation Dokkabi 1. A U.S. Army truck stopped before Captain Shim Ki-jae, the 12th Company commander, on a road while he was leading his company toward the enemy position on Nui Hon Ba. A clover mark on his shoulder showed that he was a U.S. 4th Infantry Division soldier. The American truck driver approached Captain Shim and laughed at him, tapping his helmet and shouting, "Made in the USA." He continued to boo Captain Shim, pulling a canteen, a bayonet, and a cartridge belt one by one, saying, "All made in the USA." Being felt humiliated by an American soldier, the 12th Company commander shouted that he would kill him. Still, the U.S. soldier called, "Come on! Come on!" and gestured, "Shoot if you want." Angered by his provocation, Captain Shim fired about ten shots at the sky with a Calvin rifle. Only then did the American soldier return to his truck and drive away.

Who would you complain to about your sorry appearance that a Korean army infantry company commander with 180 subordinates was humiliated by a mere American enlisted man in Vietnam? Captain Park reflects later that he wanted to return to his home country immediately because his mind to fight disappeared.[22]

## Korean attitude toward the Vietnamese people

The attitude of the Korean soldiers toward Vietnam and its people was ambivalent. While taking a leisurely tour in a rickshaw through the city streets of Nha Trang, you could enjoy the swaying palm trees lined along the long white sand beach. You could also see the Lambretta taxis waiting for guests, young girls riding motorcycles in fluttering *Ao Dai* and *Non-La* on their heads, and some Buddhist monks collecting alms in their yellow gowns. The city, thick with tropical trees, a crystal clear sky, and a blue sea, seemed to enjoy the afternoon sleep under the burning sun as if it had forgotten the war long ago. Thus, Vietnam was the land of excitement to all who knew nothing about war. And it was the land of adventure, inviting everyone to a challenge. To most KMA graduates who had no chance of going overseas, Vietnam was just a romantic land, far off to the south, with the Southern Cross twinkling in the starry nights and thickly wooded with palm trees. At that time, it was a safety valve to advance to the outside world.

However, the Korean troops' attitude toward the Vietnamese people was primarily negative. First Lieutenant Song Pil-won (Class of 1970) showed a slightly affirmative attitude toward the people but a negative one toward their military and society. Song, a White Horse's 30th Regiment platoon leader, went to a traditional market and bought a watermelon. He paid the lady shopkeeper 500 dongs (the official currency in Vietnam). Then he saw she was selling a same-sized watermelon to another Vietnamese woman for 300 dongs. Thus, Song asked her the reason. Then the shopkeeper minced no words, "You're a *Dai Han* (Korean in Vietnamese; today, *Han Quoc* is more common), and this woman is the person of our country. If you don't want it, return it to me. Here's the money you paid. Have it back." Of course, Song left the shop, taking the melon with him. And he instead considered the shopkeeper's attitude good because she was more sympathetic toward her fellow countrywoman than foreigners, which was, in his opinion, a natural phenomenon in a warring country.[23] Still, in his heart, he had a bitter smile. The same thing happened in Korea in the past. The scene overlapped with Korea's in the sixties when some taxi drivers would overcharge foreigners coming into downtown Seoul from Gimpo international airport.

The corruption of the South Vietnamese military was a well-known fact in the Vietnam War era. An American officer once remarked, "As to the Vietnamese forces, only one-third of its total strength reported to the higher headquarters is its real strength." And the remaining two-thirds of its strength was fake numbers, and the commanders of higher units embezzled their salaries and supplies. More shocking was that some South Vietnamese generals only selected the sons of the rich people among the recruits and hid them away, taking some bribe money. And they listed them in the units as if they were in their organic strength. It was said that the number of such personnel per general had been as many as 40 to 50.

A high-ranking staff officer in the Korean army command in Saigon once recalled an experience with his South Vietnamese counterpart. In a conversation with the Vietnamese colonel, he heard him saying, "The objective of both Korea and South Vietnam is to secure something from the rich country America as much as possible. The same is true of you and me. What we need to do is to take this opportunity of amassing a sizable fortune. Don't you agree?" Dumbfounded at his words, the Korean said, "I'm a soldier. I'm here because my country told me to fight. That's my only mission. I'll only do my best as a soldier while here." Looking back, South Vietnam was destined to collapse. The Korean officer couldn't understand why he had to fight for such people and their country.[24] In an American novel about the Vietnam War, *No Bugles No Drums,* a protagonist soldier also criticizes the incompetence of the South Vietnamese soldiers in this way: "South Vietnamese may be the most useless since Mussolini made privates outa all those grape pickers."[25]

In an interview with a Vietnam veteran, he said that at that time, the Korean soldiers took a racist attitude toward South Vietnam as an uncivilized nation and its people as second-rate citizens.[26] The same was true of the U.S. military's thoughts toward the Koreans and the Korean soldiers they were helping during the Korean War. Such a way of thinking was an expression of superiority toward the country's people who were receiving their help. As Caputo rightly diagnosed, in the mind of American soldiers in Vietnam, the "racist theory" proposed "that the American soldier found it easy to slaughter Asians because he did not regard them as human beings."[27] Of course, this went too far because a civilian was one thing, and a VC in combat was another. But it was evident that many Koreans and Americans felt like they were in a native or aborigine land.

Many KMA graduates write their impressions about Vietnam and its people in this vein. Captain Lee Bong-wan (Class of 1960), an intelligence officer of the White Horse Division's Engineer Battalion, says what he saw in Vietnam, "Most Vietnamese I've met seemed to be idle. Perhaps it was because they did not have to work hard to earn a living. The hot weather all year round and three harvests of rice a year may have made them what they are today. Despite the hot weather, most men and women wore black clothes with long sleeves and long trousers.

Why? The first reason was to avoid direct sunlight, and the second was that the black does not get easily dirty. Most people walk barefoot, so their trousers always look dirty. Also, when Vietnamese women married, they dyed their teeth with betel nuts. Ironically, it was a sign of purity. When a woman married a man, she made herself look ugly and tried to avoid the gaze of other men. Over 95% of married women had ugly black teeth in the sixties. Whenever children under ten saw the Korean soldiers, they swarmed around and asked for cigarettes. It made the soldiers sick because Koreans could not imagine seeing four- or five-year-old boys smoke on the street or any other place in Korea."[28]

Second Lieutenant Shin Jung (Class of 1964) found that, at that time, there were no toilets in the houses of Vietnam's rural villages, so, in the morning, young and old alike would seek excretion by a stream or in a secluded bush. Occasionally, when the Korean troops camped in a place like this, they suffered from the smell everywhere. This was also true of the American soldiers who had fought in the Korean War. The scent is commonly mentioned in American novels and memoirs about Korea. It smelled of human excretion dumped into the farming fields at the time and used as manure. The scent that wafted from all over the Korean Peninsula was excruciatingly painful for American soldiers who had never had such an experience before. Even in Webb Beech's novel *Make War in Madness* (1965), an officer meets General Kramer, who arrives in Korea on a transport plane at the airfield after a vacation and asks how the trip was. The general says, "The only way to avoid the smell is on an airplane in the sky."[29] To American soldiers at that time, Korea was uncivilized.

Jang Chang-gyu (Class of 1965), a platoon leader of the Tiger Division, compared the Vietnamese with the Koreans later in an interview. "The Vietnamese and we had many similarities in appearance and culture from the same oriental region. Therefore, they felt like Koreans were friends. But the problem was with the Korean soldiers. They had a sense of superiority over the Vietnamese people, regarding them as savages rather than their friends. Above all, there was a perception that the Vietnamese were citizens of a donated country receiving support from the Korean military. They were physically smaller than Koreans, had darker skin, and had a particularly stinky smell. At home in Vietnam, they ate a sauce called *nuoc mam* for the same purpose as our soy sauce, but that's where the smell came from. *Nuoc mam* in Vietnam is a liquid made by fermenting fish and salt stored layer by layer. Most of the areas where the Korean troops were stationed were rural areas, and since they could not obtain good fish during the war, they made it with lousy fish, which made the smell even more intense."[30]

Although the Vietnamese lived in an uncivilized state, they had a strong sense of nationalism. The Vietnamese, who have long been subject to foreign aggression and domination, were particularly resistant to foreign powers. Captain Park Jong-sik (Class of 1959), the 11th Company commander of the 28th Regiment, noted this in Vietnam. "One peculiar thing I felt while stepping on the South Vietnamese soil was that the people who welcomed us were not South Vietnamese but American soldiers. The Vietnamese almost turned away even if the Korean army waved their hands as they passed. However, when the U.S. soldiers saw us, they greeted us, gestured, and whistled. The diplomatic justification for our fighting in Vietnam was that we came to prevent South Vietnam from becoming communist at the Saigon government's request with the U.S. government's recommendation. However, the Vietnamese people seemed to think that the Korean arrival was another presence of foreign power in their country. In other words, NVA and Viet Cong actively had a head-to-head confrontation with the allies, and the South Vietnamese people passively resisted their opponents."[31]

Captain Park said on a different occasion, "We realized something big when we came to Vietnam and met Vietnamese people. Almost all Vietnamese people seemed to be fighting to drive out the French army and kick out the U.S. and Korean troops rather than fighting against their people. For example, one Christmas night, when Viet Cong attacked the local militia guarding the County office, the South Vietnamese governor unknowingly sighed in the presence of an American advisor, mumbling, "We decided not to attack each other..." Lieutenant Colonel Smith of the American advisory group said he felt pathetic when he heard this but questioned why he had to fight for these people."[32] There was a word, joking but realistic, that "you are safe if you have three sons in South Vietnam."[33] In other words, if you sent one to the South Vietnamese army, the second to the national police, and the third to Viet Cong, no one would harm your family.

Although the feeling of superiority in helping the people of a warring country made the Korean soldiers view the Vietnamese negatively, they thought they shared a common culture in the sphere of the Chinese character. Also, Koreans tried to understand the Vietnamese as the people of a country ruled by Japan and those who had fought a fratricidal war only ten or so years ago. From that perspective, General Chae Myung-shin said in an interview with Glenn C. Troelstrup, a *U.S. News & World Report* war correspondent who asked about the difference between Koreans and Americans. "Frankly speaking, in this kind of war, we are combat politicians far better than Americans and other allied forces. There are some similarities between Koreans and Vietnamese. We have no racial discrimination. We two people are mutually engaging in so many social activities. Our soldiers easily acquaint themselves with the Vietnamese people. They get along with them well. We are Orientals same as the Vietnamese. Our political situations are pretty similar. That may be why our soldiers treat them like our brethren."[34]

# PART IV

# RETURN: HOME

Chapter 1

# From Fort Hwarang to Dongjak-dong

*Fort Hwarang* is another name for the "KMA campus" in Seoul, where cadets, a modern version of ancient Hwarangs, reside. *Dongjak-dong* is where National Cemetery is located south of the Han River in Seoul. It is a symbol of sacrifice by the patriots and soldiers who have dedicated their lives to the country.

The end of war leads to the return. Whether the Korean soldiers returned alive or dead, everyone left Vietnam on March 23, 1973. The Korean military's participation in the Vietnam War lasted eight years and six months. Starting with the non-combat unit in September 1964, the Korean government began to dispatch the combat unit in October 1965. Officially, 36 KMA graduates were deceased, but the number is a little higher if those who died after returning home due to injury in Vietnam are included. These graduates, the dead or the living, fulfilled what they had chanted in the Cadet Creed: "giving their lives for the country and people." They realized what they had sung in the "KMA Anthem." Two lines of its lyrics read like this:

*The soul of KMA, though fallen a hundred times,*
*Will not return; once it goes away, never return.*

On October 27, 1967, Lieutenant Colonel Kang Jae-ryun (Class of 1955), then-President of the KMA Alumni Association and a professor of Philosophy at the Academy, wrote a brief but significant note in its

Newsletter, *Asadal,* titled, "We are on a march that will go and 'never return'":

> Our only road is from Fort Hwarang to Dongjak-dong. As a KMA graduate, if you ask about the road to Dongjak-dong, you merely confess to the mental weakness of the infant holding the razor. We have no apostles to lead us. We only have a rough road ahead. We have neither a route of retreat nor of dispersion. Although the road going up is rugged and narrow, our Blue White Line will move forward vigorously while designing a journey of the new itinerary.[1]

In these brief remarks, he defined Dongjak-dong, the National Cemetery, as the final destination for the KMA graduates to go to once they leave Fort Hwarang. He wrote this note with the graduates then fighting in Vietnam in mind. His remarks reminded the graduates how to properly embody the Korea Military Academy's spirit—Wisdom, Benevolence, and Courage (similar to West Point's Duty, Honor, and Country)--under which they had been nurtured in their cadet days.

The most moving story in the history of Korean participation in the Vietnam War is the story of Major Kang Jae-gu. Although Major Kang did not go to Vietnam alive, his death significantly moved the soldiers and the nation's people. The unit to which Major Kang belonged (the 3rd Battalion, the 1st Regiment, the Tiger division) was named after him as "Jae-gu Battalion" later and achieved tremendous combat results on the Vietnam front. Many graduates who inherited his noble spirit followed his footsteps in the jungles of Vietnam. As stated in the KMA anthem, they left their alma mater, went to the war in a faraway foreign land, never returned, and joined the ranks of the patriots buried in the National Cemetery who sacrificed themselves to defend the country.

## "So Is He Gone:" The Late Major Kang Jae-gu, a God of War

"So Is He gone" was the title of an article dedicated to the late Captain Kang Jae-gu published in *Asadal,* dated October 25, 1965.[2] It was the story about the noble sacrifice of Captain Kang Jae-gu (Class of 1960),

Cadet Kang Jae-gu (Class of 1960) in his senior year (Source: Class of 1960 Graduation Album).

the 10th Company commander, the 3rd Battalion, the First Regiment of the Tiger Division to be deployed to Vietnam.

There seems to be a sacrifice ritual before any great history is made. A "lamb for the burning offering" seems to be needed (Genesis 22:7). At 10:00 a.m. on October 4, 1965, Lieutenant Colonel Park Kyung-seok, the 3rd Battalion Commander of the First Regiment, received an urgent call from the battalion operations officer, saying, "Sir! Captain Kang Jae-gu, the 10th Company commander, died in a grenade accident." Unable to understand what the hell he was talking about, LTC. Park asked again. "Who died?" A voice on the phone repeated, "Captain Kang hugged a grenade mistakenly thrown by one of his men and exploded on his own." The battalion commander barely came to his senses.

The next moment, Park remembered Captain Kang and several of his fellow company commanders talking in the officer's dining room in the morning. Kang had been exchanging ideas about throwing a hand grenade with Captain Yong Young-il (Class of 1960), the 9th Company commander, and his classmate at KMA. Yong had been scheduled to instruct the grenade-throwing training to his men that morning. Kang had been talking loudly to Yong about his experience. "Captain Yong, you're gonna train your men to throw hand-grenade today, aren't you? You need to be careful." Suddenly, the hall became quiet when the talks focused on the hand-grenade training."[3]

Before he came here, Captain Kang had been a hand-grenade instructor at the NCO school of the 1st Corps. In the officer's dining room, he talked about his experience: "A recruit is terrified of throwing a grenade, and he's likely to drop it before himself or fly it just over his head. Then, a very critical situation takes place. Once, when I was instructing the grenade of our recruits, one of them raised the grenade over his head and threw it, but it didn't go forward and instead dropped backward. I quickly picked it up and hurled it away in a split second. At that very moment, the grenade exploded, and many soldiers were saved." A couple of hours later, Kang died of the same accident he had warned his peers about during grenade training.[4]

As a member of the Class of 1960, Captain Kang Jae-gu volunteered to join the troops to be deployed to Vietnam. Assigned to the 10th Company commander, the First Regiment, Kang instructed his men to throw a hand grenade near his unit at Hongcheon garrison on the morning of 4 October 1965. The training program was to throw a hand grenade on a hill ridge toward the bunker. Then one of his men pulled off the safety pin and tried to throw the grenade forward. But the moment he put his arms backward to obtain propelling power, the grenade slipped away from his palm. When Kang caught sight of the grenade behind the men, he tried to catch it by throwing his body toward the falling grenade.

However, as the ground was not flat, he failed to do so, and the grenade just fell into the center of the men waiting for their turns. Somebody should have kicked it out of the crowd, but no one dared. Even if anybody had done that, it wouldn't have gone so far from the men's center. Kang may have thought in such a split second that all his men would be killed if it exploded. A rugby player in his cadet days, Kang darted toward the dropped grenade and covered it with his body as if he embraced a rugby ball. At that moment, an enormous explosion erupted, and the next thing they knew was that he was transformed into a horrific figure. All men there were safe without anyone injured. But Kang's body was tragically ripped across his belly and chest.

This unbelievable incident shook the whole country with a wave of emotion. The nation's pride in dispatching the Korean military overseas for the first time in its history had peaked. That Kang saved the lives of his subordinates by covering the dropped grenade was a genuinely incredible occurrence. The people's trust and appreciation for the soldiers, especially the officers from KMA, were indescribable.

The daily newspaper *Chosun Il Bo* wrote an editorial titled "A True Soldier is Here." "During the training to go to Vietnam, Captain Kang threw himself upon the grenade that one soldier mistakenly had dropped, which saved many of his men. Upon hearing this tragic news, Captain Kang's heroic and sacrificial death, the nation's people were shocked and deeply moved. How could it be possible to decide to do such an action in an instant? It must be the expression of Kang's natural-born personality as a whole man. At the same time, we want to highly evaluate the education of Korea Military Academy that has nurtured him to be that way. After a long while, indeed, we saw a true soldier. We want to send our sincere confidence in our armed forces."[5] Many other mass media of the nation agreed with this article and expressed their condolences on Kang's righteous death.

The funeral service for Major Kang was held at the Tiger Division headquarters at Hongcheon garrison on 6 October 1965. Some days before sailing for Vietnam, all soldiers were gripped with sadness for losing a hero. In front of his coffin, they resolved to win at any cost. After that, his body was carried to his alma mater Korea Military Academy. He was temporarily laid in state at the reviewing stand of Hwarang Parade Ground amid the mourning of all 800 cadets. The next day his body was moved to the parade ground of ROK Army headquarters, where he was accorded the funeral with complete honor, participated by all the dignitaries of the National Armed Forces.

On the other hand, the Army authorities decided to posthumously promote him to a major by one rank (promotion by two levels is limited to death in combat) and award him the Hwarang Order of Military Merit, the fourth-highest medal of distinction. From the 1966 graduating class, KMA decided to present the 'Jae-Gu Award' in his honor to three graduates who had shown upright character and sacrificial spirit at the Corps of Cadets during their schooling. The KMA Alumni Association then briefly summarized Major Kang Jae-gu, the man and his short life in this way:

> To sum up the personality of the late Major Kang in a few words is almost impossible, yet, if we have to define him, perhaps he will be described as a man whose personality overflowed with a genuine love of humanity and true military spirit. Thus, his death deeply moved us in the military and the people of the public society, which had given the army an aggressive look, for they found in the personality of Major Kang, a true soldier and a man of

nobility. Major Kang's death was never a moment's action but an extension and manifestation of his solid military spirit and sincere love of humanity. We can find such personality traits in the stories of his eight-volume diaries written in his cadet days.

Major Kang had designed and built his four-year life at the Academy around two essential points, 'Make every day meaningful' and 'How much did I make an effort for others?' We can learn from these two points that he has always directed his life to pursue a righteous way and live a selfless life. Here are some aspects of his character represented in his diary: 'Man's life is definite, thus leave something worthwhile that is indefinite; make friends with sincerity and lead your men through moral virtues; live a soldier's short and thick life, making no excuse; make a military society with strict discipline and full of human love; live a life in which my presence is considered far better than my absence.' Indeed, his sacrificial death is not an accident but a result of his life as a person he has tried to make for 29 years.[6]

There was an unveiling ceremony of the late Major Kang Jae-gu's statue at the KMA campus on 2 March 1966. At the graduation ceremony for the Class of 1966, KMA dedicated the figure in the presence of then-President Park Chung-hee. Among the echoes of Jae-gu's song sung by the 100-member chorus of Soongeui Girls' high school, President Park pressed the button from the reviewing stand to unveil the statue, and its magnificent feature appeared. Major Kang's widow, Mrs. Ohn Young-soon, and her baby son were on hand. Since its erection in 1966 in the center of the KMA campus, cadet companies always pay homage to the statue with "Eyes right!" when they pass by it on their way to their quarters after completing a weekend parade. Later the Second company Major Kang belonged to in his cadet days was named the Jae-gu Company. Further, to introduce his noble spirit of sacrifice and give lessons to students, the government listed his life and deeds in the textbooks for the 6th grade of elementary school and the 2nd grade of junior high school.

A poem written by the renowned poet Lee Eun-sang was also dedicated and inscribed on the statue.

Cadet companies pay homage to the Statue of Major Kang Jae-gu with "Eyes right!" whenever they pass by it. (Source: Korea Military Academy)

Spirit and passion, like the dazzling sun and moon
Here is a torchlight blazing up endlessly.
Lo, he is leading the national armed forces.
A righteous cry to shake the earth's axis
Echoes in the hearts of you and me.

In the fragrant blood that fumed to bloom, a rose
Carve the name of an immortal man
In the history of mountains, rivers, and country.
A symbol of heroism standing high forever,
Oh! An actual living figure of the Korean people.

In later years, General Chae Myung-shin, then the Tiger Division Commander, recalled Kang's noble sacrifice and then-President Park Chung-hee's response to it in his memoir *The Vietnam War and Me*.

I received a letter from Captain Kang's immediate superior, Lieutenant Colonel Park Kyung-seok, the 3rd Battalion commander. There he stated that he would resign from the present position and be willing to get any punishment for his subordinate officer being killed and that we would keep his sacrifice for the men in our memory for good. I was deeply moved by his warm and sincere letter. Some days after the funeral, I read LTC. Park's letter and other reports by his staff again. Reading the documents repeatedly, I could not control my emotion and thus wrote a letter to President Park Chung-hee. After a while came a word from The Blue House (Presidential Office). It was the response letter from the President, who was also profoundly moved by Kang's sacrifice.

And he gave a solemn order to the Army Chief of Staff to specially promote Captain Kang to one rank higher and to award him with the Taegeuk Order of Military Merit, the highest medal of distinction, in May 1966. And he also directed the Army Chief to hold his funerary rite again as an Army Funeral. Army Funerals for the junior level officer had never been held in the history of the Korean Army, and conferring posthumous honor of the highest military medal on the person who died in the line of duty was also unprecedented. The nation was overwhelmed with a fever of love for the late Major Kang Jae-gu and pride that the nation's army would go overseas to help the oppressed people.[7]

Later the Ministry of National Defense named the 3rd Battalion of the 1st Regiment with the Tiger Division *Jae-gu Battalion* upon the request of General Chae. This battalion of the first deploying troops to Vietnam made outstanding achievements in combat and yielded the best results among all the infantry battalions in the Korean Army in Vietnam. A year after the Korean troops' deployment to Vietnam, the Jae-gu Battalion was awarded many medals, 4 Eulchi, 15 Choongmoo, and 32 Hwarang and Inheon. This number was the highest record set by a battalion throughout the entire period of Korean deployment to Vietnam. General Chae recalled, "Looking back, I feel proud that I named Major Kang's battalion 'Jae-gu Battalion.' Forty years since its naming, the 3rd Battalion of the First Regiment is still called Jae-gu Battalion in the Capital Mechanized Division (Tiger Division renamed) on the Korean front. The late Major Kang Jae-gu will live as a *god of war* in our Army forever."[8]

## The Unsung Heroes of the War: In the footsteps of the late Major Kang Jae-gu

The men of KMA are those who have pledged unconditional devotion to the country. And they take pride in it. They do not want any reward for what they have done. Neither do they want to be acknowledged by others. They only do what they are tasked to do. Many graduates fought in the Vietnam jungles following the country's orders; some lost their lives or were maimed. They are the "unsung heroes."

The first casualty among the KMA graduates in Vietnam was First Lieutenant Cho Young-shik (Class of 1963), a platoon leader of the 12th Company, the Cavalry Regiment. He was killed in an accident while placing claymore mines on 19 November 1965. Another member of the 1963 Class, First Lieutenant Kim Hyun-young, the 1st Platoon leader of the 1st Company, the Cavalry Regiment, stepped on a booby trap and was killed with his company commander while they were going on a patrol mission on 11 January 1966. Some weeks before the incident, he was on a reconnaissance patrol with only his 17 men and found a squad of the VC. After a fierce fire-fight, Kim's platoon killed six Viet Cong and captured some small arms, grenades, and documents. Thanks to this achievement, he was awarded the Inheon medal. He was honored to be invited to a party held by the then-Prime Minister of South Vietnam Nguyen Cao Ky. The death of Lieutenant Kim in Vietnam was second among the KMA graduates. He sent a letter to his parents in Seoul a few days before: "In recent days, I miss you, Mom and Dad, too much. Every day I wonder whether a letter to me is included in the package of letters delivered from the battalion headquarters once a day. I'm well aware that I receive as much love and respect from my men as I love them. Indeed, they're excellent soldiers with sincere hearts."[9]

During Operation Fierce Tiger 5 on March 23, 1966, First Lieutenant Chung Joo-young (Class of 1962) got a piercing wound on his stomach and died while raiding the enemy position. He was an executive officer of the 5th Company, the 1st Regiment of the Tiger Division. On March 23, the company commander Captain Park Dong-won (Class of 1958) attempted to destroy the enemy with his 2nd and 3rd Platoons attacking and the 1st Platoon as a reserve. While moving forward, the company faced the enemy's strong resistance from the village. The 3rd Platoon leader was killed by the enemy sniper fire.

At this, Captain Park told Lieutenant Chung, his executive officer, to take the headquarters personnel to the position of the 3rd Platoon and to attack the enemy in front. Immediately after he got this order, Lieutenant Chung led some of his personnel to the 3rd Platoon position and began to attack by moving by bounds. However, Chung was hit by an enemy bullet in his belly and fell on the spot. Lieutenant Chung was a taciturn person and an officer who was always faithful to his mission. His cremated body was transported to his alma mater Korea Military Academy, and a funeral service was held at Hwarang Parade Ground on April 23, 1966.[10] Lieutenant Chung's sacrifice partly contributed to the success of Operation Fierce Tiger V. He did not die in vain. He embodied an infantry officer's command spirit model, "Follow me."[11]

In particular, the letter he wrote to his younger sister Seon-ja, a day ahead of the operation in which he died, resonates with our hearts. It contains heartfelt words such as his pride in participating in the crusading war, deep affection for valiantly fighting subordinates, concerns about his parents, and a request to care for the younger siblings:

Dear Seon-ja

The weather is hot. Any adjectives for the burning sun seem insufficient to describe this weather. Sweat pours out like rain. Since it is a densely forested battlefield, you can avoid direct sunlight and still breathe. Otherwise, you would likely collapse because of sunstroke.

Today, we are searching for Viet Cong through the jungle and destroying them. When I wander through the thorny bushes, my face, not to mention my hands, is scratched all over from the thorns, but when I see our soldiers rushing bravely forward, I feel confident, determined, and comforted. While traveling through the land of this tragedy, I often feel rewarded for participating in this war as a crusader for world peace and freedom.

Our soldiers are proud to come here to help others, even in our poor situation. Since we are family, I sometimes have to be a father, mother, and friend. Whenever I eat, I make sure they are taken care of first. Before going to bed, I must visit them where they sleep, and I feel relieved to find they are OK. Of course, the soldiers are also respecting me with sincere hearts. I wish there were a shower. Rice is growing abundantly in the field. The whole area is only green. The mountains are thick, and the fields are lush here, but the bald mountains in my hometown always pass by in my head. I wish Korea also had such a forest. It is a peaceful rural area unless there is the sound of gunfire.

What will our hometown look like now in this season? Of course, green sprouts will come of the earth, and colorful azaleas will be embroidered on the mountains front and back. And the village will be busy preparing for farming. How are our parents? Let's do our best for the day we all meet, Seon-ja. There's little time left until I can go back. I hope you will help our father and mother and sincerely care for the younger siblings. Please say hello to the adults in the neighborhood. Your elder brother is fighting bravely. March 22, 1966.[12]

*"A platoon leader who died trying to save his man"* is the title of "Detailed Combat Report," written about a platoon leader killed in Operation Fierce Tiger 5.[13] He is First Lieutenant Kim Moo-seok (Class of 1964), the 2nd Platoon leader of the 10th Company of the First Regiment. On March 23, 1966, at 03:00 hours, an infiltration operation into the enemy line began in the area 20 km north of Quy Nhon. The 10th Company occupied Hill 78 under the control of the Viet Cong two days before the start of the attack and conducted a search operation. At 07:00 hours, it moved to a point 6 km northeast of Hill 78 to block the Viet Cong's escape. At 15:00 hours, Lieutenant Kim Moo-seok led the platoon at its head and rushed toward the target village. While moving, his platoon faced the enemy's powerful resistance, and one of his men was injured about 60 meters from the village. After carrying him to safety, Kim returned to the original position to care for two other fallen soldiers.

At this moment, Private Kim Jeong-bu was hit by the enemy's bullets in his stomach. The platoon leader hurriedly went to rescue him, but he was also hit in the head by the enemy's bullet. Seeing this scene, all his men made an all-out assault and broke through the Viet Cong's defense line, killing 28 Viet Cong and capturing many firearms.

However, Lieutenant Kim did not regain consciousness. He died during surgery at the 6th Evacuation Hospital. With a gentle yet upright personality, he has been respected and trusted by his men as if he were their father or older brother. Private Kim Jeong-bu, who had been evacuated before his platoon leader, only called his platoon leader, "Lt. Kim must not die. He got hurt because of me. Where is our platoon leader?" He kept crying, holding the surgeon while he was treating the man. Lieutenant Kim's hometown was Incheon, where his 48-year-old single mother ran a small shop.

Lieutenant Kim often had written letters to his niece Young-ae in his hometown, and a letter he wrote to Young-ae shortly before his death touched our hearts. It was a letter that showed how much he loved his men, who were loyal to him.

> My dear Young-ae: The members of our platoon you were wondering about were 40. Second Lieutenant is the lowest rank among officers, but has the highest number of men he can command directly. It is also the platoon leader's job to care for and wipe his members' noses. Everyone thinks of the platoon leader the most, but they fear him. All of my men are brave and intelligent soldiers. They follow me the best and are so loyal that they will risk their lives if I tell them to do it.[14]

According to his classmates in Vietnam, Lieutenant Kim Moo-seok was on an R&R to Saigon after receiving the Hwarang medal for his achievements in the Battle of Nam Tang before Operation Fierce Tiger 5 began. He was resting comfortably at Vung Tau's beach resort. Suddenly he said, "I'm worried about my platoon. We fought all together, but I can't be comfortable like this. Why me alone? I have to return to my platoon." If he had been in Vung Tau during his vacation, he would not have been part of Operation Fierce Tiger 5. Still, his staunch military spirit and sense of duty made him return to the unit earlier.[15]

Lieutenant Kim Gil-bu (Class of 1964) recalls later about his classmate Lieutenant Kim Moo-seok: "After arriving in Vietnam, I believe, Lieutenant Kim Moo-seok turned to the Catholic Church. In this operation, I watched him charge right before he was killed, and in all three of the charges he made, his men got up at once at his command and jumped into the enemy's massive fire. His leadership on the battlefield was so brilliant that I wondered whether he had been such a person when he was a cadet. His charging posture was dignified, and the soldiers who followed him were also brave. He was a man with a warm heart who personally wrote letters by hand to the families of each of his subordinates. He regarded his men as brothers and paid sincere attention to their personal lives. For example, when taking a bath near a stream, Kim manually washed the soldiers' backs and had a friendly conversation with them. He consistently treated his men with this unadorned human warmth. His philosophy was "first make a human and then a soldier."[16]

Among the KMA graduates who died in the Vietnam War, the Class of 1965 graduates had the most deaths, seven. Lieutenants Kim Ho-kyung and Kim Jang-woong were among them. They followed in the footsteps of Major Kang Jae-gu. Lieutenant Kim Ho-kyung achieved great success in Operation White Horse 1 in February 1967. While clearing the LZ in the second phase of the operation, one of his men stepped on a booby trap set up by the enemy. Lieutenant Kim pushed him aside and saved him, but he did not have enough time to evade the explosion. He was the second Kang Jae-gu.

The third Kang Jae-gu was Second Lieutenant Kim Jang-woong. On 18 August 1966, Second Lieutenant Kim, the 3rd Platoon leader of the 3rd Company, the 26th Regiment, the Tiger Division, died while engaging in an ambush operation in the Dhong Tung village 60 km northwest of Quy Nhon. At three o'clock at dawn, Kim set off for the location his company commander had designated, leading a squad of his platoon. Arriving at the destination at 10:30 a.m., he set up an ambush and waited for the enemy to appear. After some time, a whistling sound was heard from the forest 10 meters in front. Looking ahead, Kim found five VCs approaching through tall grass, one with an M1 rifle wearing a cowboy hat and four others with jungle knives. Kim directed his men not to fire before they came within ten meters of them. When the VCs were within distance, Kim aimed at a VC's head and killed him. At the sight of the fallen VC, all his men began to fire, and everything was in flames. Through the flames were seen some dead bodies of the VC.

Then Kim approached the dead bodies with two of his men to gather their rifles. At that moment, Kim saw a VC still alive who was about to pull off the safety pin of a hand grenade to throw at his men. In that split second, Kim shouted, "Lie down!" And with the warning, Kim jumped onto the enemy and took the grenade off his hand. Kim's action saved the two men beside him. But the grenade in his hand exploded before he threw it away, and the lethal blow ruptured his abdomen. Although he was evacuated to the 6th Evac Hospital on his men's back, his condition deteriorated, and he died from the wound some weeks later. A man of faithful personality, Lieutenant Kim was loved and respected by his men. Among the personal stuff, he left was an unopened letter that his fiancé had sent him from Seoul. Kim had no time to open it. He was an orphan child. He is survived only by his two younger brothers.[17]

First Lieutenant Chung Moon-ki (Class of 1963), the 2nd Platoon leader of the 6th Company of the Cavalry Regiment, was killed in a fire-fight during Operation Guang Ma (Crazy Horse). Conducted in two valleys in Binh Dinh Province from May 16 to June 6, 1966, the operation was to destroy the Viet Cong's 2nd Regiment, believed to be located 60 km northwest of Quy Nhon. The enemy regiment of three battalions was stationed in the heavily forested jungles of Hill 753 called Nui Suoi Ca. The 2nd battalion, to which Chung belonged, in charge of the southern part of the area of operation, launched an attack on the hill at 13:00 on June 2.

While searching the thickly wooded valley of Nui Suoi Ca, Lieutenant Chung's platoon detected the enemy and started a firefight. Then, right after the engagement report, the radio communication was suddenly cut off. Judging the situation as severe, the battalion commander committed the remaining men of the 6th and 7th Companies to the dark jungles in which Chung's platoon was supposed to be missing. After five hours of search, they found Staff Sergeant Kim of Chung's 2nd Platoon. With tears in his eyes, Kim said the platoon attacked the enemies and killed about 20. Then, when the platoon came down the hill to collect the enemy's weapons, the platoon leader and his radio operator were killed by the enemy's sniper fire from a crevice in the rocks.

War is always accompanied by death, but death in the Vietnam War, a guerrilla war, was even more painful. Death was caused by enemy snipers hiding in a cave or booby traps buried underground or hanging from ordinary trees. The thought that the mere act of walking on a road or a forested jungle trail you usually go on will take your life makes you feel a greater fear of death. Although the number of fatalities among KMA graduates was not high, the fear of death in the Vietnam War was more significant than in any other war.

Chapter 2

# The War that has Never Ended: The Living and the Dead

Memory and oblivion are opposites, yet mutual and like two sides of a coin. Memory is constantly fighting against oblivion. Even if something is forgotten, it is not to be forgotten forever. On the contrary, even if something is in the memory, it does not reside there for good. Although the Vietnam War is a war that ended some fifty years ago, it remains in our memory, sometimes as a nightmare or a sweet dream at other times. At the same time, the war remains in the memories of friends, comrades, or bereaved families, as a trauma not to be permanently forgotten.

The Vietnam veterans frequently recalled their experiences of the comradeship between the platoon leader and his men long after the war. Lieutenant Kim Hyun-young (Class of 1963), the Weapons Platoon leader of the 1st Company of the Cavalry Regiment, and his company commander were killed by a booby trap installed by the enemy. The incident occurred because of the Platoon Sergeant's mistake at Tan Nghi village, 30 kilometers northwest of Quy Nhon, on January 13, 1966. The Platoon Sergeant also suffered two eye injuries and eventually lost one eye. After receiving emergency treatment at a local field hospital in Vietnam, he was transported to Korea. Later, when he heard that two of his immediate superiors had died, he was gripped by the sense of guilt that they were killed by his own mistake and suffered trauma for a long time.

Indeed, comradeship on the battlefield was more profound than death. Many platoon leaders, such as First Lieutenant Choi Young-ha (Class of 1966) and Second Lieutenant Kwon Kyung-seok (Class of 1969), suffered the agony of losing their men in the battles. They never forgot their men's deaths, visited their hometowns, and maintained relations with the bereaved families of the dead long after the war. Also, on Memorial Day every year, they used to go to their graves at Dongjak-dong, National Cemetery in Seoul, to remember their sacrifices on the battlefield decades ago. Here are some short notes of the living--classmates, subordinates, or widows—who remember their deceased loved ones. Why does the living write about war and the dead? Such an act is to remember the dead and to keep them alive. Reflecting on the dead brings them back to life, providing comfort to the living.

On Memorial Day, ten years after the war ended, Kim Hyung-seok (Class of 1965) visited the grave of Sgt. Park Chul-jin, his platoon member, at the National Cemetery in Dongjak-dong. He met an older woman with pure white hair dressed in a white hanbok (traditional Korean clothing). She was Sgt. Park's mother. Kim Hyung-seok said, 'I'm a comrade who fought with Park in Vietnam. I'm sorry we couldn't come together alive. I feel ashamed to face you." In her words, "He was such a good boy. From the day I heard that my son was killed, my life probably also passed away with him. Though I manage to live on, there will be no meaning for me to be alive." Kim wondered what a child was to a mother at the thought of how she had spent such long years in sorrow.[1]

The highest-ranking officer among the KMA graduates who died in the Vietnam War was Lieutenant Colonel Kim Hong-sik (Class of 1955). The 1st Battalion commander of the Cavalry Regiment, LTC Kim was supervising his battalion reconnaissance team reconnoitering Route 19 across his battalion TAOR when he was ambushed by the enemy and killed during the ensuing engagement. It was around 7:30 in the morning of July 12, 1968. He was posthumously promoted to full colonel and awarded the Eulji Medal, the second-highest medal of distinction. Lieutenant Colonel Lee Dong-hee, his classmate and professor of Political Science at KMA, who mourned his death, wrote a note of condolences to *Asadal,* KMA Alumni Association Newsletter.

Kim Hong-shik, Cadet No. 13, was earnestly serving as the 1st Battalion commander of the Cavalry Regiment in Vietnam. He was killed during one of the patrols in his operational area. It is said that a ruby ring was on his left hand, and the commander's insignia was shining on his enthusiastic chest. With a childish face, Cadet Kim entered the Chinhae campus 15 years ago and soon earned a reputation as a gentleman-type, fine fellow. He was a cadet with an intelligent brain to memorize the informative speech of the Commandant of the Cadet Corps. A decent person in his cadet days, Kim deeply admired the life of Admiral Yi Sun-sin as a warrior. He was a cadet who showed off his fashion by dating a beautiful girlfriend from Sookmyung Women's University.

When he served as a commander in the field, his commanding philosophy was, "There is no hesitation and no regrets." Hong-shik, you are the incarnation of the immortal, Wisdom, Benevolence, Courage, the motto of the KMA. You have become a living saint, like our elders. And you reminded us of the image of a model soldier that we should pursue. Your loving KMA juniors and men still are fighting for their country and alma mater in the jungle of Vietnam. May you rest in a heavenly kingdom without war and death.[2]

Captain Yoon Tae-gon (Class of 1962) was the 9th Company commander of the 26th Regiment. He killed two infiltrating enemies during the night ambush 30 km north of Tuy Hoa and captured a handgun and other weapons in 1968. And while carrying out the mission with corpses left on the battlefield to lure the enemy, at midnight on July 17th, he drew the approaching enemy to within 15 meters. He threw a grenade, but the enemy's counterfire hit him in the chest. Yet he continued to command his troops until the enemy's mine killed him. He was posthumously promoted to major, and the Choongmoo Order of Military Merit, the third-highest medal of distinction, was conferred to him. First Lieutenant Kim Won-young (Class of 1967), S-3 Officer of the 3rd Battalion, the 26th Regiment, contributed an article to *Asadal* in honor of the late Captain Yoon.

The dawn moonlight was scattering obliquely among the jungle forests. After receiving a report of the appearance of a black shadow, Captain Yoon rushed forward as if he were trying to handle it all by himself. However, he was killed by the explosion of the mine buried by the enemy. In an unknown place in the forest, scorched by the war, where even the nightingale cannot cry, he closed my eyes, leaving the last words, "I loved you, men." The person who thoroughly obeyed the orders of his superiors so meticulously, who shouted that he had to pay more sweat and effort to save his subordinates and send them back to their homeland, Captain Yoon died first. Known as a typical soldier who was inflexible and stubborn when he decided he was right, he went to a distant place where he could never return. I pray for the death of an elder I admire.[3]

Second Lieutenant Chung Chan-ho (Class of 1966), a platoon leader of the 4th Company, the First Regiment, was killed in Operation Flying Tiger 10, which was carried out from September 11 to 16, 1967. Tall and handsome, he was loved by his classmates for his lively and gentle personality. He was well-known as chief of the KMA Cheering Team during his cadet days. Second Lieutenant Lim Jong-sup (later two-star general), his classmate and an artillery FO in the 4th Company of the First Regiment, wrote a note of condolences in *Asadal* in memory of Lieutenant Chung.

Chan-ho! I am sobbing silently with 177 of my classmates in deep sorrow after losing you. In that cold winter, when a blizzard pierced the sleeves of our clothes, we entered this Fort Hwarang with grand hopes and dreams in our small, pounding chests. That was already six years ago. We used to run and jump over the walls of hardship and pain under the motto, Wisdom, Benevolence, and Courage; our roaring sounds still reverberate in this Hwarang Parade Ground. How come you've gone so quickly? You volunteered for Vietnam, saying that a soldier should be on the battlefield just like a tiger needs to be in the mountains and a fish in the water. The memory of being trained together to go to Vietnam at Oeum-ri (the 7th Replacement Training Camp) makes my heart ache.

When you finished training and boarded a ship from Busan, your old mother cried, wishing you well and giving you the gimbap (cooked rice rolled with seaweed) she made with a wishing heart. Wasn't it you who heard the sad news of Choi Eui-gu's death after arriving in Vietnam and burst into tears that there should be no more casualties among our classmates? How deep was the anger in your heart the moment you had to give up the grand ambition you had had in your heart? How great was your sinful mind when you had to close your eyes without doing your filial duty as a son to your elderly parents? Chan-ho! You are gone now, but you will live forever in the hearts of our classmates. I wish you peace in Heaven, without war and death."[4]

Many widows of KMA graduates who died in Vietnam wrote letters vowing to stand firm as soldiers' wives, missing their husbands. The following is a letter contributed to *The Crusade of Hwarang Warriors* by Mrs. Yoo Hee-sook, wife of the late Major Oh Hak-hyun (Class of 1957), in memory of her deceased husband.[5]

"A Sad Recollection"

I didn't cry when I heard about you. You had told me the soldier's wife had to stand firmly on the desolate land, enduring heart-rending grief, and learn the pride of not falling even if she was lonely and not losing her reason even in the darkness of death. That's why I wandered into a dream-like reality after your funeral, but whenever I looked miserable, I smiled deeper than blood. However, when our son was confirmed to enter elementary school, I missed you so much that I struggled with tears of resentment in search of your grave. When our son was born into the world, you said that 'the loud cry makes you feel responsible as a father and that you will raise him to enjoy the happiest and most rewarding life in the world.' And you boasted that you could become a parent seven years since we married, but you left early without waiting for today's 7th anniversary.

As you left, my life ended, so I promised your soul: 'I will raise our son brightly and brilliantly as you wanted him to be. And I will proudly come by your side in the distant future.' I always remind my son. 'It's not that your dad isn't here; he just died a little early. But your dad was a better soldier than anyone else and died proudly and bravely. Your dad is now in your heart and your thoughts. What you think is right is what your dad says, and you have to do what your dad tells you to do. That's why you're always with your dad.' You left for a distant place out of my reach, but I didn't let you go. When I miss your love, I think of you as a cadet who used to walk in the Taereung forest and passionately speak of the creed of the soldier's wife. I miss you. I love you until the end of my life, and I will not forget all the memories with you.

Another heartbreaking letter is written by Mrs. Ohn Yeong-soon, the widow of Major Kang Jae-gu (Class of 1960), five months after his death. She recalls her last meeting with him while he was in training before going to Vietnam at the Hongcheon garrison.

"A Dialog at Midnight"

My darling! What are you doing now? Are you looking down on me and your son Byung–hoon from Heaven's eternal throne now? I'm crying out for your name but in vain. I miss you very much. I don't know why you make me heartbroken, especially tonight. September 23, 1965, is an unforgettable day. That morning you sent me a telegram to bring your underwear and jacket to you. Did you imagine how much joy I had when I received your message? The following day I left for Hongcheon with our baby Byung-hoon on my back. You were in full combat gear when you came to see us. You took the baby from me and returned with him in your arm to your base. That morning, I wanted to buy a chicken for you, but I could only buy a packet of boiled eggs because of a tight budget. Who knew that would be our last meeting?

Tomorrow is Easter, and we all were told to bring one candle stick for the sunrise service at Imoon-dong Church. While listening to the pastor's sermon about Jesus' resurrection, I thought of your resurrection. Darling! You cannot resurrect even just for one second, can you? Hoon's Dad! It's midnight. He is already gone to bed a few hours earlier. Hoon has now grown to be an expert in taking steps. The problem is that he is too stubborn. If you were with him, it would be no problem. What worries me is that people may talk in whispers, 'That guy behaves like a spoiled child because he has grown up without his father. Does he?'

Since you're gone, solitude has become a chronic case for me. It has been no less than death to me. While wandering in no man's land, some neighbors led me to the church. When I learned from the pastor's sermon that man's salvation comes through the substitutionary death of Jesus Christ, I was sure that your sacrifice had also been death for saving scores of your men. Such a sense of coincidence gave me an inexpressible pride that temporarily led me to forget my solitude. However, I feel like an insignificant creature once I leave the church door again.

Hoon's Dad! I don't know how I felt so bitter about your absence on September 16, when your mom passed away. That was just one week before you'd be gone. I thought you could not come though you had known of your mom's death by telegram. That night, at about 11:50 hours, however, you entered the house and burst into tears as if you could not believe your mom's death. Indeed, I am sorry now. What made me feel bitter was my thought you might not come. Though you arrived before 11 o'clock, you didn't go into the house because you wanted to return to your unit immediately if the news was inaccurate, didn't you? Now I understand how you may have felt between the noble cause to defeat the communists in Vietnam and your duty to fulfill as the eldest son of the seven family members. You've been under the heavy burden of taking all these things on your shoulder. This thought comes to me after you're gone and makes my heart ache.

Now it's midnight—time for me to go to bed. You also have to go to sleep, don't you? I am hearing that the warriors of the Tiger Division are outstandingly performing the duties you cannot fulfill. Also, the 'Jae-gu Battalion,' your Third Battalion renamed after you is said to be doing well. Every time I hear about that, I feel immersed in your warm and soft breath, as if I were with you. A new morning will dawn when tonight passes by. Like the hero of *The Myth of Sisyphus* rolling a rock up the mountain, I will do that, too. I'll do so with all my heart and mind as your wife. Good night!"[6]

The dead do not speak, but they leave many stories behind. In that sense, it is necessary to mention some stories of 1970 class graduates who served mainly as platoon leaders toward the end of the Vietnam War. Like most of their immediate elders, they did not get much exposure to such fierce fighting as their elders had done in the turbulent period of the late 1960s. There were no significant battles at the war's end except for a few, including the Battle of An Khe Pass (April 11-26, 1972), and small-scale ambush and patrol missions were the mainstays. Two of the 1970 graduates died in action, and the third died after returning home because of the aftermath of the war. But the stories of their last days touch everyone's heart.

Those two fallen soldiers were First Lieutenants Park Young-shik and Cha Byung-gyu. According to First Lieutenant Oh Hyun-koo (Class of 1970), Park, a platoon leader of the Tiger's First Regiment, stepped on the booby trap on a reconnaissance patrol one day before he finished his tour as a platoon leader. He got a lethal injury that meant both his legs should be amputated. Immediately evacuated to 106 Evacuation Hospital in Quy Nhon, Park fought a lone battle for his survival but eventually met his fate. When he got better a little after his suturing operation, he even joked, "Can I get married, eh?" to his classmates there to see him. His father had been killed in the Korean War. Both father and his only son sacrificed themselves for the country, and the family suffered when its lineage died. Later, Park's cremated body was sent home and buried at the National Cemetery at Dongjak-dong.

Some days before Park's death, Captain Oh, a surgeon at the hospital, called First Lieutenant Kim Moon-hwan (Class of 1970) to come immediately to the hospital and take down Park's will. The surgeon said a thrombosis, one of the rare surgical sequelae of serious wounds, had appeared on Park. In the hospital room, Kim said to Park, "Anything to say?" while trying to read his mind. Park, a Catholic, already seemed to know his present state and to give up everything and said, handing over a rosary on his neck to Kim, "Bury it together with me. I've neither regret nor lingering attachment to anything at all. Do your best for the country." While Kim was listening to his quiet voice, his heart felt empty. "We don't give him any medical treatment of late," the surgeon said.

Twenty years later, on Memorial Day, when Park's death was almost forgotten, Kang Hang-joong (Class of 1970) visited Park's grave and found a middle-aged woman. Asked who she was, she replied she was Park's younger sister. A young lady in her early twenties, when Kang had met with her in Busan decades ago, had changed into a woman of fifty. Although the woman married and left her home many years ago, she said she still kept in her heart the brother who had been the pillar of her family and came to the grave every Memorial Day. And also, on Memorial Day some years ago, Kim Moon-hwan visited Park's grave. There he happened to see Park's aged mother. She said, "She still keeps her son's room as it is." Kim felt his heart aching too much for Park's mother, who still missed her son.

This conversation is similar to the story of an American mother who had lost her only son in the Korean War some seventy years before. According to the heart-breaking story uploaded to an Internet site, the 96-year-old mother, Mommy Dew, has been writing a letter every day for the past 56 years, wishing for her son's safe return. Her son went missing at the Battle of Nakdong River, South Korea, in August 1950. From that day on, the mother has never gone a day without writing a letter. And the letter written on the 20,440th day was listed on the site mentioned above. In that letter, the mother says that she has kept her son's room as it is and has never moved because she was afraid that he would not find his home. She says she has been living such a hard life with the single-minded desire to see her son's face and will never give up hope until she dies. It's a heart-aching story of a mother's love that goes beyond time and space.[7]

Years later, Heo Nam-seong (Class of 1970) dedicated a poem to Park Young-shik at his grave on June 6, Memorial Day in 2015. Park was his best buddy in their Fourth Company, the Corps of Cadets.

Young-shik!
Are you still twenty-five?
I'm already sixty-nine.
Your age stopped in the jungle of Vietnam,
Only my clock kept going.

Your blue and bluish youth
And your red and redder passion
Lie here in the figure of twenty-five.
Only the frost fallen on my forehead
Races with the second hand of life.
A long way from now
When I meet with you there,
I'm afraid you won't know who I am.

Young-shik!
Between you and me
The forty-fourth dawn breaks.
Yet your face in my vacant heart,
Still grinning with delight,
Remains a twenty-five-year-old
Green bamboo tree.[8]

Tombstones of the dead officers in the Vietnam War are lined up in the National Cemetery at Dongjak-dong in Seoul. The inscriptions engraved on the stones break people's hearts.

Here is another sad story. The Class of 1970 lost one more classmate, First Lieutenant Cha Byung-gyu. His death remains a painful regret in the heart of First Lieutenant Han Chul-yong. Upon completing his tour as a platoon leader of the White Horse's 28th Regiment, Han was transferred to the regiment headquarters as a staff officer and used to recommend better positions for his classmates newly arriving in Vietnam. When Cha arrived at his regiment, Han told his senior officers that Cha was an able young man, and they assigned him to the post of helicopter officer, one of the best positions among junior officers. What good or evil angel came to Han's mind to have recommended Cha for such an assignment at the time? Some months later, Cha died in a helicopter accident. "Why had I recommended him to that position?" Han returns to the White Horse Division at Tuy Hoa in Vietnam when he is sometimes alone in a

Tombstones of the killed officers in the Vietnam War are lined up at the Dongjak-dong National Cemetery in Seoul. The inscriptions engraved on the stones make people heart-aching. (Source: ROK Army Museum)

quiet mood. He recalls, "If I had not sent him there, he'd not have died." Indeed, scars of tragic war leave the living with senses of guilt, self-reproach, remorse, and regret forever.

At the time, Cha was on his way back to his unit after he escorted his regiment commander to the meeting of the division by helicopter. Flying over the top of Nui Hon Ba, his plane suddenly crashed to the side of the rocky hill when it was about to descend to the area of the 1st Company of the White Horse's 29th Regiment at Vung Ro Bay. Cha and an American helicopter pilot were killed on the spot. Cha's horrendously maimed body was collected by some authorities in charge and kept safely at the regiment's dispensary before being evacuated to 102 Evacuation Hospital in Nha Trang. The next day he was cremated at the crematorium of the hospital. Toward the late afternoon, when the sun was going down, soldiers' dead bodies were always being cremated. And the white smoke symbolizing the souls of the dead soldiers would come out of the tall brick chimney of the building and rise to the sky.

On the tombstone of Cha at the Dongjak-dong National Cemetery is an inscription dedicated by his classmates: "Our wrestling champion, varsity forum champion! Though your body rests here, your soul will always be in our heart." What made all his classmates sadder was that his sweetheart, a school teacher somewhere in the southern part of Korea, had kept writing every day to Cha in Vietnam, not knowing of his death. It was said, however, that Kim Choong-bae (Class of 1970) had collected all her letters and returned them to her later. What would she have felt when she received all the unread letters? Rumour has it that she had been too shocked at the death of her sweetheart and lived alone for a long time. War is a tragedy to the dead on the battlefield, leaving an indelible scar on the living.

Besides Park and Cha, the Class of 1970 has a third man whose story is too sorrowful. This story may be recorded as the most heart-breaking one in the history of the Korean forces in Vietnam. The protagonist of the story is First Lieutenant Lee Wan-soo. Born of a noble family in Cheong Ju, the central part of Korea, Lee received an education in Chinese literature from early childhood and tried to maintain dignity like that of a scholar in the Joseon dynasty even when he entered the Academy. Keeping to the single path based on his principles, Lee joined the road to Vietnam after graduation from the Academy. Upon arrival in Vietnam in early 1972, he was assigned to the operational situation room of the ROKMACV. But, by a curious twist of fate, Lee's tragedy began from this rear echelon, far from the front. Lee fell in love with a South Vietnamese college student working for his office. The student, Miss Yung, was then a sophomore at the University of Saigon and hired as a typist and interpreter. However, the love of transcending national borders was destined to end in the thick of war.

As the date of his return home was approaching, Lee made her a promise that they would meet again, and he alone came back to Korea with his withdrawing unit. However, South Vietnam was taken over by North Vietnam, and there was no way to see her. Innumerable South Vietnamese became boat people and left their country, but no boats came to Korea. But if she had come to Korea, he thought, she would be in the refugee camp in Busan. So, he went down to Busan several times but in vain. He couldn't find her anywhere. In despair, he couldn't live an ordinary life. Eventually, he was hospitalized in a national mental hospital in the city of Cheongju where he had been born and grew up. Without meeting his girlfriend, Miss Yung, he went to the other world some years later, from which he would never return.

Indeed, the sad love between Lee and Miss Yung was a piece of tragic drama that happened between a man and a woman of different nationalities. This was also a by-product of the Vietnam War that had taken place in the Cold War era. The victim of the war was Lee Wan-soo. Although the young girl has now become a woman of another man, perhaps she will keep the love with the young Korean officer in her memory as a painful and unforgettable affair. In short, the Vietnam War remains a tragic war that has never ended. As Norman Mailer suggests in the title of his novel, war seems only to produce "the naked and the dead."[9]

Nha Trang citizens waving farewell flags to departing Korean troops in March 1973. (Source: ROK Army Museum)

MG Kim Young-sun (KMA 7th Class), the 9th Division Cmdr, and his staff report to President Park Chung-hee their triumphant return at a welcome-home rally in Seoul in March 1973. (Source: Institute for Military History, MND)

# Conclusion: Legacies of the Vietnam War

So far, in this book, I have examined the war experiences of KMA graduates in Vietnam over eight years and six months. Their journey from Hill 92 to Nui Hon Ba ended fifty years ago in 1973. Among 3,037 graduates from the Class of 1955 to the Class of 1971, 2,069 of them, about 70%, have been to Vietnam. The remaining 30% did not have a chance to go to Vietnam due to the assignments here in the homeland, early discharge, and the war's end, but in their minds, they all *were* in Vietnam. Although the KMA participants amounted to only 9% of the total Army officers (22,339) deployed during the entire war period, they significantly contributed to successfully conducting the war in crucial positions for each unit. Some 36 graduates, about 14% of the 259 Army officers who died, couldn't return home alive.[1]

Looking back from this vantage point, the Vietnam War was not worthwhile because South Vietnam collapsed on April 30, 1975, two years after the Korean troops withdrew. The country's fall undermined the justification for the deployment of the Korean military to help it stand. The meaning of their sacrifice and devotion faded with the defeat of South Vietnam. Was then their service and sacrifice in vain? As a military historian said, "There is not a war worthwhile to fight for, but there's a war that one should fight."[1] Perhaps the war in Vietnam was such a war for those KMA graduates. No graduates questioned why they had to fight. They had only responded to the nation's call. Especially to the second-generation KMA graduates, the Vietnam War was the war they had been destined to fight as the Korean War had been to their elders, the first-generation graduates. They put the first line of the Cadet Creed into practice: "We give our lives for our country and people."

As the commanders, leaders, and chief architects of various operations, most KMA graduates completed their assigned missions. They did their best to accomplish what they were ordered to do. Almost all the graduates returned home alive. But those who returned alive did not just return but carried the scars of war with them. In the war, which they jumped into with youthful passion and innocence, they experienced various discrepancies and contradictions. Although they saw indomitable will and courage shown by the men before desperate crises, they could also see men's cowardice and timidity in the face of death. Sometimes they were surprised by their atrocities and the extent of extremity to which they could degenerate. They underwent what human beings were like in such a harrowing situation. The war gave them a deep insight into the true nature of human beings.

The Vietnam War became an opportunity to reveal some weaknesses of the Korean military. In particular, the KMA graduates, seeing such injustice, boldly exposed the wrongdoings conducted in the military through various testimonies and memoirs. They may not allow any injustice to be practiced because of their sense of justice. This is because they have lived by the Cadet Creed and the Cadet Code of Ethics for four years at KMA. Manipulation of combat achievement and false reports, especially at the late stage of the war, were open secrets known to all Vietnam War veterans. How often did they have to bring back some parts of the dead enemy's bodies to prove their combat result? It was a barbaric act that could not exist in a civilized society. A soldier who committed a self-inflicted wound to avoid battle became a hero later and received a medal. There were cases where some small elements on night ambush chased the enemy away by making noises intentionally. These events should be deeply engraved as lessons for the next generation. All these experiences provided them with the nourishment necessary to pursue and sustain their long career in the military, as well as valuable lessons for them to keep in life.

The KMA graduates often would also become moral phi-losophers asking, "why war?" They saw the enemy as fellow human beings like them thrown into the same space of a killing field. The enemy were also human beings who had loving families waiting for them to return alive. They, too, were conscripts mobilized for their country's cause. These Viet Cong were extremely poor in weapons and equipment compared to the U.S. and other allied forces and lacked food and supplies. What they depended upon in conducting the guerrilla war was the advantage of the jungle terrain of Vietnam and the spirit of winning their revolutionary war. In particular, to avoid formidable bombing by the mighty American bombers, they had to hide deep in caves in the mountains and even accompany their families to the hideouts. They had to live there like moles. They had to endure all those hardships to survive. Many KMA graduates felt sad and even sympathetic at their skinny and pitiful appearance when dragged out of the caves. What in the world was this ideology? And yet we couldn't extend them any mercy. Because if we didn't kill them, they would kill us.

This was not only true for the Vietcong. The miserable life of the South Vietnamese people who had suffered long wars also made the KMA graduates feel more compassion. Indeed, it overlapped with the life of the Korean people during the Korean War, which might have been reflected too in the minds of American soldiers. Some graduates confess that they had let the enemy run away on some occasions. Also, after killing an enemy, they sometimes buried him with respect and thought of the family who wished him a safe return home with all their heart. The Viet Cong was also their fellow human beings with a deep love for their family. All of them, friendly or hostile, were humans.

## What did the KMA graduates learn tactically in their tours of duty to Vietnam?[2]

First, there was an essential difference between training and actual situations. The military adage "Tough training will guarantee victory in battle, and every drop of sweat during training is a drop of blood in battle" was half right and half wrong because the training situation cannot assume the situation of death. During training, you can act naturally without fear of death, but in practice, you have to face the reality of death. In fact, in Vietnam, soldiers rumored to be brave in usual times often were cowards in actual combat. The army must discover and nurture true fighting soldiers.

Second, the commanders had to display rational command and control on the battlefield. Commanders in war must have two skills: combat competence and leadership ability. No matter that life and death may be a matter of a split second, the commanders should be able to perform tasks with rational command and control with which their men can comply. In the Vietnam War, the prevailing atmosphere was to control the subordinates through coercive authority, given the specific war situation. However, such coerciveness could not elicit the men's loyalty to the platoon leader or company commander.

Third, the most crucial role of a leader on the battlefield was to ignite the soldiers' will to fight. On the battlefield, the high-sounding rhetoric for the nation did not work to rouse the soldiers' combat motivation. You had to show the example of "Follow me" at the forefront of the unit. You also needed to earn the trust of the soldiers. You had to communicate closely with them, truly understand their feelings, and touch their pain. A coercive command never made them run forward in battle. Servant leadership was needed, depending on the situation.

Fourth, the purpose of the military is to win the battle. The military needs to know how to locate good soldiers who can lead the battle to victory. The personnel management system of the army must be designed so that such soldiers are promoted. But the reality was not so. The Vietnam War was a war that made us aware of that reality. Indeed, many soldiers had made outstanding achievements in the war. In many cases, though, those soldiers were not duly recognized. Because of this, the actual war heroes felt incredible frustration, which gave the officers the scornful perception that "promotion is one thing and a combat performance another, " thus weakening their will to fight. What is the purpose of a soldier's existence in battle? It is to win. If only those officers who display good management skills and maintain good interpersonal relationships were promoted, actual fighters who know how to fight well would have no place in the military. In many cases, those who had made tremendous achievements in the Vietnam War were not promoted later in their career. This is a lesson learned by many KMA graduates who have experienced the battles of the Vietnam War.

## What did the Korean military gain by participating in the Vietnam War?

Above all, Korea made tremendous military achievements and enhanced national prestige by participating in the war. If Korea had rejected the request of the U.S. for the dispatch of troops, there might have been severe damage to the relations with the allies and the alliances with the U.S. The U.S. might have removed its two infantry divisions stationed in Korea and put them in Vietnam, which would have caused a decisive crisis in the Korean defense. This might have forced Korea to spend vast amounts of money on national security, causing a decline in national power. Also, Korea could take the military modernization path thanks to the U.S.'s firm commitment to its defense.[3]

In addition, the achievements made in the combat power of the Korean military are even more remarkable. In the Vietnam War, a great learning ground for modern warfare, mastery of aerial maneuver and assault operation by helicopters, and combat adaptation capabilities using new equipment became an opportunity for the Korean military to grow into a first-class army. The Korean military, which had been 100% dependent on American military doctrine, developed its strategic tactics, applied them to significant operations, and achieved remarkable results that significantly elevated the status of the Korean military. General William Westmoreland, Commander of the USMACV, praised the enormous achievements of the ROK Army fighting alongside U.S. forces in the Vietnam War at a joint session of the U.S. Congress on April 28, 1967, and stated as follows:

> During the Korean War in 1952, few people would have believed that the Korean military would become the world's first-class combat force. However, looking at the Korean soldiers bravely fighting in Vietnam today, there is no doubt that they are the first-level warriors and the best volunteer service corps in Vietnam.[4]

In some corners of Korean society, the success of the Korean military in Vietnam was said to have some negative influence on Korean politics. They argued that the political position and economic development of the Third Republic of Korea, which grew through the deployment of troops to Vietnam, became the background for President Park Chung-hee to adopt the three-term constitutional amendment and the Revitalizing Reforms (to extend his regime).[5]

As mentioned above, the Vietnam War veterans regret that South Vietnam's collapse diluted the justice of their participation in the war. North Vietnam, an enemy country, ended up the final victor. Moreover, the general public almost forgot the Vietnam War, and the veterans had to see Korea establishing diplomatic ties with a unified Vietnam. The today's Socialist Republic of Vietnam has a close relationship with this Republic in various fields, becoming a forward base for Korea to explore the Southeast Asian markets. Yet some parts of Korean society criticized the Vietnam War participation of the Korean military and damaged the honor of its veterans by attacking their actions in the war. In the past years, some hot issues in the Vietnam War have been the massacre of civilians and the problem of mercenaries. The issue of the massacre of civilians is being continuously raised, even today.[6]

Also, some domestic scholars have raised concerns about the mercenary issue. While the veterans were evaluated as "crusaders," "there were mixed claims that they were American mercenaries." Some scholars argued that the Korean troops could not be freed from being accused of being mercenaries because "the United States paid all expenses for the Korean military and that the dispatch would never have been made if the United States had not paid the expenses."[7] However, the issue of mercenaries is a thing of the past and is no longer debated. Regarding the massacre of civilians, the current Vietnamese government has the following position:

> There is no need to discuss the past. If anyone takes responsibility for the war, the U.S. government, which was the subject of the war, will be charged. Therefore, we don't need further discussions with the Korean government in this regard. Instead, we are grateful for the humanitarian support of the Korean government, such as building schools and hospitals in the central region where the Korean military was stationed to heal the wounds of Vietnam.[8]

Indeed, for the people of Korea, the Vietnam War became a turning point for developing Korea's national power. It was an opportunity for Korea to build a foothold and start rising as a powerful country through economic development and national reformation in the sixties and seventies. Of course, there were many painful scars of war. Many young Koreans, a total of 5,099 soldiers, died in the jungles of Vietnam. Some KMA graduates could not board the return ship alive and were boxed to be shipped home by airplane. Around 11,000 soldiers were wounded in the war. Moreover, the war left over "eleven hundred thousand victims who suffer long-term effects of exposure to defoliants."[9]

It's an irony of history that the hostility between the two countries, which had once fiercely fought against each other, has now developed into a friendly relationship. The enemy of the past has become the friend of today. The well-known adage, "There is neither a permanent friend nor foe in international relations," seems correct. Nowadays, of all the tourists visiting Vietnam, Koreans number the highest of any nationality. Once a landing site for Korean and American combat troops, Da Nang has become one of Vietnam's favorite places for many Korean tourists.

## "Vietnam, Vietnam, Vietnam, we've all been there."

A war story is usually told in two ways: one is to tell a war story written to emphasize war's brutality so that it may give the coming generation some lessons that war is evil. The other is to show that although war is a tragic human affair, it is where sacrifice and love, the highest forms of human virtue, manifest. When a war ends, tons of books are written about it. But many Vietnam veterans often feel disappointed that their experiences were not the subjects of many writers. Someone should write about how they fought and what they fought for there. If not, who will remember the horrible experiences of those who survived the valleys of death?

Looking back, some KMA graduates came home by the jumbo jet from Tan Son Nhat airport in Saigon and most by the U.S. Navy transport ships from Cam Ranh Bay. When their airplane finally took off the runway, the images of screaming men who got hit by the enemy fires in the jungles glimmered in their eyes. And when they overlooked the distant tropical scenes from the railings of a returning ship off the shore, they, at last, could have time to see the unfamiliar city of a foreign country as peaceful and beautiful. They had no bad memory of the tropical country of Vietnam and its people who lived through the long and tedious war as if it were their destiny.

A week after the KMA graduates left from Cam Ranh Bay, the ship anchored at the third dock of Busan harbor, the same place from which they had left for Vietnam, and thus ended their long journey to the war in Vietnam. As platoon leaders or junior staff officers, they fought together, and in the process, they saw their innocence and devoted love toward their comrades and superiors. Regardless of high or low military rankings, they were the same contemporary youths fighting together through life and death on the battlefields at a critical moment of their life.

Every human being has one book, an autobiography.[10] One's life is a book in itself when written down. Many KMA veterans of the Vietnam War want their war experiences to be written about--the murderous realm of jungle warfare, conflicting emotions, and terrible moral dilemmas when confronting the enemy. But not everyone can write about it. They want somebody to describe what they did on their behalf. So, when someone wrote something about their combat experience, they breathed a sigh of relief, saying, "Damn it, somebody finally said it."[11] On their part, the KMA graduates needed to record what they did in the jungles of Vietnam for those long years and let others know the brutal authenticity of the war—the untold truth.

In conclusion, the story of KMA graduates' journey to Vietnam here is the story of all the Korean officers and men, a total of 324,864 participants, who fought together in its jungles. Further, their story is a story of a Korean generation who has lived through the most turbulent period of Korean history. As Michael Herr, an American writer of Vietnam, succinctly summarized the experience of Americans in the Vietnam era, "Vietnam, Vietnam, Vietnam, we've all been there,"[12] the KMA graduates of the seventeen classes, veterans or not, all have been to Vietnam. As the graduates of regular KMA, they were the soldiers being asked for limitless sacrifice for the country. Their participation in Vietnam was not one to seek any personal reward but to come out of the firm conviction that the place where a soldier should be was on the battlefield. They showed it in person some fifty years ago. Although all those deceased cannot see today's developed country, they will believe they didn't shed their blood in vain. Their noble sacrifice significantly contributed to making Korea what it is today.

# Notes

## Introduction

1 Philip Caputo, *A Rumor of War* (New York: Ballantine Books, 1977), p. 77.

2 Choi Yong-ho, *The Vietnam War and the Korean Army in Statistics* (Seoul: Institute for Military History, MND, 2007), p. 39. (Hereafter *Statistics*)

3 In Samuel Taylor Coleridge's poem, "The Rime of the Ancient Mariner," "an albatross," considered a sign of good luck, follows a ship setting out to sea. However, the titular mariner shoots the albatross with a crossbow, an act that will curse the ship and cause it to suffer terrible mishaps. Unable to speak due to lack of water, the ship's crew let the mariner know through their glances that they blame him for their plight and tie the bird around his neck as a sign of guilt. The image of albatross around the neck as a metaphor for a burden to escape came from this context. See https://en.wikipedia.org/wiki/Albatross (metaphor).

4 Mark Baker, Nam: *The Vietnam War in the Words of the Men and Women Who Fought There* (Time Warner Book, rpt. 1992). p. 10.

5 Tim O'Brien, *The Things They Carried* (New York: Houghton Mifflin, 1990), p. 68.

6 Paul Fussell, *The Great War and Modern Memory* (Oxford University Press, 1976), p.130.

7 Ibid., p. 131.

8 Viet Thanh Nguyen, *Nothing Ever Dies: Vietnam and the Memory of War* (Cambridge: Harvard University Press, 2016), p. 4.

9 William Faulkner, *Requiem for a Nun* (1951, rpt. New York: Vintage International, 2011), p. 73.

10 James Webb, *Fields of Fire* (Englewood Cliffs, NJ: Prentice-Hall, 1979), p. 168.

11 The First Indochina War was between French forces and their Viet Minh opponents in the South after Japan's surrender in 1945. The war began in French Indochina on December 19, 1946, and lasted until July 20, 1954, with the French defeat at the battle of Dien Bien Phu.

12 In the 2003 documentary *The Fog of War*, the former United States Secretary of Defense Robert S. McNamara admitted that an attack on the USS Maddox happened on August 2, but the supposed August 4 attack, for which Washington authorized retaliation, never happened. https://en.wikipedia.org/wiki/Gulf_of_Tonkin_incident

## PART I. INITIATION: INTO THE JUNGLE

### Chapter 1. KMA as the Bastion of National Defense

1 Kwon Yong-su, "Lineage of Korea Military Academy," *Asadal*, KMA Alumni Association Newsletter, No. 29 (August 20, 1966), p. 5. (Hereafter *Asadal*)

2 It is worth noting that the students from the 1st through 9th classes were called "officer candidates," but from the 10th class, they were called "cadets." The 10th class are called "Cadet First Class." Those, who entered the Academy three weeks before the Korean War broke out, but didn't graduate, are called "Cadet Second class." The official graduates of the KMA before the war include those from the 1st class through the 10th class. But Cadets Second Class, who were not officially commissioned at KMA, are regarded as "honorary graduates."

3 Army Integrated School only lasted from August 1950 to September 1951 and commissioned 7,288 officers as the second lieutenant from the classes of first to thirty-second. The duration of training at the school had been short-term, ranging only from six to nine weeks during the war period.

4 Kwon Yong-su, op. cit., p. 5.

5 From this class of 1955, two presidents of Korea came out some thirty years later, Chun Doo-hwan and Rho Tae-woo.

6 Taereung is the place where KMA is now located. It is in the northeastern part of Seoul. In 1954, it was an administrative district of Gyeonggi Province. Taereung is originally the name of the royal tomb of Munjungwanghoo, wife of Joongjong, the eleventh king of the Joseon Dynasty who ruled over it from 1506 to 1544. People call the name of the grave the name of this area. The tomb's site is very close to the Academy's campus.

7 Designated officially by then-President Rhee Syngman on March 16, 1957, Hwarang-dae or Fort Hwarang is the place symbolizing the mind and the spirit of Hwarangdo, the organization of youths of the Silla Kingdom from circa B.C. 57 to A.D. 935 in the southern part of this peninsula.

8 *Asadal*, No. 23, December 15, 1965.

9 LTG. Park Jong-sun, "The KMA Man and the Cadet Creed," *The KMA Press*, Vol. 7(July 2015), p. 159.

### Chapter 2. The Road to Vietnam

1 Choi Yong-ho, *The Vietnam War and the Korean Army in One Book* (Seoul: Institute for Military History, MND, 2004), p. 146. The discussion on Korea's entry into the Vietnam War was mainly based on this book. (Hereafter *In One Book*)

2 Choi Yong-ho, ed., *The Vietnam War and the Korean Army through Testimonies*, Vol. I (Seoul: Institute for Military History, MND, 2001), p. 34. (Hereafter *Testimonies*, Vol. I)

3 Ibid., p. 159.

4 Ibid., p. 163.

5 Ibid., p. 187.

6 Chae Myung-shin, *The Vietnam War and Me* (Seoul: Palbokwon, 2006), pp. 48-49. (Hereafter *The Vietnam War and Me*)

7 Choi Yong-ho, *Statistics*, pp. 39-40.

8 Yoon Heung-jung, et al., *An Integrated Study of the Vietnam War* (ROKFV, 1974), p. 25.

9 Ibid., pp. 80-81.

10 Ibid., pp. 80-81.

11 We do not use the term "Class of Year" in Korea as used at West Point. Instead, we use the ordinal numbers like 11th, 12th, and 27th class.

12 Choi Yong-ho, ed., *Statistics*, p. 39.

13 KMA Alumni Association, ed., *The Crusade of Hwarang Warriors* (Seoul: Army Printing Office, 1967), pp. 60-67. (Hereafter *The Crusade of Hwarang Warriors*).

### Chapter 3. Why They Went to Vietnam

1 *A Rumor of War*, p. xi.

2 Chung Youn-son, *Memories of the Forgotten War: The Korean War in American Novels* (Seoul: Munye Publishers, 2019).

3 Kim Jin-gyu, "From Hongcheon to Ahn Ke Pass," in *The Crusade of Hwarang Warriors*, p. 262.

4 Interview with Park Jung-ki, November 16, 2021.

5 Park Jong-sik, *The Forgotten War and Dollar: a Memoir of an Infantry Company Commander* (Seoul: Dong Hwa Books, 1998), pp. 62-63. (Hereafter *The Forgotten War and Dollar*)

6 *The Crusade of Hwarang Warriors*, pp. 61-62.

7 Shin Jung, *The Wall of Jungles* (Seoul: Chunggu Publishing Company, 1967), p. 145.

8 Many writers in later years took up the soldier's remark as the title of their books, including William Crawford, *Give Me Tomorrow* (1962).

9 Captain Park Jong-sik (Class of 1959) commented on why the Koreans went to Vietnam. On his way to Busan, he saw a beggar boy following his train: "The beggar boy waving his hand toward the train slowly moving out of Daegu Station says, 'Hi, soldiers, go to Vietnam and kill a lot of Viet Cong and come back alive.' No one ordered the soldiers who saw this scene to give money to him, but everyone searched their pockets, collected the remaining Korean money, and gave it to him. And some soldiers encouraged the boy to 'grow healthy and have hope.' He kept saying 'Thank you,' bending his back repeatedly until they were invisible. We wouldn't fight the enemy; we would fight poverty." *The Forgotten War and Dollar*, p. 117.

10 Kim Hyung-seok, *Fighting Spirit: A Platoon leader of the Vietnam War*, Vol. II (Ilsan: Jisik & Gonggam, 2016), pp.89-92 (Hereafter *Fighting Spirit*).

11 Choi Young-ha, *A Rock not Staying* (Goyang: Malgeunsam, 2012), p. 79.

12 Interview with Noh Nam-seop, January 11, 2022.

13 Interview with Paik Nam-hwan, January 20, 2022.

14 Sigmund Freud, "Reflections upon war and death" in *Character and Culture* (New York: Collier Books, 1963), p. 130.

15 Interview with Oh Hyun-koo and Han Chul-yong, March 9, 2020.

16 See Harry Maihafer, *From the Hudson to the Yalu: West Point '49 in the Korean War*, Texas A&M University Press, 1993.

## Chapter 4. Organizing and Training of Combat Troops

1 See *The Vietnam War and Me*, pp. 101-103.

2 *Asadal*, No. 31, October 1966.

3 *Testimonies*, Vol. I, p. 214.

4 Ibid., pp. 199-202.

5 William Childress, "A Poet Remembers Korea," *War, Literature & the Arts*, Vol. 9, No. 2 (Fall/Winter 1997), 173-174.

6 *The Vietnam War and Me*, pp. 112-113.

7 *Testimonies*, Vol. I, pp. 203-204.

8 *The Vietnam War and Me*, p. 131.

9 Stephen Crane, *The Red Badge of Courage*, (1895, rpt. New York: Puffin, 1986), p. 15.

## Chapter 5. The Voyage: Busan to Quy Nhon

1 See *Memories of the Forgotten War: the Korean War in American Novels*, p. 124.

2 Interview with Paik Nam-hwan, April 26, 2022.

3 Lee Jae-tae, *The 11th Company of the Jae-Gu Battalion at the Vietnam Warfront* (Seoul: Uncheon Company, 2018), pp. 43-44. (Hereafter *The 11th Company*)

4 Jee Man-won, in Choi Yong-ho, ed., *The Vietnam War and the Korean Army through Testimonies*, Vol. II (Seoul: Institute for Military History, MND, 2002), p. 169. (Hereafter *Testimonies*, Vol. II)

5 *The Wall of Jungle*, pp. 33-34.

6 *The Forgotten War and Dollar*, pp. 133-134.

7 *The 11th Company*, pp. 60-61.

8 *The Crusade of Hwarang Warriors*, p. 267.

9 *The Forgotten War and Dollar*, p. 135.

10 *Fighting Spirit*, Vol. II (Ilsan: Jisk & Gonggam, 2016), pp. 15-16.

### Chapter 6. The Vietnam War: "A Formless War against a Formless Enemy"

1 See *Testimonies*, Vol. II., p. 90; Philip Caputo subtitled a chapter in his book, "Splendid little war," and defined the war inside the text, p. 89.

2 *The Vietnam War and Me*, p. 57.

3 *A Rumor of War*, p. 82.

4 *Statistics*, p. 60. Also see *The Vietnam War and Me*, p. 45.

5 *Testimonies*, Vol. I, pp. 250-251.

6 *The Crusade of Hwarang Warriors*, p. 139.

7 *The Vietnam War and Me*, p. 71.

8 *A Rumor of War*, p. 89.

9 For the Viet Cong tactics, see *The Vietnam War and Me*, pp. 69-76.

10 Ibid.

11 Ibid., pp. 464-465.

12 Park Kyung-seok, *Route 19* (Seoul: Daeyoung-sa, 1966), p. 279.

13 *Fighting Spirit*, Vol. II,p. 153.

14 *A Rumor of War*, p. 79.

15 "Lessons of Helicopter War" *Asadal*, No. 66 (January 1971), pp. 24-25.

16 Ibid, pp. 24-25.

17 Among these officers, Lt. Cha Byung-gyu was killed in a helicopter accident.

18 *The Forgotten War and Dollar*, p. 233

# PART II. ENGAGEMENT: HOW THEY FOUGHT

## Chapter 1. Class of 1955 to Class of 1957: The Pathfinders

1 Lt. Gen. James A. Van Fleet was promoted to four-star general on July 31, 1951. By the time he attended the commencement ceremony for the eleventh class (Class of 1955), he had already been retired and given congratulatory remarks in the capacity of the distinguished guest. His long military career came to an end on March 31, 1953. Here he mistakenly mentioned West Point's motto instead of the motto of KMA--Wisdom, Benevolence, and Courage.

2 *Testimonies*, Vol. I, pp. 242-248.

3 Lee Hyo, "Some Korean Army Operations and Ambush Tactics in the Densely Populated Area," *Asadal*, No. 34 (February 20, 1967), pp. 2-5. Also see *Testimonies*, Vol. I, p. 17.

4 Ibid.

5 Lee Sang-hoon, *Asadal*, No. 50 (September 1968), pp. 19-26.

6 *The Crusade of Hwarang Warriors*, p. 4. General Chae Myung-shin is a member of the KMA's 5th Class commissioned before the Korean War.

7 Ibid., pp. 24-26.

8 See *Asadal*, No. 33 (January 20, 1967), p. 19; and *The Vietnam War and Me*, pp. 346-348.

9 *Testimonies*, Vol. II, pp. 236-237.

10 *The Vietnam War and Me*, pp.383-384.

11 Kim Moo-kyung, ""Parts of the Vietnam War Seen by an Intelligence Officer of the Tiger Division," *The Crusade of Hwarang Warriors*, pp. 249-257.

12 Ibid., pp. 249-252.

13 Ibid., pp. 254-255.

14 Park Tae-deok, "The Operation of Artillery Fire in the Vietnam War," *The Crusade of Hwarang Warriors*, p. 28.

15 *Testimonies*, Vol. II, pp. 258-262

16 1Lt. Kim Hyun-young was the first victim among the KMA graduates in the Vietnam War.

17 Kim Dae-sik, "Fighting in Operation Lightning Flash 66-3," *The Crusade of Hwarang Warriors*, pp. 168-174. Also see *Testimonies*, Vol. I, pp. 301-307.

18 Kwon Yong-soo, "An Analysis of Operation White Horse 1," *Asadal*, No. 38 (June 20, 1967), pp. 15-17.

19 Koo Ja-yoon, *Asadal*, No. 49 (August 1968), pp. 13-16.

20 *Testimonies*, Vol. II, pp.468-472.

## Chapter 2. Class of 1958 and Class of 1959: Into the Fire

1 ROK Army HQs, Military History Research Office, ed., *The Combat Histories of ROK Army in the Vietnam War*, Vol. I (Seoul: Sammun Printing Co., Ltd., 1992), pp. 791-792.

2 *Testimonies*, Vol. I, pp. 307-308.

3 *Testimonies*, Vol. I, pp. 464-466.

4 *The Crusade of Hwarang Warriors*, pp. 267-268.

5 See Kim Dae-heon, "A Consideration of the Ambush Tactics," *The Crusade of Hwarang Warriors*, pp. 40-47.

6 *Testimonies*, Vol. II, pp. 460-465.

7 *Hold Back the Night* (1952) is a Korean war novel by Pat Frank. It deals with the heroic story of the 1st U.S. Marine Division withdrawing from the Changjin Reservoir in northern Korea in December 1950. It tells a beautiful story of the courage and dedication of the American soldiers in the severe winter weather under fierce attack by the massive Chinese forces.

8 The Battle of Tra Binh was fought in the Tra Binh village on February 14-15, 1967. The ROKMC 11th Company defeated a regiment-sized attack of PAVN and VC in four hours of close-quarters combat. In the battle, 246 PAVN and VC were killed, and many weapons were retrieved. Since this incident, the phrase "Myth-making Marines" of Korea began to appear.

9 Ibid., p. 201.

10 Ibid., p. 206.

11 *The Vietnam War and Me*, pp. 316-317.

12 Institute for Military History, MND, *Pictorial War History of ROK Forces to Vietnam, September 1964-June 1970* (Seoul: Gonghwa Publishing Co., 1974), p. 54.

13 *The Forgotten War and Dollar*, pp. 138-139.

14 This word reminds me of Tim O'Brien's autobiographical account of his Vietnam experience, *If I Die in a Combat Zone, Box Me Up and Ship Me Home* (1973).

15 Nuoc mam (Quang ganh) is the most well-known dipping sauce made from fish sauce. Its simplest recipe is some lime juice, or occasionally vinegar, one part fish sauce (*nước mắm*), one part sugar, and two parts water. Collect fish intestines, put them in a bowl, sprinkle salt on them, rot, and the liquid that comes out black.

16 *The Forgotten War and Dollar*, pp. 244-259.

17 *Testimonies*, Vol. I, pp. 289-290; See also Chae Myung-shin, p. 197.

18 Ibid., pp.433-435.

19 *The Combat Histories of ROK Army in the Vietnam War,* Vol. I, pp. 838-839.

## Chapter 3. Class of 1960(16th Class): Company Commanders

1 Choi Yong-ho, *Statistics,* p. 82.

2 Yong Young-il, "A Night Surprise Attack of Hoa Hip Village," *The Crusade of Hwarang Warriors,* pp. 115-117.

3 *The Vietnam War and Me,* p. 81.

4 Ibid., p. 331-334.

5 General Chae Myung-shin issued a stern order not to leave comrades' dead bodies or the injured behind. Later he wrote in his book, "I never allowed the dead bodies of comrades or comrades who had been injured to be abandoned within the enemy's sphere of influence. My firm command policy was that you should recover them at any cost. None of my comrades' corpses were left behind during my tenure." See *The Vietnam War and Me,* p. 301.

6 Lee In-soo, "To Retrieve the Bodies of his Men," *The Crusade of Hwarang Warriors,* pp. 228.

7 Ibid., p. 232.

8 *The Crusade of Hwarang Warriors,* pp. 272-274. See also *The Combat Histories of ROK Army in the Vietnam War,* Vol. I, pp. 103-110.

9 *Testimonies,* Vol. I, p. 368.

10 Ibid., pp. 365-372.

11 *The Combat Histories of ROK Army in the Vietnam War,* Vol. I, p. 340.

12 *The Vietnam War and Me,* p. 132.

13 Park Kyung-seok, *The Heritage of Jae-gu* (1984), quoted in Lee Jae-tae, *The 11th Company,* p. 132.

14 "The Heroic Story of Captain Lee Jae-tae," *The Seoul Newspaper,* dated March 26, 1966, quoted in Lee Jae-tae, *The 11th Company,* p. 432,

15 Ibid., pp. 418-419.

## Chapter 4. Class of 1961 to Class of 1963: In the Midst of Politics

1 *The Vietnam War and Me,* p. 327.

2 Interview with Kim Sung-cho, April, 6, 2022.

3 Interview with Kim Sung-cho, February 24, 2022.

4 *The Crusade of Hwarang Warriors,* pp. 134-138; *Asadal,* No. 28 (July 20, 1966), p. 21.

5 *Testimonies*, Vol. I, p. 328.

6 A coup de grâce (blow of mercy) is a death blow to end the suffering of a severely wounded person or animal. It may be a mercy killing of mortally wounded civilians or soldiers, friends or enemies, with or without the sufferer's consent.

7 Ibid., pp. 259-260.

8 *Testimonies*, Vol. I, pp. 332-335.

9 *Testimonies*, Vol. I, pp. 276-277.

10 Ibid., pp. 673-677.

11 Ibid., pp. 618-620.

## Chapter 5. Class of 1964 and Class of 1965: Platoon Leaders

1 *Asadal*, No. 41 (September 21, 1967), pp. 31-32.

2 *Testimonies*, Vol. I, pp. 339-343.

3 Ibid., p. 780.

4 *The Crusade of Hwarang Warriors*, pp. 118-121.

5 *The Combat Histories of the Korean Army in the Vietnam War*, Vol. I, pp.799-800.

6 *The Crusade of Hwarang Warriors*, pp. 175-178.

7 Ibid., pp. 213-220.

8 *Asadal*, No. 31 (October 30, 1966), p. 14.

9 *Testimonies*, Vol. II, pp. 118-120.

10 *Fighting Spirit*, pp. 55-76.

11 *Testimonies*, Vol. I, pp. 576-579.

12 *Testimonies*, Vol. II, pp. 636-644.

13 *Testimonies*, Vol. I, pp. 652-654.

14 *Testimonies*, Vol. I, pp. 728-730.

## Chapter 6. Class of 1966 to Class of 1968: In the Footsteps of their Elders

1 *Asadal*, No. 44 (January 15, 1968), p. 47.

2 Choi Young-ha, *A Rock not Staying*, pp. 78-94.

3 Lee Jong-geun, "A Memory of the Vietnam War," *The 30th Anniversary of Commission: Pictures and Anthology*, KMA Class of 1966 Alumni Association, May 1996.

4 *Testimonies*, Vol. I, pp. 668-670.

5 *Testimonies*, Vol. I, pp. 731-733.

6 *Testimonies*, Vol. II, p. 172.

7 Hong Gil-dong is a fictional character from a widespread and traditional folk tale of Korea: *The Biography of Hong Gil-dong*. This novel was written during the Joseon Dynasty. Hong Gil-dong, an illegitimate son of a nobleman and his lowborn concubine, is the story's main character. Gifted with supreme intelligence and supernatural abilities, he steals from wealthy and corrupt aristocrats, drawing comparisons to famous bandits like the English folk hero Robin Hood. Historical sources point to the existence of a bandit named Hong Gil-dong, who was arrested in 1500. The name of the operation came from this legendary figure.

8 Ibid., p. 176.

9 Interview with Yang Gil-yong, April 21, 2022.

10 Interview with Jang Pan-yong, December 30, 2022.

11 Interview with Lee Jong-gyu, April 17, 2022.

12 *The KMA Press*, May 1969.

13 *Dong A Daily Newspaper*, August 10, 1968.

14 Interview with Yang Sang-jin, May 6, 2022.

15 Cha Ki-moon, *My Journey in the Eternal Life* (Seoul: GeulNurim Publishing Co., 2020), pp. 131-141.

16 Interview with Noh Nam-seop, May 2, 2022.

17 *Asadal*, No. 60 (December 1969), p. 57.

18 *Asadal*, No. 59 (October 1969), pp. 35-36.

19 Interview with Kang Ho-tae, December 29, 2021.

20 Interview with Hwang Eui-seok, April 23, 2022.

### Chapter 7. Class of 1969 to 1971: Toward the End

1 For this matter, see Yoo Jei-hyeon, *The Vietnam War* (Seoul: Han Won, 1992), pp. 287-290.

2 *Testimonies*, Vol. II, p. 708

3 An Interview with Kwon Kyung-seok, April 26, 2022. Also see "A Wife of Public Official is also a Public Official," *The Monthly Chosun* (June 2002), pp. 262-264.

4 Interview with Paik Nam-hwan, April 25, 2022.

5 Interview with Oh Hae-jeong, March 25, 2022.

6 *Asadal*, No. 68 (May 1971), p. 21.

7 *Testimonies*, Vol. I, pp. 745-747.

8 *Testimonies*, Vol. I, P. 304.

9 *Testimonies*, Vol. II, p. 470.

10 *Testimonies*, Vol. II, pp. 740, 744.

11 *Testimonies*, Vol. II, pp. 788-790.

12 Interview with Chang Yang-woon, January 15, 2020.

13 Interview with Kim Pil-soo, January 17, 2020.

14 Interview with Oh Hyun-gu, January 20, 2020.

15 Interview with Kim Soon-shin, January 22, 2020.

16 Interview with Park Heung-keun, January 25, 2020.

17 Interview with Kang Kwang-seok, January 27, 2020.

18 Interview with Kim Won-gyung, January, 31, 2020.

19 Interview with Han Chul-yong, February 3, 2020.

20 Interview with Noh Kyung-hwan, March 21, 2020.

21 For this matter, see Yoo Jei-hyeon, *The Vietnam War*, pp. 296-298.

22 Interview with Moon Dhu-shiek, May 20, 2022.

# PART III. CONSIDERATION: ISSUES

## Chapter 1. The Vietnam War as a war of Platoon Leaders

1 *Statistics*, p. 81.

2 Lee Sang-hoon, "Ambush Operations in Vietnam," *Asadal*, No. 50 (September 1968), p. 20.

3 *Statistics*, p. 39.

4 *The Wall of Jungles*, p. 169.

5 Lee Sang-hoon, op.cit. p. 20.

6 *Fighting Spirit*, Vol. II, 2016.

7 *Testimonies*, Vol. II, p. 755.

8 *Testimonies*, Vol. I, p. 747.

9 Ibid., p. 619.

10 *The Things They Carried*, 1990.

11 Interview with Kwon Kyung-seok, April 26, 2022.

12 *Testimonies*, Vol. II, p. 740.

13 Michael Herr, *Dispatches* (Avon books, NY, 1978), p.64.

14 *The Vietnam War and Me*, p. 71.

15 Ibid., pp. 71-72.

16 *The Crusade of Hwarang Warriors*, p. 141.

17 *The Vietnam War and Me*, p. 276.

18 For the concept of a company tactical base, see *The Vietnam War and Me*, pp. 184-185.

## Chapter 2. Combat Motivation: What Makes Them Fight?

1 Sigmund Freud, "Reflections upon War and Death" in *Character and Culture* (NY: Collier Books 1963), p. 130.

2 *The Red Badge of Courage*, pp. 183, 192, 193.

3 *The Things They Carried*, p. 141.

4 *The Wall of Jungles*, p. 95.

5 *The 11th Company*, p. 212.

6 Ibid., p. 165.

7 *Testimonies*, Vol. II, p. 395.

8 *Testimonies*, Vol. I, p. 872.

9 James Webb, *Fields of Fire*, p. 29.

10 *The Wall of Jungles*, pp. 100-101.

11 *A Rumor of War*, pp. xviii-xix.

12 *Testimonies*, Vol. I, p. 263.

13 *Testimonies*, Vol. II, pp. 642-643.

14 *Fighting Spirit*, pp. 118-119.

15 Ibid., p. 152.

16 *The Crusade of Hwarang Warriors*, p. 39.

17 *Dispatches*, p. 145.

18 Michael Lynch, *An American Soldier* (Boston: Little, Brown and Company, 1969), p. 211.

19 Kim Hyung-seok, pp. 204-205. Also see Mark Twain, "The Man that Corrupted Hadleyburg," 1899.

**Chapter 3. Dehumanization of War**

1 *Dispatches*, p. 275.

2 *A Rumor of War*, p. 235.

3 Peter L. Stromberg, "A Long War's Writing: American Novels about the Fighting in Vietnam Written While American Fought," Diss. Cornell University, 1974, p. 209.

4 *Dispatches*, p. 29.

5 Robert Jay Lifton, "Victims and Executioners," in Richard A. Falk et al. eds., *Crimes of War* (NY: Random House, 1971), P. 423.

6 *A Rumor of War*, p. 309.

7 The title of James Jones' novel refers to the heroic efforts of the Charlie company during the fighting at Guadalcanal in World War II. The book's title can be easily deduced from the motto of James Jones' book: 'A Thin Red Line of Heroes,' quoted from Rudyard Kipling's poem "Tommy." It reports to an old Midwestern saying: "There is only a thin red line between the sane and the mad."

8 Gustav Hasford, *The Short-Timers* (NY: Bantam Books, 1979), p. 176.

9 Interview with Koh Tae-soon, February 22, 2020.

10 *The 11th Company*, p. 397.

11 *The Forgotten War and Dollar*, p. 209.

12 *Fighting Spirit*, pp. 252-259.

13 Interview with Hwang Won-tak, February 20, 2022.

14 *Testimonies*, Vol. II, p. 230.

15 *The 11th Company*, p. 368.

16 Interview with Noh Nam-seop, February 20, 2022.

17 William Tecumseh Sherman, *Memoirs* (1875), pp. 119-120, quoted in Michael Walzer, *Just and Unjust Wars* (New York: Basic Books, 1977), p. 32.

18 Michael Walzer, p. 32.

19 Ibid., p. 230.

20 *Fighting Spirit*, p. 120.

21 *The 11th Company*, p. 324.

22 The position of the Vietnamese government about the allegation of the massacre of civilians is as follows: "There is no need to discuss the past between Korea and Vietnam now. We must cover up the matter of the past and work together for the future. If anyone needs to be held accountable, the U.S. government is the one. We will discuss it with the U.S. We will not discuss the issue further with the Korean government." See Choi Yong-ho, *In one Book*, pp. 431-432.

23 *The Forgotten War and Dollar*, p. 246.

24 Ibid, "Preface"

25 *The Forgotten War and Dollar*, p. 220.

26 *Ganh Gao* is a traditional carrier used in Vietnam. It consists of two baskets hanging from a wide and long bamboo tree from both ends. It is a means to transport goods, similar to A-frame in Korea.

27 *A Rumor of War*, p. xx.

### Chapter 4. Appearances and Realities in the Vietnam War

1 The ancient Greek playwright Aeschylus is credited for the dictum.

2 Francis Pollini, *Night*, 1961.

3 *The Things They Carried*, 1990.

4 *The Crusade of Hwarang Warriors*, p. 264.

5 Michael Walzer, *Just and Unjust Wars*, 1977, p. 195.

6 Michael Maclear, *The Ten Thousand Day War, Vietnam: 1945-1975* (New York: St. Martain's Press, 1981), p. 164.

7 For further discussion of these matters, see Chung Youn-son, *American War Novel: from the Civil War to the Vietnam War* (Seoul: Seoul National University Press, 2002), pp. 311-330.

8 *Testimonies*, Vol. I, p. 251.

9 *A Rumor of War*, p. xix.

10 *Testimonies*, Vol. I, p. 89

11 *A Rumor of War*, p. 229.
12 Ibid., p. 230.
13 *The Vietnam War and Me*, pp. 211-212.
14 *The Wall of Jungles*, pp. 163-164.
15 *Testimonies*, Vol. II, p. 296.
16 *Testimonies*, Vol. II, p. 221.
17 *Fighting Spirit*, pp. 162-164.
18 Ibid., p. 202.
19 *Asadal*, No. 39 (July 23, 1967), p. 22.

## Chapter 5. War and Humanism: To Humanize an Inhuman War

1 *A Rumor of War*, p. 229.
2 Archibald MacLeish, "Lines for an Interment," *New Republic*, LXXVI (Sept. 20, 1933), p. 159.
3 *The Things They Carried*, p. 80.
4 *The 11th Company*, p. 397.
5 Interview with Moon Dhu-shiek, January 22, 2022.
6 *The 11th Company*, p. 221.
7 *Fighting Spirit*, p. 280.
8 *A Rumor of War*, pp. 116-117.
9 Ibid., p. 117.
10 *Testimonies*, Vol. II, p. 516.
11 Ibid., pp. 275-277.
12 *The Combat Histories of Korean Army in the Vietnam War*, Vol. I, pp. 822-825.
13 William March, *Company K* (New York: Harrison Smith, 1933), p. 53.
14 Park Kyung-seok, *Route 19* (Seoul: Dae Young Sa, 1966), pp. 171-181. Also see Lee Jae-tae, pp. 70-71.

## Chapter 6. The Korean Soldiers Reflected upon the Americans and the Vietnamese, and Vice Versa

1 *A Rumor of War*, p. xviii.
2 *The Vietnam War and Me*, p. 241.
3 *The Forgotten War and Dollar*, p. 241. Also see *The Vietnam War and Me*, p. 406.
4 *Testimonies*, Vol. 1, pp. 920, 924.

5 *The Vietnam War and Me*, p. 241.

6 Lee Bong-wan, "Various Aspects of Vietnam," *Asadal* No. 41 (September 21. 1967), p. 34.

7 Choi Yong-ho, *Statistics*, p. 140.

8 *The Saigon Post*, March 20, 1968.

9 Chesly J. Manly, *Chicago Tribune*, March 13, 1966.

10 The name of Ojakgyo Bridge comes from the legendary bridge on the Milky Way. According to the legend in Korean culture, Gyeonwoo, a male star, and Jiknyeo, a female star, meet at a bridge named Ojakgyo on the Milky Way on July 7, once every year by the lunar calendar.

11 *The Vietnam War and Me*, pp. 418-419.

12 Ibid., pp. 413-414.

13 Ibid., p. 418.

14 Ibid., p. 418.

15 *The Later Han*, Vol. 85; *Chronicles of Eastern Ethnics*, Vol. 75; and *Eastern Ok Jeo*. All quoted in Park Jung-ki, *Wake Up, Korea* (Sejong: Poetry and Essay, 2022), p. 182.

16 *Classic of the Mountains and Seas*, quoted in *Wake Up, Korea*, p. 185.

17 *Wake Up Korea*, pp. 185.

18 *The Vietnam War and Me*, p. 508.

19 Ibid., p. 276.

20 Charles Durden, *No Bugles, No Drums* (1976. Rpt. Avon Books, 1984), p. 204.

21 *The Vietnam War and Me*, pp. 52-53.

22 *The Forgotten War and Dollar*, p. 178.

23 Interview with Song Pil-won, January 20, 2020.

24 *The Vietnam War and Me*, p. 169.

25 Charles Durden, p. 67.

26 Interview with Hwang Won-tak, February 22, 2022.

27 *A Rumor of War*, p. xviii.

28 Lee Bong-wan, "Various Aspects of Vietnam," *Asadal*, No. 41 (September 21, 1967), pp. 33-34.

29 Webb Beech, *Make War in Madness* (NY: Fawcett Publications, Inc., 1965), p. 35.

30 *Testimonies*, Vol. I, p. 578.

31 *The Forgotten War and Dollar*, p. 135.

32 Ibid., p. 193.

33 *Testimonies*, Vol. II, p. 592.

34 Chae Myung-shin interviewed by Glenn C. Troelstrup, *US News & World Report* in *Asadal*, No. 39 (July 23, 1967), pp. 20-23.

# PART IV. RETURN: HOME

## Chapter 1. From Fort Hwarang to Dongjak-dong

1 Kang Jae-ryun, *Asadal*, No. 42, October 27, 1967.
2 *Asadal*, Vol. 22, October 25, 1965, p.18.
3 Park Kyung-seok, *Route 19*, pp. 18-19.
4 Ibid., pp. 19-20.
5 Quoted in *Asadal*, No. 22 (October 25, 1965), p.19.
6 Ibid., p. 19.
7 *The Vietnam War and Me*, pp. 123-124.
8 Ibid., p. 127.
9 *Asadal*, No. 23 (January 20, 1966), p. 22.
10 *Asadal*, No. 26 (April 22, 1966), p. 20.
11 *The Combat Histories of Korean Army in the Vietnam War*, Vol. I, pp.374-375.
12 *The Crusade of Hwarang Warriors*, P. 279.
13 *The Combat Histories of Korean Army in the Vietnam War*, Vol. I, pp. 795-796.
14 *Testimonies*, Vol. I, p. 346.
15 Ibid., p. 347.
16 Ibid., p. 345.
17 *Asadal*, No. 30 (September 15, 1966), p. 15.

## Chapter 2. The War that has Never Ended: The Living and the Dead

1 Kim Hyung-seok, *Combat Diary: A Platoon Leader in the Vietnam War* (Seoul: Knowledge & Sympathy, 2022), p. 430.
2 Lee Dong-hee, *Asadal*, No. 49 (August 1968), p. 44.
3 Kim Won-young, *Asadal*, No. 49 (August 1968), p. 45.
4 Lim Jong-sup, *Asadal*, No. 43 (November 27, 1967), pp. 42-43.
5 *The Crusade of Hwarang Warriors*, pp. 285-286.
6 Ibid., pp. 283-284.
7 See Chung Youn-son, *The Memories of the Forgotten War*, 2019.
8 From *The 50th Anniversary Collection of KMA's Class of 1970*, Seoul, 2020.
9 Norman Mailer, *The Naked and the Dead*, Rinehart & Company, 1948

## Conclusion: Legacies of the Vietnam War

1 Donald Konx, *The Korean War; Uncertain Victory* (Harcourt Brace Jovanovich, 1991), pp. 506-507

2 Kim Hyung-seok, *Combat Diary: A Platoon Leader in the Vietnam War* (Seoul: Knowledge & Sympathy, 2022), pp. 431-455.

3 *The Vietnam War and Me*, pp. 505-506.

4 Ibid., p. 507.

5 Choi Yong-ho, *In one Book*, pp. 395-396.

6 On August 17, 2022, a Vietnamese lady named Nguyen Ti Tanh (62), then an eight-year-old girl, submitted to the Korean court the case of the incident of the massacre allegedly committed by the Korean Marine Corps in the Pong Ni village of Quang Nam Province on February 12, 1968. She applied for compensation and testified for the killings of about 70 residents. Indeed, the Vietnam War did not end yet.

7 Choi Yong-ho, *In one Book*, p. 425.

8 Interview with Nguyen Duy Quy, Director of the Vietnamese Institute of Humanities and Social Sciences, who visited the Institute for Military History of MND, on May 7, 2002. Quoted in Choi Yong-ho, *In one Book*, pp. 431-432.

9 Shin Jong-tae, "The Defoliant Patients of Vietnam Veterans and Directions for Promoting Veterans Affairs Policy (in Korean)," *Journal of Studies on Military Development*, Vol. 7, No. 2 (Dec. 2013), pp. 147-163.

10 Susan Rubin Suleiman, "War Memories," *New Literary History*, p. 563.

11 James Webb, *Fields of Fire*, New York: Bantam, 1978.

12 *Dispatches*, p. 278.

# Glossary

**AK:** Automatic Kalashnikov.

**AK-47:** The standard rifle used by the North Vietnamese and Viet Cong.

**AKA:** Also Known As.

**AP:** Advance Party.

**AOR:** Area of Responsibility.

**APC:** Armored Personnel Carrier

**AR:** Automatic Rifle.

**Arc light:** B-52 bomber strike, shaking earth for ten miles from target area.

**ARVN:** Army of the Republic of Vietnam.

**Bird:** Any aircraft, but usually referring to helicopters.

**B40:** Bazooka 40 (an 80-mm unguided heat rocket up to 150 m).

**BMNT:** Beginning Morning Nautical Twilight.

**BN:** Battalion

**Boonierat:** a combat infantryman

**Bush:** An infantry term for the field or jungle.

**CCL:** Civilian Control Line.

**CG:** Commanding General

**Chinook:** A supply and transport helicopter.

**Claymore:** Anti-personnel mine carried by infantry.

**CKS:** Chiang Kai-Shek (rifle).

**CMDR:** Commander.

**CO:** Company; Command officer.

**DEROS:** Date of Expected Return from Overseas.

**Dust-off:** Medical evacuation by helicopter or a helicopter used for fast evacuation of wounded from the battlefield.

**EENT:** End Evening Nautical Twilight.

**FAAO:** Field Artillery Forward Observer.

**FDC:** Fire Direction Center.

**FDO:** Fire Direction Officer.

**FFE:** Fire for Effective.

**FNG:** Fucking New Guy.

**FO:** Forward Observer.

**FTX:** Field Training Exercise.

**IO:** Interpretation Officer.

**JCS:** Joint Chiefs of Staff.

**KIA:** Killed in Action.

**KMA:** Korea Military Academy.

**KMAG:** Korea Military Assistance Group.

**KOICA:** Korea International Cooperation Agency

**LMG:** Light Machine Gun.

**LP:** Listening Post.

**LAW:** Light Anti-Tank Weapon (M-72 Rocket Launcher).

**LD:** Line of Departure.

**LST:** Landing Ship, Tank.

**LZ:** Landing Zone.

**MASH:** Mobile Army Surgical Hospital

**MDL:** Military Demarcation Line.

**ME:** Meeting Engagement.

**MGR:** Mere Gook Rule.

**MIA:** Missing in Action.

**MIG:** Military Intelligence Group.

**MND:** Ministry of National Defense (Republic of Korea)

**M-1:** World War II vintage American rifle.

**M-16:** The standard American rifle used in Vietnam after 1966.

**M-79:** Single-barreled grenade launcher used by infantry.

**MP:** Military Police

**MRE:** Meal Ready to Eat.

**MSTS:** Military Sea Transportation Service.

**NCO:** Noncommissioned Officer, usually a squad leader or platoon sergeant.

**NOE:** Nap of the Earth; A type of very low-altitude flight course used by military aircraft to avoid enemy detection and attack in a high-threat environment.

**NVA:** North Vietnamese Army.

**O-1:** Observation plane; **OH-1:** Observation helicopter

**OAC:** Officers' Advance Course.

**OBC:** Officers' Basic Course.

**O/O:** on order

**PAVN:** People's Army of Vietnam.

**PFC:** Private First Class.

**Point:** The forward man or element on a combat mission.

**POW:** Prisoner of War.

**Prep fire:** preparation fire before beginning the attack

**PT:** Platoon

**PZ:** Pick-up Zone.

**R&R:** Rest and Recuperation.

**RCT:** Regiment Combat Team.

**RECONDO:** Reconnaissance Commando.

**ROK:** Republic of Korea.

**RGT:** Regiment

**ROKFV:** Republic of Korea Forces in Vietnam.

**RPG:** Rocket-Propelled Grenade.

**RPD:** Russian-made 7.62 mm machinegun.

**RR:** Recoilless Rifle

**RTO:** Radio Telephone Operator.

**Sapper:** A VC or NVA commando, usually armed with explosives.

**SGT:** Sergeant

**SSGT:** Staff Sergeant

**Short-timer:** Soldier nearing the end of his tour in Vietnam.

**SME:** Small and Medium-sized Enterprise.

**SMG:** Submachine Gun.

**SNU:** Seoul National University.

**TACCP:** Tactical Command Post.

**TAOR:** Tactical Area of Responsibility.

**TDY:** Temporary Duty.

**Tet Offensive:** a major uprising of the NLF, their sympathizers, and NVA characterized by a series of coordinated attacks against military installations and provincial capitals throughout Vietnam. It occurred during the lunar New Year at the end of January, 1968.

**TNT:** Trinitrotoluene, Explosives.

**TOG:** They are Only Gooks.

**TOT:** Time on Target.

**USMA:** United States Military Academy.

**USMACV:** United States Military Assistance Command Vietnam.

**VC:** Viet Cong.

**WIA:** Wounded in Action.

**XO:** Executive Officer, the second in command of a military unit.

# Selected Bibliography

## Books and Articles


Atkinson, Rick. *The Long Gray Line: The American Journey of West Point's Class of 1966.* Houghton Mifflin Company, 1989.

Baker, Mark, *Nam: The Vietnam War in the Words of the Men and Women Who Fought There.* Time Warner Book, rpt. 1992.

Basker, Lawrence and William A. Strauss. *Chance and Circumstance: The Draft, the War and the Vietnam Generation.* NY: Vantage Books, 1978.

Beech, Webb. *Make War in Madness.* NY: Fawcett Publications, Inc., 1965.

Caputo, Philip. *A Rumor of War.* New York: Ballantine Books, 1977.

Cha, Ki-moon, *My Journey in the Eternal Life* (in Korean). Seoul: GeulNurim Publishing Co., 2020.

Chae, Myung-shin, *The Vietnam War and Me* (in Korean). Seoul: Palbokwon, 2006.

Childress, William. "A Poet Remembers Korea," *War, Literature & the Arts,* Vol. 9, No. 2 (Fall/Winter 1997)

Choi, Yong-ho, ed. *The Vietnam War and the Korean Army through Testimonies,* Vol. I (in Korean). Seoul: Institute for Military History, MND, 2001.

__________. *The Vietnam War and the Korean Army through Testimonies,* Vol. II (in Korean). Seoul: Institute for Military History, MND, 2002.

__________. *The Vietnam War and the Korean Army in One Book* (in Korean). Seoul: Institute for Military History, MND, 2004.

__________. *The Vietnam War and the Korean Army in Statistics* (in Korean). Seoul: Institute for Military History, MND, 2007.

Choi, Young-ha, *A Rock not Staying* (in Korean). Goyang: Malgeunsam, 2012.

Chung, Youn-son, *The Memories of the Forgotten War: The Korean War in American Novels* (in Korean). Seoul: Munye Publishers, 2019.

__________. *American War Novel: From the Civil War to the Vietnam War* (in Korean). Seoul: Seoul National University Press, 2002.

Coleridge, Samuel Taylor. "The Rime of the Ancient Mariner," https://en.wikipedia.org/wiki/Albatross (metaphor).

Crane, Stephen. *The Red Badge of Courage,* 1895. Rpt. New York: Puffin, 1986.

Durden, Charles. *No Bugles, No Drums.* NY: Charter Books, 1976. Rpt. Avon Books, 1984.

Emerson, Gloria. *Winners and Losers: Battles, Retreats, Gains, Losses and Ruins from the Vietnam War*. New York Harcourt Brace, 1976.

Falk, Richard A. et. al. eds., *Crimes of War*. NY: Random House, 1971.

Faulkner, William. *Requiem for a Nun* (1951), rpt. New York: Vintage International, 2011.

Fitzgerald, Frances. *Fire in the Lake: The Vietnamese and the Americans in Vietnam*. New York: Bantam Books, 1972.

Freud, Sigmund. "Reflections upon war and death" in *Character and Culture*. New York: Collier Books, 1963.

Frye, Northrop. *Anatomy of Criticism*. Princeton: Princeton Univ. Press, 1957.

Fussell, Paul. *The Great War and Modern Memory*. Oxford University Press, 1976.

Hasford, Gustav. *The Short-Timers*. NY: Bantam Books, 1979.

Herr, Michael. *Dispatches*. NY: Avon books, 1978.

Hickey, James. *Chrysanthemum in the Snow*. Crown Publishers, 1990.

Hirsh, Marianne. "The Generation of Postmemory," *Poetics Today*, 29:1 (March 2008), pp. 108-128.

____________. "Connective Histories in Vulnerable Times," *PMLA*, 129-3 (2014), pp. 330-348.

Institute for Military History, MND, *Pictorial War History of ROK Forces to Vietnam*, September 1964-June 1970, Seoul: Gonghwa Publishing Co., 1974.

Kim, Hyung-seok. *Fighting Spirit: A Platoon leader of the Vietnam War*, Vol. I, II (in Korean). Ilsan: Jisik & Gonggam, 2016.

____________. *Combat Diary: A Platoon Leader in the Vietnam War.* Seoul: Knowledge & Sympathy, 2022.

Kwon, Yong-su, "Lineage of Korea Military Academy (in Korean)," *Asadal* Vol. 29 (August 20, 1966).

KMA Alumni Association, ed., *The Crusade of Hwarang Warriors* (in Korean). Seoul: Army Printing Office, 1967.

Lee, Jae-tae. *The 11th Company of the Jae-Gu Battalion at the Vietnam Warfront* (in Korean). Seoul: Uncheon Company, 2018.

Lifton, Robert Jay. *Home From the War: Vietnam Veterans, Neither Victims nor Executioners.* NY: Simon & Schuster, 1973.

Lynch, Michael. *An American Soldier.* Boston: Little, Brown and Company, 1969.

Maclear, Michael. *The Ten Thousand Day War, Vietnam: 1945-1975*. New York: St. Martain's Press, 1981.

MacLeish, Archibald. "Lines for an Interment," *New Republic*, LXXVI (Sept. 20, 1933).

Maihafer, Harry J. *From the Hudson to the Yalu: West Point '49 in the Korean War.* Williams-

Ford Texas A&M University, 1993.

Mailer, Norman. *The Naked and the Dead.* NY: Rinehart & Company, 1948.

March, William. *Company K.* New York: Harrison Smith, 1933.

Moore, Harold G. and Joseph L. Calloway. *We Were Soldiers Once and Young.* NY: Random House, 1993.

Nguyen, Viet Thanh. *Nothing Ever Dies: Vietnam and the Memory of War.* Cambridge: Harvard University Press, 2016.

O'Brien, Tim. *The Things They Carried.* New York: Houghton Mifflin, 1990.

Park, Jong-sik, *The Forgotten War and Dollar: A Memoir of an Infantry Company Commander* (in Korean). Seoul: Dong Hwa Books, 1998.

Park, Jung-ki, *Wake up, Korea*! Sejong: Quarterly Poetry, 2022.

Park, Kyung-seok, *Route 19* (in Korean). Seoul: Daeyoung-sa, 1966.

__________. *The Heritage of Jae-gu* (in Korean). 1984.

Pollini, Francis. *Night*, 1961.

ROK Army HQs, Military History Research Office, ed., *The Combat Histories of Korean Army in the Vietnam War*, Vol. I (in Korean). Seoul: Sammun Printing Co., Ltd., 1992.

Santoli, Al. *Everything We Had: An Oral History of the Vietnam War by Thirty-Three American Soldiers Who Fought It.* New York: Ballantine Books, 1982.

Shin, Jung, *The Wall of Jungles* (in Korean). Seoul: Chunggu Publishing Company, 1967.

Stromberg, Peter L. "A Long War's Writing: American Novels about the Fighting in Vietnam Written While American Fought," Ph.D. Diss. Cornell University, 1974.

Suleiman, Susan Rubin. "War Memories," *New Literary History*, 24 (1993).

Twain, Mark. "The Man that Corrupted Hadleyburg," 1899.

Walzer, Michael. *Just and Unjust Wars.* New York: Basic Books, 1977.

Webb, James. *Fields of Fire.* Englewood Cliffs, NJ: Prentice-Hall, 1979.

Yoo, Jei-hyeon, *The Vietnam War* (in Korean). Seoul: Han Won, 1992.

Yoon, Heung-jung, et. al. *An Integrated Study of the Vietnam War* (in Korean). ROKFV, 1974.

## Periodicals, Newspapers, Study Reports, and Internet sites

KMA Alumni Association Newsletter, *Asadal* (in Korean), 1964-1971.

*The Korea Military Academy Press* (The KMA Press) (Monthly in Korean)

*Korean Armed Forces Comrade Newspaper*

*The 30th Anniversary of Graduation: Pictures and Anthology* (in Korean), KMA Class of 1966, May 1996.

*The 50th Anniversary Collection of KMA Class of 1970* (in Korean), Seoul, 2020.

*Chicago Tribune*

*The 50th Anniversary of KMA History* (in Korean), Seoul, 1996.

*Dong A Daily Newspaper* (in Korean)

*The Monthly Chosun* (in Korean)

*The Post-Report on the Withdrawal of ROK Forces from Vietnam* (in Korean), ROKFV, 1973.

*New Republic*

*The Saigon Post*

*US News & World Report*

https://www.earticle.net/Article/A228331

https://en.wikipedia.org/wiki/Gulf_of_Tonkin_incident

# Index


A
Ahn Byung-moo 238
Ahn Choon-saeng 27
Ahn Hyun-tae 154
Ahn Jae-seok 101
Ahn Joong-geun 27
Ahn Seung-bong 48, 58
*An American Soldier* 287, 404, 415
An Khe Pass 39, 40, 232, 260, 262, 263, 264, 323, 330, 376
Ao Dai 346
*Apocalypse Now* 85
Area of Responsibility 36, 410, 412
Ark-light 111
Army Integrated School 26, 56, 57, 393
*A Rumor of War* 14, 45, 85, 271, 283, 311, 326, 336, 392, 394, 396, 404, 405, 406, 407, 414
*Asadal* 59, 98, 198, 224, 225, 244, 356, 371, 372, 373, 393, 395, 396, 397, 398, 399, 400, 401, 403, 406, 407, 408, 415, 417
Attrition Rate 42, 109, 124, 153, 175, 197, 210, 224, 235, 247, 261, 310

B
Bae In-ki 58, 79, 274, 286
Bae Jung-do 56
Bang Seo-nam 56, 58, 61, 124, 125
Baptism of Fire 140, 191, 240, 252, 292
The Battle of Binh Tan 339
The Battle of Changjin Reservoir 49
The battle of Chi Hoa 149
The Battle of Duc Co 122, 126, 185, 248
The Battle of Phung Son 139
The Battle of Re Wen Village 240
Beach, Dwight 36
The Blue White Line 29, 30, 32
Body Count 307
Boule, Pierre 345
Brown, Winthrop 37
Byun Gil-nam 198, 205
Byun Hong 195, 196, 271

C
The Cadet Code of Ethics 306
Cadet Creed 30, 31, 32, 33, 48, 68, 306, 355, 384, 393
Cadet First Class 25, 26, 393
Cadet Second Class 26, 52, 56
The Capturing of Korean Soldiers Alive 337
Caputo, Philip 45, 80, 85, 271, 283, 289, 308, 311, 318, 326, 336, 392, 396
*Catch-22* 291
Cha Byung-gyu 86, 376, 379, 396
Chae Hee-seon 135, 169
Chae Myung-shin 36, 38, 49, 56, 62, 76, 82, 83, 99, 100, 102, 127, 129, 135, 138, 139, 149, 151, 176, 192, 198, 275, 276, 277, 302, 306, 309, 315, 337, 338, 351, 361, 394, 397, 398, 399, 407
Cha Jun-hoe 125
Cha Ki-moon 220, 401
Chang Yang-woon 248, 249, 251, 402
*Chicago Tribune* 339, 407, 417
Childress, William 414
Cho Geum-joo 246
Choi Bong-ryong 155
Choi Byung-soo 57
Choi Eui-gu 373
Choi Ung 107
Choi Young-chang 251
Choi Young-gu 57
Choi Young-ha 51, 199, 371, 395, 400, 414
Cho Seung-ok 262
Cho Young-shik 363
*Chrysanthemum in the Snow* 321, 415
Chun Doo-hwan 44, 94, 115, 123, 163, 193, 393
Chung Chan-ho 373
Chung Il-seok 251
Chung Joo-young 117, 118, 167, 363
Chung Man-gil 96
Chung Moon-ki 148
Chung Poong-jeon 136
Chung Pyoung-yang 162
Chun Gwan 210, 418
Chung Yeon-soo 229
Chung Yong-woo 113
Chung Young-moo 198
Chung Yu-hee 155
Chun Yong-min 251
*Classic of Mountains and Seas* 343
Clavell, James 345

Code of Battlefield Ethics 318
*Combat Diary* 415
*Company K* 331, 406, 416
Company Tactical Base 95, 96, 97, 111, 119, 253, 276, 277, 321, 403
*Comrade Newspaper* 194, 417
Comradeship 278, 282
*Confessions of a Humanist* 161
Coppola, Francis 85
Counterinsurgency 78, 272, 276, 306
Coup de Grâce 170, 400
Crane, Stephen 414
*The Crusade of Hwarang Warriors* 99, 103, 106, 109, 306, 374, 394, 396, 397, 398, 399, 400, 403, 404, 405, 408, 415
Cu Mong Pass 139, 280, 332, 333, 334, 335

D

Daihan 240, 316
The Democratic Republic of Vietnam 17
Dien Bien Phu 15, 17, 392
*Dispatches* 273, 287, 289, 403, 404, 409, 415
Domino Theory 15
Dongjak-dong 240, 355, 356, 371, 377, 379, 380, 408
Dove Force 35, 39, 100, 154
Duty, Honor, Country 94

E

*Eastern Ok Jeo* 342, 407
Einstein, Albert 318
*Endless Journey* 184
Engagement 16, 138, 307, 411
Enlisted Men 24, 42, 59, 60, 310
Estimated Kills 313

F

False Reporting 312
Faulkner, William 415
Fear of Death 48, 52, 214, 228, 231, 232, 241, 250, 255, 256, 278, 281, 286, 289, 290, 310, 319, 321, 369, 385
*Fields of Fire* 17, 282, 291, 392, 404, 409, 416
*Fighting Spirit: A Platoon leader of the Vietnam War* 395, 415
Finger Mountain 41
The First Generation 13, 24
The Five Rules to Observe in Life 32
Fort Hwarang 28, 31, 32, 355, 356, 373, 393, 408
Frank, Pat 398
Free Fire Zone 307, 308
Freud, Sigmund 415

G

*Gaffey* 67
Geneva Conventions 318
Go Boi plains 143, 144, 170, 178, 218, 219, 231, 254, 272, 277, 282, 293
*Going After Cacciato* 271
The Graduates after the Korean War 27
The Graduates before the Korean War 24
*Green Berets* 54
Guerrilla War 16, 45, 61, 75, 76, 77, 78, 79, 123, 138, 182, 206, 220, 221, 272, 273, 275, 276, 292, 299, 308, 344, 369, 385

H

Hahm Duk-sun 180
Han Chul-yong 53, 256, 310, 379, 395, 402
Han Il-soo 103, 113
Han Kwang-duk 128, 129, 183
Harkins, Paul D. 18
Hasford, Gustav 291, 404
Helicopter Liaison Officer 84
The Helicopter War 82, 396
Heller, Joseph 291
Heo Nam-seong 378
Heo Sam-soo 155
*The Heritage of Jae-gu* 151, 399, 416
Herr, Michael 415
Hickey, James 415
Hill 67 175
Hill 92 6, 175, 383
Hit-and-run Tactics 79
Ho Chi Minh Trail 18, 76
Hodge, John R. 24
*Hold Back the Night* 125, 398
The Honor Code 306
Huangpu Military Academy of China 24
100 Day-jinks 311
Hwang Eui-seok 225, 232, 401
Hwang Won-tak 164, 195, 296, 404, 407
Hwarang 28, 29, 31, 32, 33, 93, 99, 103, 106, 109, 117, 121, 122, 130, 152, 164, 167, 195, 210, 211, 213, 250, 306, 312, 355, 356, 359, 362, 364, 366, 373, 374, 393, 394, 396, 397, 398, 399, 400, 403, 404, 405, 408, 415
Hwarang-dae 28, 29, 32, 393

J

Jae-gu Battalion 125, 140, 151, 158, 185, 254, 356, 362, 376
Jae-gu Village 134, 135
Jang Chang-gyu 191, 349
Jang Joon-ik 121
Jang Pan-yong 211, 212, 401
Jang Sei-dong 139, 176, 193, 329
Jang Yang-hee 120
Jee Man-won 70, 198, 208, 396
Jeon Sang-woon 182
Johnson, Lyndon B. 235
James Jones 291, 404
Jung Rae-hiuk 244
*Just and Unjust Wars* 405, 416

K

Kang Deok-hwang 160
Kang Eui-yeong 120
Kang Gwang-seok 255
Kang Hang-joong 251, 331, 377
Kang Ho-tae 225, 228, 401
Kang Jae-gu 56, 58, 134, 137, 151, 195, 248, 356, 357, 358, 359, 360, 361, 362, 363, 367, 375
Kang Jae-ryun 31, 355, 408
Kang Myung-oh 155
Kang Sei-ho 127, 129
Killing Zone 80, 81, 223, 277
Kim Chan-bok 56
Kim Chang-ho 251
Kim Chan-sam 184
Kim Choong-bae 380
Kim Dae-heon 123, 398
Kim Dae-sik 109, 397
Kim Gil-bu 158, 185, 366
Kim Gu-woong 210
Kim Ho-kyung 367
Kim Hong-sik 371
Kim Hyung-seok 49, 63, 68, 73, 83, 179, 188, 191, 270, 285, 286, 288, 294, 300, 314, 325, 326, 371, 395, 404, 408, 409
Kim Hyun-young 109, 363, 370, 397
Kim Jae-gyu 155
Kim Jang-woong 367
Kim Jeong-jo 192
Kim Jin-gyu 46, 73, 121, 128, 306, 394
Kim Jin-seon 172, 206
Kim Jin-wook 155
Kim Jin-young 154
Kim Ji-young 328, 329
Kim Joi-soo 103
Kim Jong-hwan 198
Kim Joo-cheon 170
Kim Jung-woon 55
Kim Ki-taik 100, 101
Kim Kye-doo 113
Kim Kyei-won 143
Kim Moo-kyung 103, 104, 105, 397
Kim Moon-hwan 377
Kim Moo-seok 365, 366
Kim Myung-sei 214, 215
Kim Pil-soo 250, 251, 402
Kim Sang-gu 125
Kim Sang-soon 248
Kim Seong-eun 34
Kim Seong-hoi 327
Kim Seong-ki 155
Kim Soon-shin 251, 253, 283, 402
Kim Sung-cho 155, 159, 162, 399
Kim Tae-geun 251
Kim Tae-seop 154
Kim Tae-woong 207, 251, 321
Kim Won-gyung 255, 256, 402
Kim Won-young 372, 408
Kim Yeon-gak 155
Kim Yong-bae 62
Kim Yong-jin 56
Kim Young-won 210
*King Rat* 345
KMA Alumni Association 31, 59, 98, 198, 224, 258, 355, 359, 371, 393, 394, 415, 417
The KMA Anthem 30, 356
KMA Man 32, 33, 393
*The KMA Press* 32, 214, 393, 401, 417
KMA's Return to Seoul 28
Koh Kun-young 127
Koh Tae-soon 282, 292, 404
Koh Young-bok 218
Koo Bon-geol 135
Koo Ja-yoon 110, 112, 398
Korea Military Assistance Group 27, 411
Korean Liberation Army 24
Korean Military Aid Team 34
The Korean War 13, 14, 23, 24, 26, 27, 28, 33, 38, 42, 45, 49, 51, 53, 55, 56, 57, 60, 61, 66, 70, 75, 85, 152, 165, 182, 240, 287, 310, 321, 339, 345, 348, 349, 377, 383, 385, 387, 393, 394, 395, 397, 409, 414, 415

Kwon Joon-taik 56
Kwon Kyung-seok 236, 239, 240, 269, 271, 371, 401, 403
Kwon Seung-joon 185, 186
Kwon Wu-heon 58
Kwon Yong-soo 110, 397
Kwon Young-dae 120

L
Laird, Melvin 344
Larson, Stanley R. 125
*The Later Han* 342, 407
Lee Bong-wan 337, 348, 407
Lee Bum-cheon 218
Lee Byung-gan 232
Lee Byung-gu 58, 94, 95, 96
Lee Byung-tae 155
Lee Cheol-ho 58, 239
Lee Chun-geun 124, 125, 127, 129, 184, 248
Lee Chun-shik 128
Lee Dong-hee 214, 371, 408
Lee Eui-ho 198, 206, 271
Lee Eun-sang 360
Lee Gang-eon 247, 421
Lee Gyu-sup 261
Lee Gyu-wan 176
Lee Hak-gun 96
Lee Hee-sung 196
Lee Heung-yeol 251
Lee Hyo 58, 94, 96, 97, 98, 397
Lee Hyun-woo 154
Lee In-soo 144, 399
Lee Jae-kwan 187
Lee Jae-tae 58, 61, 67, 69, 72, 149, 151, 152, 249, 279, 293, 297, 300, 320, 324, 333, 339, 396, 399, 406
Lee Jong-geun 202, 203, 400
Lee Jong-gyu 213, 401
Lee Jong-koo 119
Lee Jun 172, 173
Lee Jung-lin 155, 158
Lee Kwon-bok 49, 145, 146, 171
Lee Moon-seok 154
Lee Nam-joo 57
Lee Pil-jo 56
Lee Sang-don 261
Lee Sang-ho 98, 154, 225, 270, 397, 403
Lee Sang-hong 225
Lee Sang-hoon 98, 270, 397, 403
Lee Sang-tae 261
Lee Sei-ho 164, 192, 244, 259
Lee Seok-myung 44, 94
Lee Seung-gwan 86, 258
Lee So-dong 37, 341
Lee Sung-woo 86
Lee Wan-soo 380, 381
Lee Won-yang 125, 136
Lee Yong-su 123, 124
Lee Yoon-gu 176
Lim Chung-soo 165
Lim In-jo 154
Lim Jong-sup 373, 408
Lim Joong-bok 86, 258
The Logistics Support Command 41
The Long Gray Line 29
Lynch, Michael 415

M
Maclear, Michael 415
MacLeish, Archibald 415
Mai Shao 217
Mao Zedong 78
Maruta 292
MASH 34, 35, 411
McNamara, Robert 15, 82
Meeting Engagement 194, 269
Mere Gook Rule 309, 411
"Me Too" Menu 69
Military Deity 137
Min Chan-ki 208
Min Kyung-gu 154
Min Pyung-shik 101
Model Hwarang 117, 164, 195
Moon Dhu-shiek 262, 264, 323, 402, 406
Moon Hee-heun 86
Myung San-ok 176

N
Na In-hwan 225
National Liberation Front 17
Ngo Dinh Diem 17
Nguyen Cao Ky 35, 72, 363
Nixon, Richard 235, 260
Noh Kyung-hwan 259, 402
Noh Nam-seop 51, 221, 297, 395, 401, 405
Noh Yoon-sup 258
Non-La Hat 72
*Nothing Ever Dies: Vietnam and the Memory of War*

16, 392, 416
Nothing Happens to Me 52, 278
Nui Hon Ba 2, 6, 7, 41, 84, 101, 111, 155, 193, 217, 222, 226, 245, 256, 258, 314, 346, 380, 383
Nui Phu Cat 7, 39, 40, 95, 104, 105, 142, 143, 144, 156, 158, 173, 185, 218, 285, 298, 321

O

O'Brien, Tim 416
Oh Hae-jeong 241, 401
Oh Hak-hyun 374
Oh Hyung-geun 198
Oh Hyun-guk 165
Oh Hyun-koo 52, 86, 252, 376, 395
Oh Kil-hwan 225
Oh Yong-geun 261
One Medal for One Gun 296
One Shot One Kill 64
Operation Advance 3 119, 146
Operation Bat 26 124, 193, 194, 285
Operation Crazy Horse 147, 149
Operation Dokkabi 1 107, 108, 113, 346
Operation Dokkabi 6 112, 114, 212, 213
Operation Dokkabi 14 107, 108
Operation Dokkabi 21 88, 247, 281
Operation Dokkabi (Goblin) 2 101
Operation Dongbo 17 246
Operation Eagle 70-1 205, 244
Operation Eagle 71-1 124, 327
Operation Fierce Tiger 5 105, 107, 117, 118, 119, 144, 167, 170, 178, 248, 343, 363, 365, 366
Operation Fierce Tiger 6 103, 104, 107, 121, 135, 143, 155, 156, 157, 158, 169, 185, 298
Operation Fierce Tiger 9 155, 202, 203
Operation Fierce Tiger 15 249
Operation Fierce Tiger 16 195, 207, 244, 251, 253, 271
Operation Fierce Tiger 17 207, 230, 251, 313
Operation Hong Gil-dong 189, 208, 209
Operation Horse Head 2 199
Operation Lightning Flash 3 119
Operation Lightning Flash 5 96
Operation Lightning Flash 66-3 109, 110, 397
Operation Ojakgyo Bridge 41, 89, 131, 195, 301, 336, 340, 341, 342
Operation Plum Blossom 304
Operation Sacred Horse 72-1 221, 232
Operation Tiger 169, 194, 205, 237, 240, 242, 285
Operation Typhoon 7-4 165
Operation White Horse 2 223
Operation White Horse 9 83, 155, 160, 162, 234
Operation White Horse 10 107, 162, 211

P

Paik Myung-yoon 114, 246
Paik Nam-hwan 51, 68, 238, 239, 395, 401
Paik Un-taek 192
Park Chang-sik 125
Park Chan-wook 58
Park Chung-hee 25, 34, 62, 65, 149, 151, 152, 163, 307, 360, 361, 362, 382, 388
Park Dong-won 78, 117, 167, 248, 308, 333, 363
Park Gap-yong 94
Park Han-young 56
Park Heung-keun 254, 283, 402
Park Jae-hwa 154, 155, 156
Park Jong-nam 193
Park Jong-sik 48, 63, 72, 73, 88, 125, 130, 131, 294, 301, 346, 350, 394, 395
Park Jong-sun 32, 147, 393
Park Jong-sung 147
Park Jung-ki 5, 46, 47, 342, 394, 407
Park Kyung-seok 52, 56, 83, 134, 151, 334, 357, 361, 396, 399, 406, 408
Park Kyung-won 154
Park Man-joo 74
Park Min-soon 225
Park Pan-gil 203
Park Sei-jik 102
Park Seong-hwan 171
Park Sung-joon 258
Park Tae-deok 106, 108, 397
Park Young-il 194
Park Young-shik 376, 378
The People's Army of Vietnam 19
Period System 310
*Platoon* 271
Point System 310
Policy of Vietnamization 235
Psalm 23 289, 290

R

Racist Theory 348
RECONDO 216, 217, 218, 219, 412
*Requiem for a Nun* 16, 392, 415
Rho Young-chul 150
ROK Army Creed in Vietnam 38
ROKFV 36, 37, 38, 40, 41, 44, 78, 100, 126, 155,

164, 189, 192, 205, 208, 221, 244, 259, 296, 306, 394, 412, 416, 417
ROK West Point 93
Rotation Policy 307
Route 1 39, 40, 41, 42, 52, 76, 83, 95, 96, 111, 115, 122, 126, 136, 140, 149, 195, 260, 262, 263, 294, 330, 331, 333, 340, 341, 343, 371, 396, 406, 408, 416
Route 19 39, 40, 52, 76, 83, 95, 96, 122, 126, 136, 260, 262, 263, 330, 331, 343, 371, 396, 406, 408, 416
Rules of Engagement 307
Ryu Byung-hyun 82, 143, 341

S
Sadler, Barry 54
*Sands of Iwo Jima* 54
Search and Destroy 38, 89, 95, 110, 111, 114, 119, 149, 151, 164, 274, 308, 313
The Second Generation 27
Seo Yong-won 236
The 7th Replacement Training Camp 65, 236, 271, 373
Sherman, William Tecumseh 299, 405
Shin Hyun-soo 55
Shin In-sun 87, 251
Shin Jung 49, 58, 71, 176, 181, 270, 279, 282, 312, 349, 394
Shin Sang-gil 68, 320
Shin Sun-gyun 327
Silla Kingdom 31, 393
Sixteen-Letter Tactics 78
Snatching Combat Achievements 313
Sohn Tae-ik 155
Sohn Yong-nam 176, 178
Sohn Young-gil 47, 94
Song Pil-won 232, 347, 407
Song Sung-han 44
Song Yeol-jae 251
Southern Cross 41, 54, 100, 249, 347
South Joseon National Defense Constabulary Academy 24, 26
Stromberg, Peter L. 416
Suh Soo-geun 204, 285
Sun Tzu 78, 286

T
Taegeuk Order of Military Merit 362
Taereung 28, 109, 116, 124, 374, 393
TAOR 36, 39, 40, 41, 95, 96, 99, 104, 111, 123, 135, 138, 139, 140, 143, 178, 192, 212, 232, 272, 274, 275, 333, 340, 371, 412
Temerity 284
Tet Offensive 38, 164, 202, 224, 235, 412
*The Art of War* 286
*The Bridge on the River Kwai* 345
*The Forgotten War and Dollar* 301, 394, 395, 396, 398, 404, 405, 406, 407, 416
*The Naked and the Dead* 381, 408, 416
*The Red Badge of Courage* 65, 278, 284, 395, 403, 414
*The Saigon Post* 338, 341, 407, 417
*The Short-Timers* 291, 404, 415
*The Things They Carried* 271, 278, 305, 319, 392, 403, 405, 406, 416
*The Thin Red Line* 291
*The Times* 343
*The Vietnam War and Me* 12, 361, 394, 395, 396, 397, 399, 403, 406, 407, 408, 409, 414, 423
Tomalin, Nicholas 343
Tonkin Gulf Incident 15
Troelstrup, Glenn C. 316, 351, 407

U
Uhm Myoung-chul 251
Unit 731 291
U.S. 7th Marine Regiment 36
U.S. 101st Airborne Division 36, 95, 140, 342
U.S. Army Military Government 24
U.S. Army Special Forces School 134
U.S. Infantry School at Fort Benning 57, 71
USMACV 18, 35, 36, 39, 40, 100, 211, 260, 342, 387, 413
U.S. Military Academy at West Point 23, 25
U.S. Military Advisory Group 25
U.S. Military Assistance Command Vietnam 18
*U.S. News and World Report* 316
*USS Barret* 67, 69
*USS Enterprise* 67
*USS General Leroy Eltinge* 66, 71
*USS Maddox* 15, 18, 392
*USS Upsher* 243
*USS Victoria* 70

V
Valley of Death 144, 145
Van Fleet, James A. 27, 28, 94, 397
Vietminh 15

Viet Thanh Nguyen 16, 392
Vo Nguyen Giap 78, 80, 81

W

Wake up, Korea 416
Walzer, Michael 416
*The Wall of Jungles* 49, 176, 270, 279, 394, 403, 404, 406, 416
War is Hell 299, 304, 319
Wayne, John 54, 256, 257
Webb, James 17, 282, 291, 392, 404, 409, 416
Westmoreland, William 18, 35, 36, 85, 100, 342, 387
Winning and Fighting Well 318
Winning Hearts and Minds 307
Wisdom, Benevolence, Courage 29, 372
Won Jong-gwan 243, 270

Y

Yamamoto Musashi 67
Yang Chul-seok 311
Yang Gil-yong 211, 401
Yang Sang-jin 216
Yang Seung-bok 155
Yong Young-il 56, 58, 59, 103, 142, 143, 149, 158, 339, 357, 399
Yoo Gwang-sun 236
Yoo Gyo-jun 245
Yoon Hang-joong 236, 237
Yoon Seok-tae 258
Yoon Tae-gon 372

# 『92고지에서 혼바산까지』
## 육사 졸업생들의 베트남전 참전기

저자: 정연선 (Chung Youn-son)

육군사관학교 (26기)
서울대 영문과 및 동 대학원 학사/석사
미국 에모리 (Emory)대 미국학 박사
육사 영어과 교수 (현 명예교수)
한국영어영문학회 회장
한국아메리카학회 회장

주요 저서 및 역서
『미국전쟁소설: 남북전쟁으로부터 월남전까지』
『잊혀진 전쟁의 기억: 미국소설로 읽는 한국전쟁』
『영미 전쟁시선』
『한국에서의 미국학』(공저)
『미국학의 이론과 실제』(공역)
*Beyond the Limits of Human Ability* (영역)
*Stories of Leadership* (영역)